Economic Dynamics and General Equilibrium

Springer
Berlin
Heidelberg
New York
Hong Kong
London
Milan
Paris
Tokyo

Anders Borglin

Economic Dynamics and General Equilibrium

Time and Uncertainty

With 49 Figures
and 26 Tables

Springer

Prof. Anders Borglin
Lund University
Department of Economics
22007 Lund
Sweden
anders.borglin@nek.lu.se

ISBN 3-540-00265-0 Springer-Verlag Berlin Heidelberg New York

Cataloging-in-Publication Data applied for
A catalog record for this book is available from the Library of Congress.
Bibliographic information published by Die Deutsche Bibliothek
Die Deutsche Bibliothek lists this publication in the Deutsche Nationalbibliografie; detailed bibliographic data is available in the Internet at <http://dnb.ddb.de>.

Springer-Verlag is a part of Springer Science+Business Media

springeronline.com

Printed in Germany

SPIN 10905012 43/3130/DK-5 4 3 2 1 0 – Printed on acid-free paper

PREFACE

This book grew out of the teaching and lecture notes for a course in "Dynamic Economics" given at Copenhagen University where it has been part of a combined study program in economics and mathematics since 1986. The subjects in the course have varied over time but a common theme has been the study of economies over time and under uncertainty in the form of incomplete markets and the overlapping generations model. When I took up a position at Lund University in 1996 Mich Tvede took over responsibility for the course and together we developed a large part of the material, which formed the basis for the present book. He has generously allowed me to use all of our common material.

There is a considerable overlap between general equilibrium theory with incomplete markets (GEI), finance and macroeconomic theory, which stems from the considerations of time and uncertainty, as can be seen from the splendid textbooks by LeRoy and Werner [2001], Pliska [1997] and the seminal work on GEI by Magill and Quinzee [1996], in particular its "Historical Remarks" which offer a brilliant history of modern economic theory.

The exposition here aims at making some of the results of GEI and on overlapping generations (OG) economies readily available to a reader acquainted with economic theory at an upper undergraduate level.

The restricted possibilities to transfer income between dates or across states is the distinguishing feature of GEI. By initially focusing on economies over time the interpretation is simplified. The discussion of assumptions and terminology for economies with uncertainty is then be pursued at length in a separate chapter. A main role is played by the (generalized) indirect utility function which depicts the consumer's preferences over net income vectors with spot-prices given. The indirect utility function makes it possible to decompose the consumer's problem into a financial decision of choosing a net income vector and the spot-market problems of choosing a goods bundle at each spot-market, given a net income vector. The results of GEI are illustrated by explicit examples. No proofs are given for the general results but there is a thorough discussion of the assumptions and concepts involved.

The exposition of OG economies is from the outset restricted to simple OG economies; a single good at each date and a single consumer, who acts at two dates, born at every date. Time is taken to extend into the indefinite future and past.
Different notions of optimality are studied and a simple version of Cass' celebrated efficiency criterion is derived for reduced models. From Chapter 9 and onwards the economy is assumed to be stationary so that all consumers are alike. Using support functions and approximations of upper contour sets it is shown that Cass' efficiency criterion applies to stationary economies.

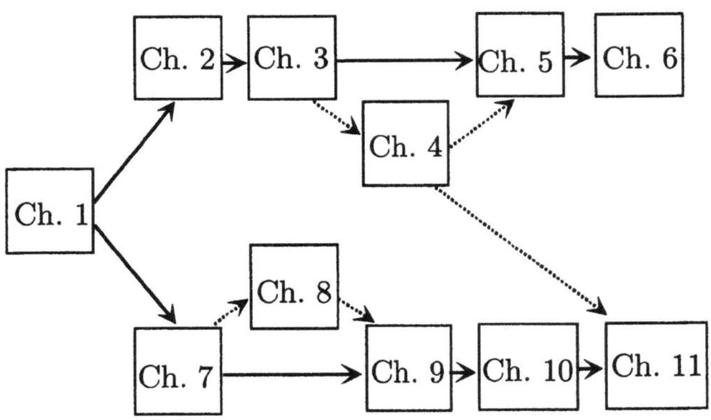

Figure A: *Solid arrows indicate major dependence. Dashed arrows indicate minor dependence*

The equilibrium locus summarizes the pair of relative prices resulting in market balance. Each stationary economy has a maximal set of relative prices which can be extended to an equilibrium price system; a maximal invariant set. From the characterization of the invariant set it is immediately seen, on the one hand, that there is in general a continuum of equilibria and, on the other hand, that each equilibrium for a no-trade economy or an economy close to such economy is either a steady state equilibrium or converges to such an equilibrium both in forward and backward time. Once the nominal equilibrium of the economy involves a higher intensity of trade this simplicity is lost and the possibility of cyclical equilibria and "sun-spot" equilibria occur.

Organization of the book: Figure A indicates the dependence among chapters. The book is divided into 11 chapters and an Appendix. Chapters, apart from the appendix chapter, are divided into numbered *sections*. *Subsections* are distinguished by headings with capital initial letters, and *subsubsections* are distinguished by headings with small letters.

In each of the chapters, except the appendix chapter, all the Definitions, Lemmata, Corollaries, Propositions and Theorems are numbered alphanumerically within each section. Thus, for example, "3.1.C" refers to the third item in Chapter 3, Section 1 and the item may belong to any of the mentioned categories. *Figures* are separately numbered alphanumerically within each section and so are *Tables*.

The appendix chapter is divided into four *sections;* Appendix A, B, C and D and within each of these, items are numbered alphanumerically but also in the appendix chapter, figures are numbered separately. In the text, items in the appendix chapter are always referred to by the appendix section and the alphanumeral within that appendix section.

The symbol □ is used to signal the end of a proof or an example unless the proof or example is followed a subsub, sub or section heading, in which case it is dropped. Section or subsections with a * can be skipped at first reading.

Acknowledgement. My first and foremost gratitude is to Mich Tvede for all the interesting and enlightening discussions over the years. Danish and Swedish students have provided valuable inspiration and feedback on the numerous versions of the lecture notes. The references to published papers fail to convey the major influence and enthusiasm which numerous visitors to Copenhagen and Lund have provided through their brilliant lectures.
My colleagues; Tommy Andersson, Ola Jönsson, Bo Larsson and Birger

Nilsson have to assume full responsibility for any missing misprints and lack of incorrect grammar while remaining shortcomings, as usual, rest with the author

Anders Bäckstrand provided the initial document style and Joakim Ekstrand prevented the time needed for reading manuals to Latex packages to extend into the indefinite future. Valuable support has also been given by the staff at Springer Verlag, in particular, Martina Bihn has provided helpful comments and encouragement from the outset.

Financial support from the Swedish Insurance Association, Trygg Hansa Research Foundation and the Craaford Foundation is gratefully acknowledged.

Lund 19 October 2003
Anders Borglin

Contents

Chapter 1

CONSUMERS AND ECONOMIES - A REVIEW

Introduction

The purpose of this chapter is twofold. On the one hand, we will introduce consumers and economies and recall a few results, which are probably well known to the reader. On the other hand, we give a set of assumptions regarding the consumer to be used in the sequel; the *Maintained Assumptions.*

Although the results may be well known to the reader the methods used in the proofs may be unfamiliar. We do not prove the results under the weakest assumptions possible but rather apply methods which will turn out to be useful in the subsequent exposition. We stress the differentiable point of view which allow us to compare the consumers' subjective evaluations and the market evaluation as given by the prices.

In contrast to following chapters we refrain here from giving any interpretation of the studied economies. Exchange among the consumers can be thought of as taking place at a single date and at a single location.

The chapter is organized as follows. In Section 1.1 we describe consumers and introduce the Maintained Assumptions on consumers. We also give some properties of solutions to the *Walrasian Consumer Problem* and derive the *Walrasian demand function.*

An economy is a (finite) set of consumers. Economies are introduced in Section 1.2, where we also define the notion of a *Walras equilibrium* and of a *Pareto optimal allocation.* The section ends with a discussion of the First and Second Theorem of Welfare Economics. The proof of the Second Theorem highlights the equalization of subjective consumer evaluations at a Pareto optimal allocation.

Often we will be concerned with a single consumer and it is then convenient to drop the index for the consumer. It will not always be pointed out when we do so.

1.1 Consumers

The purpose of this section is to introduce a set of assumptions on consumers, to be used in the throughout the book, and to give a review of the consumer's decision problem. The *Maintained Assumptions* will, implicitly, be assumed to hold in all proofs, but in some examples one or more of the Maintained Assumptions may fail to be satisfied. We will point out when this is the case.

Assume that there are n commodities so that the *commodity space* is $\mathbb{R}^n$. A consumer (in a private ownership, pure exchange economy) is characterized by a *consumption set* C, which is a non-empty subset of the commodity space $\mathbb{R}^n$, a *utility function* $u : C \longrightarrow \mathbb{R}$ and an *initial endowment* e which is a vector in the commodity space. Hence, for the purpose of the theory a consumer is triple (C, u, e).

In the interpretation, the consumption set includes the feasible consumptions taking into account "non-economic" restrictions, that is, restrictions not related to prices or income. The consumption set describes the needs of the consumer whereas the utility function depicts the consumer's taste. The consumer has to satisfy her needs and among the consumption plans which achieve this the consumer chooses on the basis of her taste. The initial endowment describes contracts for deliveries to (and from) the consumer which the consumer has entered into prior to our study of the economy.

Maintained Assumptions on Consumers

We now list the set of Maintained Assumptions. These may not be familiar to the reader but Lemma 1.1.A below states that they imply some well known properties of the utility function. In the sequel we will assume that any consumer satisfies the Maintained Assumptions except occasionally in examples or problems. Recall that $\mathbb{R}^n_{++} = \{c \in \mathbb{R}^n \mid c_k > 0 \text{ for } k = 1, \ldots, n\}$ and $\mathbb{C}^2(C, \mathbb{R})$ is the set of functions with domain C taking values in $\mathbb{R}$ which are twice differentiable, or equivalently, have partial derivatives of first and second order which are continuous functions.

Maintained Assumptions: *A* **consumer**, (C, u, e), *is assumed to satisfy:*

(C1) $C = \mathbb{R}^n_{++}$

(C2) $e \in \mathbb{R}^n_{++}$

(C3) $u \in \mathbb{C}^2(C, \mathbb{R})$ *and each contour set of* u *is a closed subset of* $\mathbb{R}^n$

(C4) $Du(c) \in \mathbb{R}^n_{++}$ *for each* $c \in C$

(C5) *for* $c \in C$; $h^t D^2u(c)h < 0$ *for* $h \in \mathbb{R}^n \backslash \{0\}$ *such that* $Du(c)h = 0$

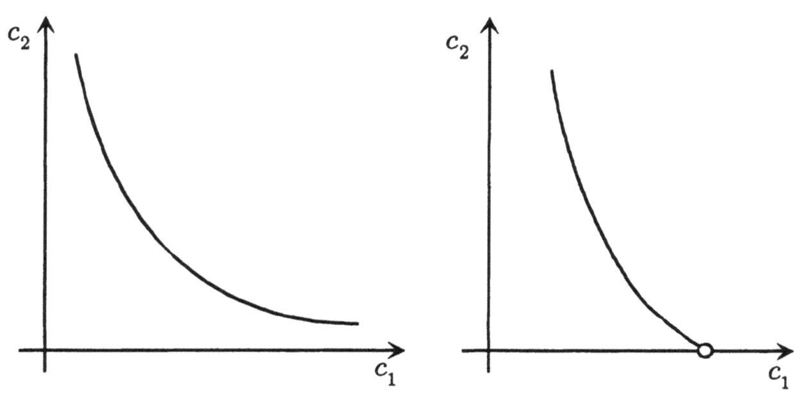

Figure 1.1.A: *On the left a contour set of a utility function satisfying assumption* (C3) *and on the right a contour set of a utility function violating assumption* (C3)

The function $Du : C \longrightarrow \mathbb{R}^n$ defined by

$$Du(c) = (D_1u(c), D_2u(c), ..., D_nu(c)) = (\frac{\partial u(c)}{\partial c_1}, \frac{\partial u(c)}{\partial c_2}, \dots, \frac{\partial u(c)}{\partial c_n})$$

takes the point c to the vector of partial derivatives, evaluated at c. The vector $Du(c)$ can be identified with the linear form defined on $\mathbb{R}^n$ taking $h \in \mathbb{R}^n$ to the number $Du(c)h$. The vector $Du(c)$ will also be denoted $\operatorname{grad} u(c)$ and referred to as the *gradient* of u evaluated at the point c.

$D^2u(c)$ is the *Hessian* of u evaluated at the point c. It is a symmetric $n \times n$ matrix whose entry at position (i, j) is

$$D^2_{ij}u(c) = \frac{\partial^2 u(c)}{\partial c_i \partial c_j}$$

Assumptions (C1) and (C2) are easy to interpret. In particular they imply that the initial endowment is a possible consumption for the consumer.

A function is twice differentiable if and only if it has continuous first and second order partial derivatives. Hence (C3) asserts that the utility function, u, has partial derivatives of first and second order which are continuous functions. (C3) also excludes that a contour set for u has a limit point belonging to the boundary of $\mathbb{R}^n_{++}$. An alternative way of stating this part of the assumption is to say that the closure (in $\mathbb{R}^n$) of each contour set is contained in $\mathbb{R}^n_{++}$. Cf. Figure 1.1.A where the left panel shows a contour set satisfying assumption (C3) and the right panel shows a contour set which has a boundary point not belonging to $\mathbb{R}^2_{++}$. Thus the panel on the right illustrates a contour set of a utility function which does not satisfy assumption (C3).

(C4) is a differentiable version of the assumption that the utility function is increasing in each component. It implies that preferences are strongly monotone; $c' \geq c, c' \neq c$ implies $u(c') \geq u(c)$, but is a strengthening of that assumption since no partial derivative can take the value 0 at any point. Cf. Exercise 1.C.

Recall that u is strictly quasi-concave if:

$$u(c) \geq u(c'), c \neq c', \alpha \in]0,1[\quad \text{implies} \quad u(\alpha c + (1-\alpha)c') > u(c')$$

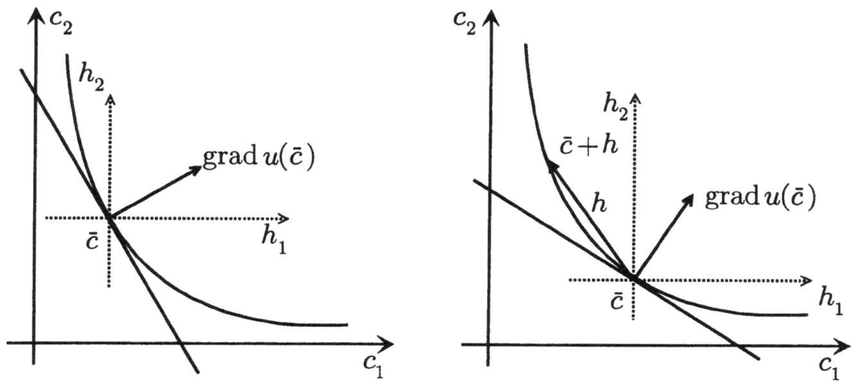

Figure 1.1.B: *On the left is shown the hyperplane defined by the gradient at $\bar{c}$ on which the quadratic form defined by $D^2u(\bar{c})$ is assumed negative definite;* (C5). *On the right is shown how* grad $u(\bar{c})$ *defines a supporting hyperplane to the upper contour set which has only the point $\bar{c}$ in common with the plane*

(C5) implies that the utility function is strictly quasi-concave. It is a somewhat stronger assumption than strict quasi-concavity since strict quasi-concavity would only imply that the quadratic form, given by $D^2u(c)$, is negative *semi*definite on the linear subspace given by $Du(c)h = 0$. This subspace may be identified with the tangent hyperplane (suitably translated) to the contour set of u at $\bar{c}$ as depicted in the left panel of Figure 1.1.B.

We will at times say that the utility function satisfies the Maintained Assumptions which is then taken to mean that (C3)-(C5) are satisfied.

Implications of the Maintained Assumptions

The Maintained Assumptions are useful since they allow us to use differential calculus in a rigorous way. They are somewhat stronger than needed for some arguments and often we can use the implied properties of the utility function, given in Lemma 1.1.A below, which may be more familiar to the reader. A more thorough account of the implications of the Maintained Assumptions can be found in Balasko [1988] or Mas-Colell [1985].

Lemma 1.1.A *Let (C, u, e) be a consumer and let $\bar{c}$ be any point in C. Then*

(a) *u is a strictly quasi-concave function*

(b) *u is increasing in each argument*

(c) *$u(\bar{c}+h) \geq u(\bar{c})$ and $h \neq 0$ implies $Du(\bar{c})h > 0$*

(d) *$\{c \in C \mid u(c) \geq u(\bar{c})\}$ is a closed, strictly convex subset of $\mathbb{R}^n$*

(e) $$\det \left[\begin{array}{c|c} D^2u(\bar{c}) & \begin{array}{c} D_1u(\bar{c}) \\ D_2u(\bar{c}) \\ \vdots \\ D_nu(\bar{c}) \end{array} \\ \hline D_1u(\bar{c}) \quad D_2u(\bar{c}) \quad \cdots \quad D_nu(\bar{c}) & 0 \end{array} \right] \neq 0$$

There is an alternative way to state (c) of Lemma 1.1.A. Let $u(\hat{c}) \geq u(\bar{c})$ and put $h = \hat{c} - \bar{c}$ so that $\hat{c} = \bar{c} + (\hat{c} - \bar{c}) = \bar{c} + h$. Then $Du(\bar{c})(\hat{c} - \bar{c}) \geq 0$ with equality only if $\hat{c} = \bar{c}$. This is shown in the right panel of Figure 1.1.B where it is seen that the gradient of u evaluated at $\bar{c}$ defines a supporting hyperplane owning only the point $\bar{c}$ from the set of consumptions at least as good as $\bar{c}$. For the sake of completeness we give a proof of Lemma 1.1.A in the next section.

Proof of Lemma 1.1.A*

Let $\mathbb{R}^n_+$ denote the nonnegative vectors in $\mathbb{R}^n$.

(a) We will prove that (C5) implies that u is strictly quasi-concave by proving the converse: if u is not strictly quasi-concave then (C5) is not true.

Assume that u is not strictly quasi-concave. Then there are $c \in \mathbb{R}^n_{++}$, $h \neq 0$ and $t \in]0,1[$ such that

$$u(c+h) \geq u(c) \quad \text{and} \quad u(c+th) \leq u(c)$$

Consider the function $H : [0,1] \longrightarrow \mathbb{R}$ defined by $H(t) = u(c+th)$. H is constant and equal to $u(c)$ for $t \in [0,1]$ or H attains a minimum at some interior point of $[0,1]$. In either case, for some $t^0 \in \,]0,1[$,

$$u(c+t^0h) \;\le\; u(c+th) \quad \text{for } t \in [0,1]$$

Then, since $t^0 \in \,]0,1[$ and $H(t^0) \le H(t)$ for $t \in [0,1]$,

$$\frac{dH(t^0)}{dt} \;=\; Du(c+t^0h)h \;=\; 0 \quad \text{and} \quad \frac{d^2H(t^0)}{dt^2} \;=\; h^tD^2u(c+t^0h)h \;\ge\; 0$$

which implies that the quadratic form induced by $D^2u(c+t^0h)$ is not negative definite on the homogenous hyperplane with normal $Du(c+t^0h)$. Hence (C5) is not true as was to be proved.

(b) Assume, in order at arrive at a contradiction, that, for some i, $u(c+a\tau^i) < u(c)$ for $\tau^i = (0,\ldots,1,\ldots,0)$ where "1" occurs at the i'th position and $a > 0$. Consider the function $H : [0,a] \longrightarrow \mathbb{R}$ where $H(\alpha) = u(c+\alpha\tau^i)$. By the mean value theorem for derivatives, there is a point $\hat{\alpha}$ such that $DH(c+\hat{\alpha}t^i) = u(c+a\tau^i) - u(c)$, which implies $DH(c+\hat{\alpha}t^i) < 0$. Since $DH(c+\hat{\alpha}t^i) = D_iu(c+\hat{\alpha}t^i)$ we get a contradiction to (C4). It follows that $u(c+h) > u(c)$ for $c \in C, h \in \mathbb{R}^n_+\backslash\{0\}$.

(c) By **(a)** u is strictly quasi-concave. Let

$$\bar{c} \in C, h \ne 0 \quad \text{and} \quad u(\bar{c}+h) \;\ge\; u(\bar{c})$$

Then $u(\bar{c}+th) > u(\bar{c})$ for $t \in \,]0,1[$. Thus, on the one hand, the limit

$$\lim_{t\to 0} \frac{1}{t}[u(\bar{c}+th) - u(\bar{c})]$$

is nonnegative and, on the other hand, this limit equals $Du(\bar{c})h$. It remains to prove that $Du(\bar{c})h > 0$ for $h \ne 0$.

Assume, in order to arrive at a contradiction, that there is $\bar{c} \in C$, $h \ne 0$ such that $u(\bar{c}) \le u(\bar{c}+h) \ge u(\bar{c})$ and $Du(\bar{c})h = 0$. Since u is strictly quasi-concave we get

$$u(\bar{c}+\frac{1}{2}h) \;>\; u(\bar{c}) \quad \text{and} \quad Du(\bar{c})(\frac{1}{2}h) \;=\; 0$$

The set

$$\Gamma \;=\; \{h' \in \mathbb{R}^n \mid u(\bar{c}+h') \;>\; u(\bar{c})\}$$

is an open set in $\mathbb{R}^n$ owning the vector $(1/2)h$. Since $Du(\bar{c})(1/2)h = 0$ there is a vector $\hat{h}$ belonging to Γ such that $Du(\bar{c})\hat{h} < 0$. However, from the conclusion above we have $Du(\bar{c})\hat{h} \geq 0$; a contradiction. Hence $u(\bar{c}+h) \geq u(\bar{c})$, $h \neq 0$ implies $Du(\bar{c})h > 0$.

(d) Since u is a continuous function the set $\{c \in C \mid u(c) > u(\bar{c})\}$ is an open set in C for each $\bar{c} \in C$. Its boundary is the contour set for the value $u(\bar{c})$ which by assumption is contained in C. Hence $\{c \in C \mid u(c) \geq u(\bar{c})\}$ is a closed subset of C and also a closed subset of $\mathbb{R}^n$. Since u is a strictly quasi-concave function its upper contour sets are strictly convex sets.

(e) Theorem A in Appendix B contains a general result from which (e) follows.

Solutions to the Consumer Problem

The consumer problem with given wealth

Let us begin by studying the consumer's decision problem for some externally given wealth, w.

For $p \in \mathbb{R}^n_{++}$ and $w > 0$ let

$$B(p,w) = \{c \in C \mid pc = p_1c_1 + p_2c_2 + \ldots + p_nc_n \leq w\}$$

be the *budget set*, for prices p and wealth w. $B(p,w)$ is a non-empty set since $\lambda p \in B(p,w)$ for $0 < \lambda \leq w/||p||^2$. The set of consumptions satisfying the inequality in the definition of the budget set with equality is the *budget hyperplane*.

Proposition 1.1.B asserts that there exists a unique solution to the consumer decision problem and gives some of the properties of this solution.

Proposition 1.1.B *Let $p \in \mathbb{R}^n_{++}$ and $w > 0$ and consider the Consumer Problem*

$$\text{Max}_{\,c}\, u(c) \quad s.\ to\ c \in B(p,w) \qquad \textbf{1.1.A}$$

Then

(a) *there is a unique solution $\bar{c}$ to the Consumer Problem* 1.1.A *and $p\bar{c} = w$*

(b) *$\bar{c}$ is the solution to* 1.1.A *if and only if, for some $\lambda > 0$*

$$\left\{ \begin{aligned} grad\, u(\bar{c}) - \lambda p &= 0 \\ p\bar{c} &= w \end{aligned} \right. \qquad \mathbf{1.1.B}$$

(c) *the function $\hat{f} : \mathbb{R}^n_{++} \times \mathbb{R}_{++} \longrightarrow C$, which maps (p, w) to the (unique) solution of* 1.1.A*, is differentiable, hence continuous, at each $(p, w) \in \mathbb{R}^n_{++} \times \mathbb{R}_{++}$. Furthermore, $\hat{f}$ is homogenous of degree* 0. *The function $\hat{f}$ is referred to as the Walrasian demand function.*

Proof: To prove **(a)** note that since $B(p, w)$ is not empty there is a $\tilde{c}$ belonging to $B(p, w)$. Using Lemma 1.1.A it is seen that the set

$$A = \{c \in C \mid u(c) \geq u(\tilde{c})\} \cap \{c \in \mathbb{R}^n \mid pc \leq w\}$$

is a non-empty, compact set. A is also a convex set since A is the intersection of two convex sets . Since u is a continuous function we can apply Theorem A in Appendix A which asserts that there is a $\bar{c} \in A$ such that $u(\bar{c}) \geq u(c)$ for $c \in A$. Then $u(\bar{c}) \geq u(c)$ also for $c \in B(p, w)$ so that $\bar{c}$ is a solution to 1.1.A.

Assume, in order to get a contradiction, that $p\bar{c} < w$. Then $w - p\bar{c}$ is a positive number and there is a vector $h \in \mathbb{R}^n_{++}$ such that $ph \leq w - p\bar{c}$. From (b) of Lemma 1.1.A follows that $u(\bar{c}+h) > u(\bar{c})$. Since $p(\bar{c}+h) \leq w$ this contradicts that $\bar{c}$ is a solution to (1.1.A). Hence $p\bar{c} = w$.

Since $B(p, w)$ is a convex set and u is a strictly quasi-concave function there can not be more than one solution to 1.1.A.

(b) Let $\bar{c}$ be the solution to 1.1.A. From **(a)** we get that $\bar{c}$ is a solution to

$$\text{Max}_c\; u(c) \quad \text{s. to } c \in C \text{ and } \quad pc = p_1 c_1 + p_2 c_2 + \ldots + p_n c_n = w$$

Since $\bar{c} \in \mathbb{R}^n_{++}$ we can apply a Lagrange's theorem, Theorem B in Appendix A, which asserts that there exists a $\lambda \in \mathbb{R}$ such that 1.1.B is satisfied. Since $p \in \mathbb{R}^n_{++}$ (C4) implies $\lambda > 0$.

On the other hand, let $\bar{c} \in C$ satisfy 1.1.B for some $\lambda > 0$. Let $\hat{c} \in C$ be such that $u(\hat{c}) > u(\bar{c})$. Since u is a strictly quasi-concave, differentiable function and $\hat{c} = \bar{c} + (\hat{c} - \bar{c})$, we get by applying (c) of Lemma 1.1.A (with $h = (\hat{c} - \bar{c})$) that $\operatorname{grad} u(\bar{c})(\hat{c} - \bar{c}) > 0$. Using $\operatorname{grad} u(\bar{c}) = \lambda p$ we get $p\hat{c} > p\bar{c}$. The equality $p\bar{c} = w$ now implies $p\hat{c} > w$; contradicting that $\hat{c}$ satisfies the last relation of 1.1.B. Hence there does not exist $\hat{c} \in B(p, w)$ such that $u(\hat{c}) > u(\bar{c})$ and it follows that $\bar{c}$ is a solution to 1.1.A.

(c) Define $\hat{f} : \mathbb{R}^n_{++} \times \mathbb{R}_{++} \longrightarrow C$ where $\hat{f}(p, w)$ is the unique solution to 1.1.A so that

$$u(\hat{f}(p, w)) > u(c) \quad \text{for } c \neq \hat{f}(p, w),\ c \in B(p, w)$$

Let $F : C \times \mathbb{R}_{++} \times \mathbb{R}^n_{++} \times \mathbb{R}_{++} \longrightarrow \mathbb{R}^n \times \mathbb{R}$ be defined by

$$F(c, \lambda, p, w) = \begin{cases} \operatorname{grad} u(c) - \lambda p \\ pc - w \end{cases}$$

or, written out in more detail,

$$\begin{aligned} F_1(c, \lambda, p, w) &= D_1 u(c) - \lambda p_1 \\ F_2(c, \lambda, p, w) &= D_2 u(c) - \lambda p_2 \\ &\vdots \\ F_n(c, \lambda, p, w) &= D_n u(c) - \lambda p_n \\ F_{n+1}(c, \lambda, p, w) &= p_1 c_1 + p_2 c_2 + \ldots + p_n c_n - w \end{aligned}$$

Let $(p, w) \in \mathbb{R}^n_{++} \times \mathbb{R}_{++}$ be given. Then there is a unique (c, λ) such that $F(c, \lambda, p, w) = 0$, namely, $c = \hat{f}(p, w)$ and $\lambda = D_1 u(\hat{f}(p, w))/p_1$

Let $\hat{h} : \mathbb{R}^n_{++} \times \mathbb{R}_{++} \longrightarrow \mathbb{R}_{++}$ be defined by $\hat{h}(p, w) = D_1 u(\hat{f}(p, w))/p_1$. Then

$$(\hat{f}(p, w), h(p, w)) = (c, \lambda) \text{ where } F(c, \lambda, p, w) = 0 \qquad \textbf{1.1.C}$$

so that $(\hat{f}, h)$ satisfies $F(\hat{f}(p, w), \hat{h}(p, w), p, w) = 0$ for $(p, w) \in \mathbb{R}^n_{++} \times \mathbb{R}_{++}$

To prove that $\hat{f}$ is a differentiable function, let $(\bar{p}, \bar{w}) \in \mathbb{R}^n_{++} \times \mathbb{R}_{++}$ and let $F(\bar{c}, \bar{\lambda}, \bar{p}, \bar{w}) = 0$. Calculate the Jacobian, $\hat{D}$, of F with respect to (c, λ) and evaluate it at the point $(\bar{c}, \bar{\lambda}, \bar{p}, \bar{w})$,

$$\hat{D} = D_{(c,\lambda)}F(\bar{c},\bar{\lambda},\bar{p},\bar{w}) = \left[\begin{array}{cccc|c} & & & & -\bar{p}_1 \\ & & D^2u(\bar{c}) & & -\bar{p}_2 \\ & & & & \vdots \\ & & & & -\bar{p}_n \\ \hline \bar{p}_1 & \bar{p}_2 & \cdots & \bar{p}_n & 0 \end{array}\right]$$

Let D be the matrix obtained from $\hat{D}$ by changing the sign of the last column and then multiplying the last row and column by λ. From $F(\bar{c},\bar{\lambda},\bar{p},\bar{w}) = 0$ follows that the last row and column of the derived matrix equal grad $u(\bar{c})$ and from (e) of Lemma 1.1.A follows that $\det D \neq 0$. Since $\lambda \neq 0$ also $\det \hat{D} \neq 0$.

From the Implicit Function Theorem, Theorem C in Appendix A, follows that there is an open set $U_{(\bar{p},\bar{w})} \subset \mathbb{R}^n_{++} \times \mathbb{R}_{++}$, owning $(\bar{p},\bar{w})$, and a differentiable function

$$(\tilde{f},\tilde{h}) : U_{(\bar{p},\bar{w})} \longrightarrow \mathbb{R}^{n+1}$$

such that $F(\tilde{f}(p,w),\tilde{h}(p,w),p,w) = 0$ for $(p,w) \in U_{(\bar{p},\bar{w})}$. From relation 1.1.C follows

$$(\hat{f},\hat{h}) = (\tilde{f},\tilde{h}) \quad \text{for } (p,w) \in U_{(\bar{p},\bar{w})}$$

which implies that $\hat{f}$ is a differentiable function at $(\bar{p},\bar{w})$. Since $(\bar{p},\bar{w})$ was an arbitrary point of $\mathbb{R}^n_{++} \times \mathbb{R}_{++}$ it follows that $\hat{f}$ is a differentiable function.

The relation $B(p,w) = B(\alpha p, \alpha w)$ for $\alpha > 0$ implies $\hat{f}(\alpha p, \alpha w) = \hat{f}(p,w)$ for $\alpha > 0$.

□

We have illustrated the relations 1.1.B in Figure 1.1.C. At the solution, $\bar{c}$, to the Consumer Problem the price vector p, reflecting the "market evaluation" of the commodities is proportional to the gradient of the consumer. The gradient reflects the consumer's subjective evaluation. The decision problem of a consumer will be denoted "the Consumer Problem". Since we consider different market structures its meaning will often be determined by the context.

The value of the consumer's Walrasian demand function, at the price-wealth pair (p,w) is also referred to as the consumer's *Marshallian demand.* Unfortunately the terminology in use does not distinguish between demand as a function of prices and wealth and demand as a function solely of prices, with wealth given by the initial endowment.

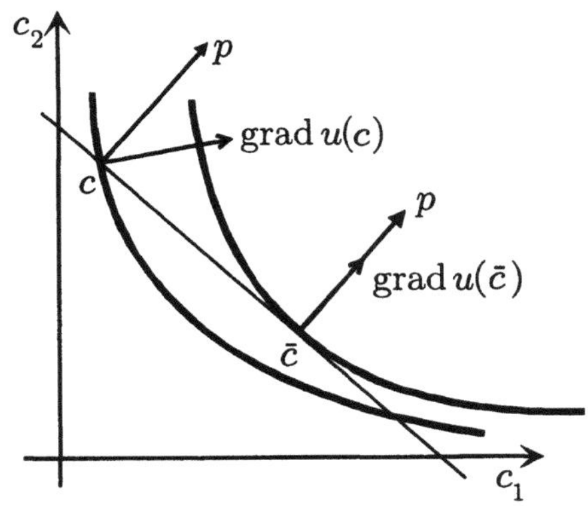

Figure 1.1.C: *A consumption, c, is a solution to the Consumer Problem if and only if c belongs to the budget hyperplane and* grad $u(c)$ *is proportional to the price vector p. The consumption* $\bar{c}$ *is a solution to the Consumer Problem*

Wealth given by the value of the endowment

The solution to the Consumer Problem, with an externally given wealth, can be used to derive some properties of the solution to the Consumer Problem, where the consumer's wealth is given by the value of the initial endowment so that $w = pe$.

Corollary 1.1.C *Let* $p \in \mathbb{R}^n_{++}$. *The Consumer Problem*

$$Max_c\, u(c) \quad s.\ to\ c \in B(p, pe) \qquad \textbf{1.1.D}$$

has a unique solution $\hat{f}(p, pe)$ *where* $\hat{f}$ *is the function from Proposition* 1.1.B. *The solution is a differentiable function of p.*

Proof: Existence and uniqueness of a solution follows by applying Proposition 1.1.B with $w = pe$. The mapping $p \longrightarrow \hat{f}(p, pe)$ is the composition of the linear, hence differentiable, mapping $p \longrightarrow (p, pe)$ and the differentiable mapping $(p, w) \longrightarrow \hat{f}(p, w)$ so that we have

$$p \longrightarrow (p, pe) \longrightarrow \hat{f}(p, pe)$$

Since the composition of differentiable mappings is a differentiable mapping, the mapping $p \longrightarrow \hat{f}(p, pe)$ is a differentiable mapping.

□

The value of the function $\hat{f}$ at prices p, $\hat{f}(p, pe)$, is the consumer's Walrasian demand. We should add that the Walrasian demand function has further properties. These are thoroughly discussed in most expositions on microeconomic theory; for example Mas-Colell et al. [1995], Balasko [1988] or Mas-Colell [1985].

1.2 Walras Equilibrium and Welfare Economics

Economies and allocations

Consider a set of consumers, $\mathbb{I} = \{1, 2, ..., I\}$, making up an economy. In the definition below the initial endowment for the economy is assigned to the individual agents so we have what is referred to as an economy with private ownership. Since we study only economies with private ownership we refer to them simply as "economies" with no mention of the "private ownership".

Definition 1.2.A *An* **economy** $\mathcal{E}$ *is a tuple* $(C^i, u^i, e^i)_{i \in \mathbb{I}}$ *where, for* $i \in \mathbb{I}$, (C^i, u^i, e^i) *is a consumer.*

Occasionally we study economies with production. Then we include in the definition of the economy a description of the production possibilities and of the ownership, among the consumers, of the producers. We postpone the introduction of these concepts until they are needed.

In the definition of an allocation for an economy $\mathcal{E}=(C^i, u^i, e^i)_{i \in \mathbb{I}}$ we include among the constraints that the consumptions making up the allocation should be individually feasible and that there should be equality between what is used up and what is available in the economy.

Definition 1.2.B *Let* $\mathcal{E} = (C^i, u^i, e^i)_{i \in \mathbb{I}}$ *be an economy. An* **allocation**, *for* $\mathcal{E}$, *is an* I*-tuple of vectors* $(c^i)_{i \in \mathbb{I}}$ *where,*

(a) *for* $i \in \mathbb{I}$, $c^i \in C^i$

(b) $c^1 + c^2 + \ldots + c^I = e^1 + e^2 + \ldots + e^I$

Walras equilibrium

Now we can define a Walras equilibrium. It is a price system and a set of "best" actions relative to the price system with the property that the actions are also compatible with the total initial endowment in the economy.

Definition 1.2.C *A* **Walras equilibrium,** *for* $\mathcal{E} = (C^i, u^i, e^i)_{i \in \mathbb{I}}$, *is a pair* $((\bar{c}^i)_{i \in \mathbb{I}}, p)$ *where* $p \in \mathbb{R}^n_{++}$ *such that*

(a) *for each* $i \in I$, $\bar{c}^i$ *is a solution to:* $Max_c\, u^i(c)$ *s. to* $c \in B^i(p, pe^i)$

(b) $(\bar{c}^i)_{i \in \mathbb{I}}$ *is an allocation*

If $((\bar{c}^i)_{i \in \mathbb{I}}, p)$ is a Walras equilibrium then $(\bar{c}^i)_{i \in I}$ is the (Walras) *equilibrium allocation* and p the (Walras) *equilibrium price system.*

More generally, given an economy, $\mathcal{E}$, an allocation $(\bar{c}^i)_{i \in \mathbb{I}}$ is a (Walras) equilibrium allocation if there exists a price system p such that $((\bar{c}^i)_{i \in \mathbb{I}}, p)$ is a Walras equilibrium and p is a (Walras) equilibrium price system if there exists an allocation, $(\bar{c}^i)_{i \in \mathbb{I}}$, such that $((\bar{c}^i)_{i \in \mathbb{I}}, p)$ is a (Walras) equilibrium. We will use a similar terminology for spot-market equilibria. The context and the equilibrium concept used will determine the meaning of "equilibrium allocation" and "equilibrium price system". We have the following

Theorem 1.2.D (Existence of a Walras Equilibrium) *Let* $\mathcal{E} = (C^i, u^i, e^i)_{i \in \mathbb{I}}$ *be an economy. There exists a Walras equilibrium* $((\bar{c}^i)_{i \in \mathbb{I}}, p)$ *for* $\mathcal{E}$.

If p is an equilibrium price system then $[(\hat{f}^i(p, pe^i))_{i \in \mathbb{I}}, p]$ is a Walras equilibrium. A proof of existence of a Walras equilibrium would take us to far afield and we refer the reader to Mas-Colell et al. [1995]. Here we merely note that the above theorem asserts that the Maintained Assumptions are sufficient to ensure the existence of a Walras equilibrium.

The First and Second Theorem of Welfare Economics

Pareto optimal allocations

The First and Second Theorem of Welfare Economics relate Walras equilibrium allocations and Pareto optimal allocations.

Definition 1.2.E *Let* $\mathcal{E} = (C^i, u^i, e^i)_{i \in \mathbb{I}}$ *be an economy and let* $(\bar{c}^i)_{i \in \mathbb{I}}$ *and* $(\hat{c}^i)_{i \in \mathbb{I}}$ *be allocations for* $\mathcal{E}$. *The allocation* $(\hat{c}^i)_{i \in \mathbb{I}}$ **Pareto dominates** $(\bar{c}^i)_{i \in \mathbb{I}}$ *if*

$$\text{for each } i \in \mathbb{I},\ u^i(\hat{c}^i) \geq u^i(\bar{c}^i) \text{ and for some } i \in \mathbb{I},\ u^i(\hat{c}^i) > u^i(\bar{c}^i)$$

An allocation $(\bar{c}^i)_{i \in \mathbb{I}}$ *is a* **Pareto optimal allocation** *for* $\mathcal{E}$ *if there is no allocation for* $\mathcal{E}$ *that Pareto dominates* $(\bar{c}^i)_{i \in \mathbb{I}}$.

The first theorem of welfare economics

The First Theorem of Welfare Economics states that the (Walras) equilibrium allocations have at least one desirable property; they are Pareto optimal allocations. Hence there is at an equilibrium allocation a conflict between the satisfaction of different consumers. The satisfaction of one consumer can be increased only at the expense of the satisfaction of some other consumer.

Theorem 1.2.F (The First Theorem of Welfare Economics) *Let* $\mathcal{E} = (C^i, u^i, e^i)_{i \in \mathbb{I}}$ *be an economy and* $((\bar{c}^i)_{i \in \mathbb{I}}, p)$ *a Walras equilibrium for* $\mathcal{E}$. *Then the equilibrium allocation* $(\bar{c}^i)_{i \in \mathbb{I}}$ *is a Pareto optimal allocation for* $\mathcal{E}$.

Proof: Consider a consumer $i \in \mathbb{I}$. From (a) of Proposition 1.1.B we have $p\bar{c}^i = pe^i$.
Choose any c^i such that $u^i(c^i) \geq u^i(\bar{c}^i)$. By (c) of Lemma 1.1.A we have $Du^i(\bar{c}^i)c^i \geq Du^i(\bar{c}^i)\bar{c}^i$ with strict inequality if $c^i \neq \bar{c}^i$. By (b) of Proposition 1.1.B $Du^i(\bar{c}^i)$ is proportional to p. Hence

$$u^i(c^i) \geq u^i(\bar{c}^i) \text{ implies } pc^i \geq p\bar{c}^i \quad \text{and}$$

$$u^i(c^i) > u^i(\bar{c}^i) \text{ implies } pc^i > p\bar{c}^i$$

Assume, in order to arrive at a contradiction, that $(\bar{c}^i)_{i\in\mathbb{I}}$ is not a Pareto optimal allocation for $\mathcal{E}$. Then there is an allocation, $(\hat{c}^i)_{i\in\mathbb{I}}$, for $\mathcal{E}$, which Pareto dominates $(\bar{c}^i)_{i\in\mathbb{I}}$. Without loss of generality we may assume

$$\begin{aligned} u^1(\hat{c}^1) &> u^1(\bar{c}^1) \quad \text{and} \\ u^i(\hat{c}^i) &\geq u^i(\bar{c}^i) \quad \text{for } i \in \mathbb{I} \end{aligned}$$

It follows that

$$\begin{aligned} p\hat{c}^1 &> p\bar{c}^1 = pe^1 \quad \text{and} \\ p\hat{c}^i &\geq p\bar{c}^i = pe^i \quad \text{for } i = 2, 3, ..., I \end{aligned}$$

Summing these inequalities we get

$$\sum_{i\in\mathbb{I}} p\hat{c}^i > \sum_{i\in\mathbb{I}} pe^i$$

On the other hand, since $(\hat{c}^i)_{i\in\mathbb{I}}$ is an allocation for $\mathcal{E}$ we have

$$\sum_{i\in\mathbb{I}} \hat{c}^i = \sum_{i\in\mathbb{I}} e^i$$

which implies

$$\sum_{i\in\mathbb{I}} p\hat{c}^i = \sum_{i\in\mathbb{I}} pe^i$$

This contradicts the inequality above and it follows that our hypothesis is false. Hence $(\bar{c}^i)_{i\in\mathbb{I}}$ is a Pareto optimal allocation.

The second theorem of welfare economics

Note that for an equilibrium allocation, $(\bar{c}^i)_{i\in\mathbb{I}}$, we have by Proposition 1.1.B, for $i \in \mathbb{I}$, with $w^i = pe^i$ and some $\lambda^i > 0$

$$\begin{cases} \operatorname{grad} u^i(\bar{c}^i) - \lambda^i p = 0 \\ p\bar{c}^i - pe^i = 0 \end{cases}$$

The gradient shows a consumer's subjective (marginal) evaluation of the commodities. The prices can be interpreted as the "market evaluation". In equilibrium each of the subjective evaluations are equalized and equal to the "market evaluation". This is no surprise since if a consumer's subjective evaluation did not agree with the "market evaluation" the consumer could gain by varying her consumption slightly, contradicting that we had an equilibrium.

The Second Theorem of Welfare Economics is concerned with an economy, which uses some *unspecified* institutions to allocate commodities among consumers. The point of departure is a set of institutions which results in an "equilibrium" allocation which is "good" in the sense that it is a Pareto optimal allocation. Then, as the conclusion of the theorem asserts, the allocation could also be achieved by a redistribution of the initial resources and the use of a price system. This is illustrated in Figure 1.2.A.

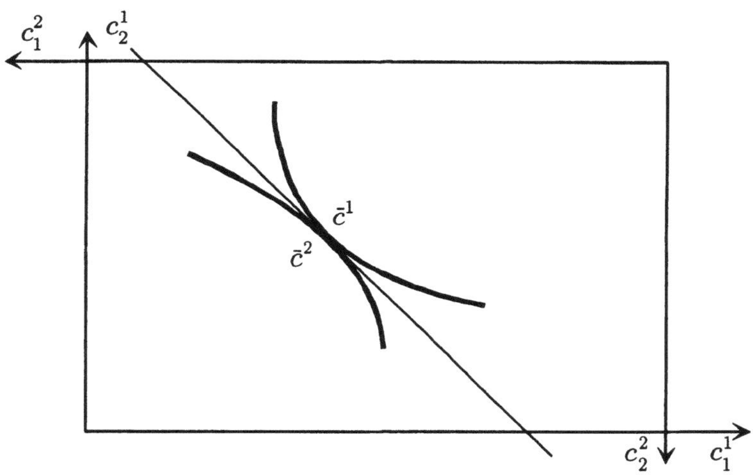

Figure 1.2.A: *$(\bar{c}^1, \bar{c}^2)$ is a Walras equilibrium allocation for any economy which has the initial endowment on the (common) budget hyperplane*

Theorem 1.2.G (The Second Theorem of Welfare Economics) *Let $\mathcal{E} = (C^i, u^i, e^i)_{i \in \mathbb{I}}$ be an economy and $(\bar{c}^i)_{i \in \mathbb{I}}$ a Pareto optimal allocation for $\mathcal{E}$. Then there is $p \in \mathbb{R}^n_{++}$ such that $((\bar{c}^i)_{i \in \mathbb{I}}, p)$ is a Walras equilibrium for $\mathcal{E}' = (C^i, u^i, \bar{c}^i)_{i \in \mathbb{I}}$.*

Proof: Let $(u^i(\bar{c}^i))_{i \in \mathbb{I}}$ be the utility vector of the given Pareto optimal allocation. Since the allocation $(\bar{c}^i)_{i \in \mathbb{I}}$ is a Pareto optimal allocation it follows that $(\bar{c}^i)_{i \in \mathbb{I}}$ is a solution to the following constrained maximization problem (on the far right we have given the Lagrange multipliers to be introduced below).

$$\text{Max}_{(c^i)_{i\in\mathbb{I}}} \quad u^1(c^1) \quad \text{s. to } c^i \in C^i \quad \text{and} \qquad\qquad \text{Multiplier}$$

$$\begin{cases} u^i(c^i) - u^i(\bar{c}^i) = 0 \quad \text{for } i = 2,\ldots,I & \alpha^i \\ \Sigma_{i\in\mathbb{I}}\, c^i_l - \Sigma_{i\in\mathbb{I}} e^i_l = 0 \quad \text{for } l = 1,2,...,n & \lambda_l \end{cases}$$

Lagrange's theorem, Theorem B in Appendix A, asserts that there are multipliers α^i, for $i = 2,..,I$, and a vector $\lambda = (\lambda_1, \lambda_2, \ldots, \lambda_n)$ such that

$$\begin{cases} \operatorname{grad} u^1(\bar{c}^1) - \lambda = 0 \\ \alpha^i \operatorname{grad} u^i(\bar{c}^i) - \lambda = 0 \quad \text{for } i = 2,\ldots,I \end{cases}$$

From the first relation follows that $\lambda \in \mathbb{R}^n_{++}$, since the value of each of the gradients is a positive vector by (C3) of the Maintained Assumptions. This in turn implies that α^i, for $i = 2,\ldots,I$, are positive numbers. The vector λ is *independent* of the consumer considered. This suggests that we could use as prices $p \in \mathbb{R}^n_{++}$ the vector of Lagrange multipliers λ. Put $p = \lambda$ to get

$$\begin{cases} \operatorname{grad} u^1(\bar{c}^1) - p = 0 \\ \operatorname{grad} u^i(\bar{c}^1) - \dfrac{1}{\alpha^i} p = 0 \quad \text{for } i = 2,\ldots,I \end{cases} \qquad \textbf{1.2.A}$$

We now use the sufficiency of the marginal conditions 1.1.B from (b) of Proposition 1.1.B. Assign to consumer $i \in \mathbb{I}$, $\bar{c}^i$ as initial endowment. Consumer i's wealth is then given by $w = p\bar{c}^i$ and it follows from 1.2.A that $\bar{c}^i$ is a solution to consumer i's problem, at (p, w). Since $(\bar{c}^i)_{i\in\mathbb{I}}$ is an allocation for the economy $\mathcal{E}' = (C^i, u^i, \bar{c}^i)_{i\in\mathbb{I}}$ it follows that $((\bar{c}^i)_{i\in\mathbb{I}}, p)$ is a Walras equilibrium for $\mathcal{E}'$.

□

Figure 1.2.A illustrates that the Pareto optimal allocation $(\bar{c}^i)_{i\in\mathbb{I}}$ is a Walras equilibrium allocation for the economy $\mathcal{E}' = (C^i, u^i, \bar{c}^i)_{i\in\mathbb{I}}$. In fact it is easy to see that $(\bar{c}^i)_{i\in\mathbb{I}}$ is a Walras equilibrium allocation for any economy where the initial endowments are chosen on the budget hyperplane(s) induced by $(\bar{c}^i)_{i\in\mathbb{I}}$ and the prices associated with that allocation.

The assumptions used in the proof are far from the weakest possible but the proof highlights the equalization of the subjective evaluations,as given by the consumers' gradients, and the "market evaluation", as given by the prices. In the proof the "market evaluation" appears as the Lagrange multipliers for the market balance conditions. The following corollary is useful.

Corollary 1.2.H *Let* $\mathcal{E} = (C^i, u^i, e^i)_{i \in \mathbb{I}}$ *be an economy and* $(\bar{c}^i)_{i \in \mathbb{I}}$ *an allocation for* $\mathcal{E}$. *The allocation* $(\bar{c}^i)_{i \in \mathbb{I}}$ *is a Pareto optimal allocation if and only if each of the gradients of the utility functions, evaluated at* $(\bar{c}^i)_{i \in \mathbb{I}}$, *are proportional to a common vector* $p \in \mathbb{R}^n_{++}$, *that is, there are numbers* $\alpha^i \in \mathbb{R}$, $i \in \mathbb{I}$, *such that*

$$\text{grad } u^i(\bar{c}^i) = \alpha^i p \qquad \textbf{1.2.B}$$

The relation 1.2.B follows from the First and Second Theorem of Welfare Economics. If $(\bar{c}^i)_{i \in \mathbb{I}}$ is a Pareto optimal allocation for $\mathcal{E} = (C^i, u^i, e^i)_{i \in \mathbb{I}}$, then, by the Second Theorem, there is a $\hat{p} \in (\mathbb{R}^L_{++})^{T+1}$ such that $((\bar{c}^i)_{i \in \mathbb{I}}, \hat{p})$ is a Walras equilibrium for $\mathcal{E}' = (C^i, u^i, \bar{c}^i)_{i \in \mathbb{I}}$. Then 1.2.B is satisfied with $p = \hat{p}$ since each consumption is a solution to the Consumer Problem.

Conversely, if 1.2.B is satisfied by some p and α^i, $i \in \mathbb{I}$, then $((\bar{c}^i)_{i \in \mathbb{I}}, p)$ is a Walras equilibrium, for $\mathcal{E}' = (C^i, u^i, \bar{c}^i)_{i \in \mathbb{I}}$, and by the First Theorem of Welfare Economics, $(\bar{c}^i)_{i \in \mathbb{I}}$ is a Pareto optimal allocation.

Summary

We have introduced consumers and given a set of Maintained Assumptions on consumers to be used in the sequel. The assumptions were easy to interpret, apart perhaps from (C5), which amounted to a slight strengthening of the strict quasi-concavity assumption.

The Maintained Assumptions ensured the existence of a unique solution to the Walrasian Consumer Problem and that the induced function, the Walrasian demand function, was a differentiable function of the prices. Thus one of the merits of the assumptions introduced was the justification of the differentiability of demand. The Walrasian demand function has other properties which we have not discussed.

Having shown that individual consumer demand was well behaved we turned to a study of Walras equilibria and Pareto optimal allocations. We noted that the Maintained Assumptions were strong enough to ensure that each economy had at least one Walras equilibrium. Each Walras equilibrium allocation was seen to be a Pareto optimal allocation and each Pareto optimal allocation could be realized as a Walras equilibrium allocation, for an economy where the initial endowments were redistributed in a suitable way.

A consequence of the First and Second Theorem of Welfare Economics was a characterization of Pareto optimal allocations. An allocation was seen to be a Pareto optimal allocation if and only if the subjective consumer evaluations, as given by the gradients, were equalized at the allocation.

Exercises

(**NB!** In the exercises below we do not assume that consumers necessarily satisfy the Maintained Assumptions.)

Exercise 1.A Consider a consumer with consumption set $\mathbb{R}^2_{++}$ and utility function $u : \mathbb{R}^2_{++} \longrightarrow \mathbb{R}$ given by $u(c_1, c_2) = \ln c_1 + \ln c_2$ and initial endowment $e = (e_1, e_2) \in \mathbb{R}^2_{++}$.

(a) Which of the Maintained Assumptions are satisfied?

(b) Calculate the gradient at the point $\bar{c} = (2, 1)$ and find the tangent hyperplane to the upper contour set at that point.

(c) Calculate the Hessian, $D^2u(\bar{c})$, and show that $D^2u(\bar{c})$ defines a quadratic form which is negative definite (and hence negative definite on the homogenous hyperplane defined by $Du(\bar{c})$).

Exercise 1.B Consider a consumer, $(\mathbb{R}^2_{++}, u, e)$ with utility function given by $u(x_1, x_2) = (ax_1^\rho + bx_2^\rho)^{\frac{1}{\rho}}$ where $\rho < 1$, $\rho \neq 0$.

(a) Let $a = b = 1$ and $0 < \rho$. Are the Maintained Assumptions satisfied? (*Hint*: The utility function is homothetic. Consider the indifference class corresponding to utility 1.)

(b) Let $a = b = 1$ and $\rho = -\frac{1}{4}$. Are the Maintained Assumptions satisfied?

Exercise 1.C Consider a consumer, (C, u, e), satisfying the Maintained Assumptions and assume that $u(\mathbb{R}^2_{++}) = \mathbb{R}$. Let $\varphi : \mathbb{R} \longrightarrow \mathbb{R}$ be a strictly increasing, twice differentiable function.

(a) Assume that $\varphi'(x) \geq 0$ for $x \in \mathbb{R}$ and let the preference relation $\succsim$ on C be defined by $c \succsim c'$ if and only if $u(c) \geq u(c')$. Show that the composition of u and φ, denoted $\varphi \circ u$, induces the same preference relation as u.

(b) Assume that $\varphi'(x) \geq 0$ for $x \in \mathbb{R}$ and $\varphi'(0) = \varphi''(0) = 0$ (for example, the function with values $\varphi(x) = x^3$). Does the consumer $(C, \varphi \circ u, e)$ satisfy the Maintained Assumptions?

(c) Assume that $\varphi'(x) > 0$ for $x \in \mathbb{R}$. Show that the consumer $(C, \varphi \circ u, e)$ satisfies the Maintained Assumptions.

Exercise 1.D Consider a consumer, (C, u, e), with $C = \mathbb{R}^2_{++}$, $e \in \mathbb{R}^2_{++}$ and $u(c_1, c_2) = c_1^\alpha c_2^\beta$ where $\alpha > 1$ and $\beta > 1$.

(a) Show that $D^2u(1,1)$ does not define a negative definite quadratic form on $\mathbb{R}^2$.

(b) Show that $D^2u(1,1)$ defines a negative definite quadratic form on $H = \{h \in \mathbb{R}^2 \mid Du(1,1)h = 0\}$.

(c) Show that the function $\ln \circ u$ defines the same preferences as u and that $\ln \circ u$ is strictly concave. Does the matrix of second order derivatives of $\ln \circ u$ induce a negative definite quadratic form (and hence a quadratic form negative definite on the homogenous hyperplane defined by $D(\ln \circ u(1,1))$?

Exercise 1.E Consider a consumer, $(\mathbb{R}^2_{++}, u, e)$ with utility function given by $u(c_1, c_2) = \alpha \ln c_1 + \beta \ln c_2$ where α, β are positive numbers and $e \in \mathbb{R}^2_{++}$.

(a) Find demand as a function of prices and wealth.

(b) Find demand as a function of prices when wealth is given by the value of the initial endowment.

(c) Find the desired net trade of the consumer given prices. Find prices such that the desired net trade is equal to 0.

(d) Find the gradient of u at e. Compare with the prices found in (d).

Exercise 1.F Consider an economy, $\mathcal{E}$, with two consumers, a and b, satisfying the Maintained Assumptions with utility functions

$$\begin{aligned} u^a(c_1, c_2) &= \alpha \ln c_1 + (1-\alpha) \ln c_2 \\ u^b(c_1, c_2) &= (1-\alpha) \ln c_1 + \alpha \ln c_2 \end{aligned}$$

where $0 < \alpha < 1$, and initial endowments $e^a = (2,1)$ and $e^b = (1,2)$.

(a) Find the aggregate excess demand for the economy.

(b) Show that there is (apart from normalization) a single equilibrium price system for the economy. What is the equilibrium allocation?

(c) Find all the initial endowments, $\hat{e}^a$ and $\hat{e}^b$, such that the equilibrium found in (b) is an equilibrium for the economy differing from $\mathcal{E}$ only in that the initial endowments are $\hat{e}^a$ and $\hat{e}^b$.

(d) Find the set of Pareto optimal allocations for the economy.

Exercise 1.G The following is an example from Mas-Colell et al. [1995, p 521] of an economy with three Walras equilibria. It will be used as a starting point for an example in Chapter 6.
Let $\mathcal{E} = (C^i, u^i, e^i)_{i \in \{a,b\}}$ be an economy with commodity space $\mathbb{R}^2$ where

$$\begin{aligned} C^a &= \mathbb{R}^2_{++}, \quad e^a = (2, r) \quad \text{and} \quad u^a(c_1, c_2) = c_1 - \frac{1}{8} c_2^{-\frac{1}{8}} \\ C^b &= \mathbb{R}^2_{++}, \quad e^b = (r, 2) \quad \text{and} \quad u^b(c_1, c_2) = c_2 - \frac{1}{8} c_1^{-\frac{1}{8}} \end{aligned}$$

where $r = 2^{8/9} - 2^{1/9}$.

(a) Do the consumers satisfy the Maintained Assumptions?

(b) Show that u^a is a concave function but not a strictly concave function. Is u a strictly quasi-concave function?

(c) Find the gradients of u^a and u^b.

(d) Show that the economy has the following three Walras equilibria

Price system		Equilibrium allocation	
		Good 1	Good 2
$p = (1,1)$	$\bar{c}^a =$	$1 + r$	1
	$\bar{c}^b =$	1	$1 + r$
$p = (2,1)$	$\bar{c}^a =$	$2 + 2^{-1/9} - 2^{1/9}$	$2^{1/9}$
	$\bar{c}^b =$	$2^{-1/9}$	$2 + 2^{17/9} - 2^{10/9} - 2^{8/9}$
$p = (1,2)$	$\bar{c}^a =$	$2 + 2^{17/9} - 2^{10/9} - 2^{8/9}$	$2^{-1/9}$
	$\bar{c}^b =$	$2^{1/9}$	$2 + 2^{-1/9} - 2^{1/9}$

Hint: Show (i) that the market balance conditions are satisfied, (ii) that the gradients are proportional to the price system at the allocations given above and (iii) that the wealths of consumer a and b agree with the value of the initial endowments.

Chapter 2

ECONOMIES OVER TIME

Introduction

In Chapter 1 we gave a review of some of the results for the Arrow-Debreu economy, but did not enter into the interpretation when deliveries of goods could occur at different dates. In this chapter we study economies over time and consider the interpretation of the model in this case. The reasoning, definitions and results of this chapter carry over to economies with uncertainty, which are studied in Chapter 4 to 6. Thus the delimitation of this chapter and Chapter 4 to 6 is somewhat arbitrary.

The *naive* interpretation of the Arrow-Debreu economy studied in Chapter 1 involves no time. We can imagine the consumers to meet, at a certain date and at a certain location, to exchange goods. Equilibrium is characterized by a price system and a set of consumptions which are "best" at the equilibrium prices with the property that the consumptions are compatible, in total, with the endowments.

Economists have for long been working with allocation problems over time, in space and under uncertainty. In the course of this work new and more sophisticated interpretations of the Arrow-Debreu model have arisen. Many of the results for the model are given in a *formal model* as results concerning mathematical objects. For example, taste is in the theory represented by a preordering which to a mathematician is a relation with certain properties. Often this predordering is in turn represented by a utility function. The interest in the formal results stems ultimately from some agreement between the formal objects of the theory and their empirical counterparts.

Modelling the Arrow-Debreu economy as a theory about mathematical objects means that new interpretations can be accommodated in the same formal model. Simply by *reinterpreting* what a "commodity" is we get a theory of resource allocation over time, a theory of location or a theory of resource allocation under uncertainty. We discuss the interpretation of the commodity concept in Section 2.1. New interpretations force us to *reconsider the assumptions* of the theory. For example, convexity assumptions regarding production possibilities may seem unreasonable when goods are distinguished according to their place of delivery. The theory also uses a particular notion of equilibrium; Walras equilibrium, to give an explanation of relative prices or exchange rates between goods. With new interpretations the equilibrium concept may seem less convincing.

The Arrow-Debreu model is not explicit about the *institutions* used in the economy and a Walras equilibrium may arise from different sets of institutions. We give two examples of institutions leading to a Walras equilibrium over time.

Firstly, in Section 2.2 we consider the case of *complete (commodity) markets at date 0.* Here each commodity can be bought and sold at date 0 for delivery at any date, at a price determined in 0-crowns. Thus buying and selling at date 0 amounts to entering into contracts for deliveries at different dates.

Secondly, in Section 2.3 we study the case where there is a *spot-market for goods* at each date and a market for *spot income at date 0; the current date.* Hence the consumers enter into contracts about deliveries of t-crowns for each of the future dates and there is for each future date a date 0 price for income to be delivered at that date. The price at the current date for income to be delivered at a future date is denominated in 0-crowns. The consumers are assumed to have common expectations which amount to perfect foresight about the goods prices that will prevail at the future dates. They can at the current date consume and trade income to be delivered at future dates and *plan* their consumption for the future dates. If price expectations are correct then the expected spot prices will be realized in the future spot-markets and each consumer will be able to carry out the consumption plan decided on at date 0.

These interpretations point to the implicit assumption in the model of unlimited possibilities of income transfers between dates but under the first interpretation there is no need to transfer income between dates and the assumption is implicit. Under the second interpretation there are maximal possibilities of income transfers. A consumer can, disregarding the date 0 cost, buy any future net income pattern. In both cases the consumer can act restricted by a single budget constraint defined by prices or discounted prices and her *wealth.*

Indeed it will be seen that when the possibilities of income transfers are maximal then there is an implicit Walras equilibrium, induced by a spot-market equilibrium. At the other extreme is a pure spot-market equilibrium where there are no possibilities of income transfers

In the theory of general equilibrium with *incomplete (asset) markets*, situations are studied where the possibilities of income transfers are more or less restricted. To do so we have to be explicit about the institutions which are used in the economy to make transfers of "purchasing power" possible. In this chapter we approach the subject of income transfers by allowing the agents to trade at the current date spot income to be used at future dates but we also explicitly allow for the case when there are restrictions on the exchange of spot income at date 0.

It turns out that in such a situation it is useful at first to study consumer choice in the extreme case where there are no possibilities of income transfers. Given net income vector and spot prices we define in Section 2.4 the *spot-market demand.* which gives the planned commodity demand at each date.

Using the spot-market demand function we define the *indirect utility function* which for given prices depicts the consumer's preferences regarding net income vectors. We then show in Section 2.6 how to *decompose* the consumer's decision problem into a choice of a net income vector at the current date and the (planned) choice of a goods bundle in each of the spot-markets. The restricted possibilities to transfer income between dates are reflected in restrictions pertinent to the choice of a net income vector at the current date.

The decomposition of the consumer problem suggests a definition of a *spot-market equilibrium relative to a set of income transfers*, given in Section 2.7. Central to this definition is the set of net income vectors available to the consumers. If this set owns only the zero vector we get a pure spot market equilibrium. On the other hand if the possibilities of income transfers are maximal then the spot-market equilibrium generates a Walras equilibrium.

In Chapter 3 and 4 we will specify in more detail the institutions used to allow the agents to make income transfers, between dates or "states of Nature". There the income transfers stem from the choice of a *portfolio*, that is, a bundle of assets paying dividends at future dates. These are the means available to obtain a suitable net income vector.

The restrictions on the choice of net income vectors is at the heart of the theory of incomplete markets. In view of this we thoroughly discuss in Section 2.8 how the more or less limited trading possibilities at date 0 may prevent the equalization of the consumers' relative evaluations of income at different dates. When the set of available net income vectors is a linear subspace then, in an equilibrium, the relative evaluations all belong to the subspace orthogonal to the space of available income transfers. We also show that it is precisely the limitations on exchange of future net income that may give rise to equilibrium allocations which are not Pareto optimal allocations.

2.1 Goods and Commodities

Physical characteristics, date and place of delivery

It is useful to make a distinction between a *good* and a *commodity*. A good is defined by its physical characteristics. The description of these characteristics may be more or less detailed. At times it is very important to have a careful description of the physical characteristics. For example, when there is trade in forward delivery buyer and seller may disagree, at the date of delivery, about whether a contract is met or not. In particular, they may disagree about whether the delivered good has the physical characteristics agreed upon at the time of contracting.

We assume that there is only a finite number of physical descriptions, that is, goods.

To a consumer or producer matters not only the physical characteristics of the objects which are exchanged. The *time (date)* of delivery and the *location* of delivery is usually of importance. For a producer it is evidently impossible to use a good delivered at some date as input at an earlier date. To a consumer a pizza delivered at her home, at some date, may be more valuable than a pizza delivered at some other location at the same date.

We assume that there is a finite number of dates and locations.

Commodities

The examples show that economic agents are concerned not only with the physical properties of the goods but also with the date when they are delivered and the location of delivery. A *commodity* is a specific good delivered at a certain date and at a certain location. The number of commodities will thus be the product of the number of goods, the number of dates and the number of locations for delivery. (In Chapter 4 and onwards delivery may also be contingent on the "state of Nature" and this state will then be regarded as part of the description of a commodity.)

The assumption that there is a finite number of goods, dates and locations for delivery implies that the number of commodities is finite. Since we are not be interested in location theory we assume that there is a single location for delivery. In this case, a commodity is (uncertainty disregarded) a good delivered at a specific date. Thus in Chapter 2 and 3, the number of commodities is the product of the number of goods and the number of dates.

When we consider economies without production it makes sense to think of the goods as perishable so that they can not be stored between dates. In fact, storage is a kind of production activity, for example using a good as input at date t to produce the same good as output at date $t+1$. One of the reasons we confine the exposition mainly to exchange economies is to avoid the additional complications introduced by considering economies with production.

The attributes of a commodity, as described above, do not always suffice for the purpose at hand. If the commodity concept captures the essential characteristics each consumer should be indifferent between any two units of a given commodity. At times it would be natural to extend or modify the commodity concept. A case in point is "green electricity". Although a consumer may not perceive of any difference in the physical characteristics between "green electricity" and conventional electricity, these are not necessarily perfect substitutes to the consumer. A consumer may prefer one unit of "green electricity" to one unit of conventional electricity. In such a situation we may want to include in the description of a commodity something about how it was produced or by whom it was delivered.

2.2 A Market for all Commodities at Date 0

Goods and dates

We now give a first interpretation of the general equilibrium model over time. This interpretation considers an economy with institutions far removed from what is observed in everyday life but it is very useful as a starting point for discussing economies over time.

We assume that there are dates $0, 1, 2, \ldots, T$ and we let

- $\mathbb{T} = \{0, 1, 2, \ldots, T\}$ be the *set of dates* and
- $\mathbb{T}_1 = \{1, 2, \ldots, T\}$ be the *set of future dates.*

It is convenient to refer to date 0 also as the *current date.* At each date there are L goods, where $L \geq 1$, and we denote by

- $\mathbb{L} = \{1, 2, \ldots, L\}$ the *set of goods.*

Since we are assuming that there is only one location we have $L(T+1)$ commodities and the commodity space is $(\mathbb{R}^L)^{T+1} = \mathbb{R}^{L(T+1)}$. We have introduced some of the notation in Table 2.2.A.

Table 2.2.A: *Notation for the spot market interpretation*

	Endowment	Consumption	Date 0 prices	Discount factor	Spot prices
Date					
0	$e(0)$	$c(0)$	$P(0)$	1	$p(0)$
1	$e(1)$	$c(1)$	$P(1)$	$\beta(1)$	$p(1)$
2	$e(2)$	$c(2)$	$P(2)$	$\beta(2)$	$p(2)$
$\vdots$	$\vdots$	$\vdots$	$\vdots$	$\vdots$	
T	$e(T)$	$c(T)$	$P(T)$	$\beta(T)$	$p(T)$

The consumer problem with a complete commodity market

At date 0 there is a complete market for commodities. Each good may be bought or sold at date 0 for delivery at each date $t \in \mathbb{T}$. This is often described by saying that we have (complete) forward markets.

The date 0 prices of the commodities can be thought of as being denominated in 0-crowns. Since the economy does not use money, or any other medium of exchange, 0-crowns is the unit of account used to debit or credit the consumers as they buy or sell commodities at date 0.

In order to obtain one unit of good $l \in \mathbb{L}$ at date $t \in \mathbb{T}$ the consumer has to give up $P_l(t)$ 0-crowns. For the promise to deliver one unit of good $l \in \mathbb{L}$ at date $t \in \mathbb{T}$ the consumer receives $P_l(t)$ 0-crowns.

Since there is a complete market for *all* commodities, at date 0, the consumer can calculate her *gross wealth* in 0-crowns. This is the amount of 0-crowns the consumer can obtain at date 0 by selling her initial endowment at date 0 and the future dates. Let w denote the consumer's gross wealth. Then

$$w = P(0)e(0) + P(1)e(1) + \ldots + P(T)e(T)$$

where

$$P(t)e(t) = \sum_{l \in \mathbb{L}} P_l(t)e_l(t) \quad \text{for } t \in \mathbb{T}$$

The consumer uses her gross wealth to secure deliveries of the commodities at date 0 to date T. The cost, at date 0 in 0-crowns, of the *consump-*

tion $(c(0), c(1), \ldots, c(T))$ is

$$P(0)c(0) \;+\; P(1)c(1) \;+\; \ldots \;+\; P(T)c(T)$$

The consumer is subject to a single *budget restriction.* The consumer is restricted to choose a consumption whose date 0 cost is no larger than her wealth. Hence c has to satisfy

$$P(0)c(0) \;+\; P(1)c(1) \;+\; \ldots \;+\; P(T)c(T) \;\leq\; w$$

In case w is not generated by the initial endowment then w may be too small for to be any consumption which satisfies the consumer's needs, that is, a consumption $c \in C$, which also satisfies the budget restriction. For the case considered here when w is given by the value of the initial endowment the budget restriction will, however, always be satisfied by the initial endowment which by the Maintained Assumptions belongs to C.

If, on the other hand, w is large enough then there may be many consumptions satisfying the consumer's needs and the budget restriction. The consumer's choice is in this case determined by her taste as expressed by her utility function. Thus the Consumer Problem is

$$\begin{aligned} \text{Max}_c \quad & u(c(0), c(1), \ldots, c(T)) \quad \text{s. to } c \in C \text{ and} \\ & P(0)c(0) \;+\; P(1)c(1) \;+\; \ldots \;+\; P(T)c(T) \;\leq\; w \end{aligned} \qquad \textbf{2.2.A}$$

The consumer problem as a choice of net trade

We have assumed that the consumer sells *all* of her initial endowment and uses the proceeds to buy a consumption. In fact he will then be buying back some of the commodities which he sold earlier. With the initial endowment $e = (e(0), e(1), \ldots, e(T))$ and the choice of consumption $c = (c(0), c(1), \ldots, c(T))$ the *net trade* of the consumer is $z = c - e = (c(t) - e(t))_{t \in \mathbb{T}} = (z(t))_{t \in \mathbb{T}}$. If, for some $t \in \mathbb{T}$ and $l \in \mathbb{L}$, $z_l(t) = c_l(t) - e_l(t) > 0$ then the planned consumption of good l at date t is larger than her initial endowment of that commodity, and there will be a (net) delivery to the consumer at date t of good l. If $z_l(t) = c_l(t) - e_l(t) < 0$ the consumer will consume less than the initial endowment of the commodity and there is a (net) delivery from the consumer. Thus positive coordinates, in the net trade, correspond to deliveries *to* the consumer and negative coordinates correspond to deliveries *from* the consumer. Cf. Figure 2.2.A.

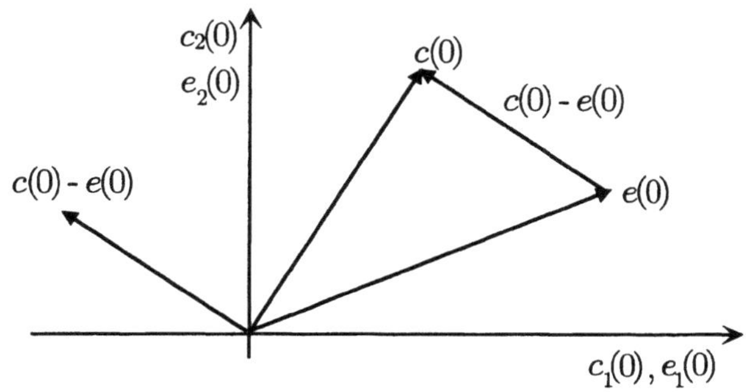

Figure 2.2.A: *The vector* $c(0) - e(0)$ *is the net trade at date 0. It gives the deliveries to and from the consumer*

If the consumer's wealth is given by the value of the initial endowment then using the net trade the budget restriction can be rewritten

$$P(0)z(0) + P(1)z(1) + \ldots + P(T)z(T) \leq 0 \qquad \textbf{2.2.B}$$

and the Consumer Problem can be reformulated as a problem of choosing a net trade. Thus

$$\text{Max}_z \; u([e(t) + z(t)]_{t\in\mathbb{T}}) \quad \text{s. to } z \in C - \{e\} \text{ and } \textstyle\sum_{t\in\mathbb{T}} P(t)z(t) \leq 0$$

Note that the net trade, $z = c - e$, belongs to $C - \{e\}$ if and only if the associated consumption, $c = z + e$, belongs to C.

Walras equilibrium

Let us consider a group of consumers, who make up an economy, $\mathcal{E} = (C^i, u^i, e^i)_{i\in\mathbb{I}}$. What is a Walras equilibrium for this economy? By the definition we should have prices $(P(0), P(1), \ldots, P(T))$ and solutions to each Consumer Problem, at these prices and with the wealth given by the value of the initial endowment. The solutions to the Consumer Problem are given by the Walrasian demand functions $\hat{f}^i$, $i \in \mathbb{I}$, from Corollary 1.1.C. Let

$$\vec{c}^i = \hat{f}^i \left(P(0), P(1), \ldots, P(T), \textstyle\sum_{t\in\mathbb{T}} P(t)e^i(t)\right)$$

The prices make up an equilibrium price system if the solutions to the consumers' problems make up an allocation so that in particular we have market balance. The market balance condition is a condition of equality between two vectors in $(\mathbb{R}^L)^{T+1}$

$$\sum_{i\in\mathbb{I}} \bar{c}^i = \sum_{i\in\mathbb{I}} e^i$$

or written out in more detail, as $T+1$ equalities between vectors in $\mathbb{R}^L$

$$\begin{aligned} \sum_{i\in\mathbb{I}} \bar{c}^i(0) &= \sum_{i\in\mathbb{I}} e^i(0) \\ \sum_{i\in\mathbb{I}} \bar{c}^i(1) &= \sum_{i\in\mathbb{I}} e^i(1) \\ &\vdots \\ \sum_{i\in\mathbb{I}} \bar{c}^i(T) &= \sum_{i\in\mathbb{I}} e^i(T) \end{aligned}$$

If the price system and the consumptions satisfy the conditions given above then

$$((\bar{c}^i)_{i\in\mathbb{I}}, P) = ((\bar{c}^i(t)_{t\in\mathbb{T}})_{i\in\mathbb{I}}, (P(t))_{t\in\mathbb{T}})$$

is a Walras equilibrium. What we have is just an instance of the general definition; we have only added more structure to prices and consumptions.

2.3 A Spot-Market at Each Date and a Market for Spot Income at Date 0

The interpretation of the Arrow-Debreu model in Section 2.2 is simple but also far removed from institutions actually observed. We observe markets for future deliveries only for a few goods. These markets are used mainly by producers to reduce risk.

In this section we consider an alternative interpretation of the Arrow-Debreu model. The interpretation is closer to the working of the economies we observe but we still have to make strong assumptions regarding expectations.

Spot-markets and expectations

Assume that at the current date it is known that there will be a spot-market for the L goods at the current date and each of the future dates.

Goods bought and sold on the market at date t are to be paid in t-crowns which is the unit of account used for purchases and sales at date t. The consumer now has to meet $T+1$ budget restrictions; one for each date. If there were no means of transferring purchasing power between dates the consumer would be constrained to an expenditure at each date agreeing with the value of the initial endowment at the same date.

To plan at date 0 the consumers have to form expectations about the prices that will prevail in the future spot-markets. We make the highly simplifying assumption that the consumers have common and correct expectations about future spot prices. Thus consumers plan at date 0 against the expected future spot prices perceiving only a single vector of spot prices to be possible at each future date. In equilibrium these plans give market balance at date 0. At date 1 there is the possibility for the consumers to reconsider their plans. If they hold on to their original expectations then the realized spot prices at date 1 will be equal to what was expected, at date 0, and the prices and expected prices will give market balance at date 1. Furthermore, the consumers' date 1 plans for dates $2, \ldots, T$ will coincide with their original plans for these dates. Let the expected spot prices be $p(t) \in \mathbb{R}^L_{++}$ for date $t \in \mathbb{T}_1$.

A market for spot income at date 0

Assume that at date 0 there is besides the goods market a market for buying and selling t-crowns for delivery at date t, $t \in \mathbb{T}_1$. Let $\beta(t), t \in \mathbb{T}_1$, be the price, in 0-crowns, of 1 t-crown to be delivered at date t. These prices are often referred to as *discount factors.*

Assume that the consumer at date 0 chooses the amounts $(r(1), \ldots, r(T))$ of t-crowns, for $t \in \mathbb{T}_1$. Here $r(t) > 0$ implies a delivery of t-crowns *to* the consumer and $r(t) < 0$ a delivery of t-crowns *from* the consumer at date t. To be able to deliver t-crowns the consumer must plan to sell some of her initial endowment of goods on the spot-market at date t.

The consumer has to finance the cost of the future net income vector $(r(1), \ldots, r(T))$ at the current date. The date 0 cost of $(r(1), \ldots, r(T))$

in 0-crowns is

$$\beta(1)r(1) + \beta(2)r(2) + \ldots + \beta(T)r(T)$$

If this is a positive number, the consumer must make a corresponding delivery of 0-crowns, $r(0) < 0$, at the current date. If it is a negative number then the consumer receives the amount $r(0) > 0$ at the current date. Careful consideration of the sign conventions for deliveries to and from the consumer shows that the consumer is bound at date 0 by the restriction

$$\beta(0)r(0) + \beta(1)r(1) + \beta(2)r(2) + \ldots + \beta(T)r(T) = 0$$

where $\beta(0) = 1$, in her choice of a *net income vector*, $r = (r(0), r(1), \ldots, r(T))$. Note that this restriction can be written $r \in H$, where H is the homogenous hyperplane

$$\{r \in \mathbb{R}^{T+1} \mid \beta r = 0\}$$

and where β is the vector $(\beta(0), \beta(1), \beta(2), \ldots, \beta(T)) \in \mathbb{R}^{T+1}_{++}$ with $\beta(0) = 1$.

The consumer problem with spot-markets

The Consumer Problem when there are spot-markets for goods at date $t \in \mathbb{T}$ and a market for spot income at date 0 can now be stated.

$$\begin{aligned} \text{Max}_{(c,r)} \quad u(c(0), c(1), \ldots, c(T)) \quad \text{s. to } c \in C \text{ and} & \\ p(0)(c(0) - e(0)) &\leq r(0) \\ p(1)(c(1) - e(1)) &\leq r(1) \\ &\vdots \\ p(T)(c(T) - e(T)) &\leq r(T) \\ r(0) + \beta(1)r(1) + \beta(2)r(2) + \ldots + \beta(T)r(T) &= 0 \end{aligned} \qquad \textbf{2.3.A}$$

Here $p(0)$ and $\beta = (1, \beta(1), \beta(2), \ldots, \beta(T)$ are known at the current date, whereas the commodity prices $p(t)$, $t \in \mathbb{T}_1$, are expected prices at future dates. For every market there is a corresponding budget restriction. At a solution, $(\bar{c}, \bar{r})$, to the Consumer Problem the *net expenditure vector*, $[p(t)(\bar{c}(t) - e(t))]_{t\in\mathbb{T}}$, equals the net income vector, $[\bar{r}(t)]_{t\in\mathbb{T}}$ by the monotonicity of the utility function.

We would like to compare the consumption choices open to a consumer in the case where each commodity can be traded for forward delivery at the current date and the case with spot-markets and trade in spot income at date 0. We have

Proposition 2.3.A *For* $p \in (\mathbb{R}^L_{++})^{T+1}$ *and* $\beta = (\beta(0), \beta(1), \beta(2), \ldots,$

$\beta(T)) \in \mathbb{R}^{T+1}_{++}$ *with* $\beta(0) = 1$ *let*

$$A = \{c \in (\mathbb{R}^L)^{T+1} \mid c \textit{ satisfies } 2.2.\text{A } \textit{with } P(t) = \beta(t)p(t)\} \quad \textit{and}$$

$$B = \{c \in (\mathbb{R}^L)^{T+1} \mid \textit{there exists } r \in \mathbb{R}^{T+1} \textit{ s.t } (c, r) \textit{ satisfies } 2.3.\text{A }\}$$

Then $A = B$.

Proof: First we show that $B \subset A$. Let c satisfy 2.3.A with $r \in \mathbb{R}^{T+1}$. Then $\sum_{i\in\mathbb{I}} \beta(t)r(t) = 0$ and multiplying the budget restriction in 2.3.A for date t with $\beta(t)$, for $t \in \mathbb{T}$ and adding the budget restrictions we get

$$\sum_{t\in\mathbb{T}} \beta(t)p(t)[c(t) - e(t)] \leq \sum_{t\in\mathbb{T}} \beta(t)r(t)$$

By 2.3.A, $\sum_{t\in\mathbb{T}} \beta(t)r(t) = 0$ and it follows that c satisfies 2.2.B with $P(t) = \beta(t)p(t)$.

Next we show that $A \subset B$. Let c satisfy 2.2.B with $P(t) = \beta(t)p(t)$. For $t \in \mathbb{T}$, define $\hat{r}(t)$ through the relations

$$\beta(t)\hat{r}(t) = \beta(t)p(t)[c(t) - e(t)]$$

Then $p(t)(c(t) - e(t)) = \hat{r}(t)$, for $t \in \mathbb{T}$, and by 2.2.B we have

$$\sum_{t\in\mathbb{T}} \beta(t)p(t)[c(t) - e(t)] \leq 0$$

and thus $\sum_{t\in\mathbb{T}} \beta(t)\hat{r}(t) \leq 0$. Put

$$\begin{cases} r(0) = \hat{r}(0) - \sum_{t\in\mathbb{T}} \beta(t)\hat{r}(t) & \text{and} \\ r(t) = \hat{r}(t) & \text{for } t \in \mathbb{T}_1 \end{cases}$$

Then, $r(0) \geq \hat{r}(0)$ and $\sum_{t\in\mathbb{T}} \beta(t)r(t) = 0$. Using the definition of $(\hat{r}(t))_{t\in\mathbb{T}}$ and $(r(t))_{t\in\mathbb{T}}$ is easy to see that (c, r) satisfies 2.3.A with $r = (r(t))_{t\in\mathbb{T}}$.

Spot-market equilibrium

A spot-market equilibrium is an allocation, net income vectors, spot prices for commodities and prices for spot income, such that the consumptions making up the allocation are solutions to the Consumer Problems and the market for spot income at date 0 balances. In particular, we have market balance for the goods at each date in a spot-market equilibrium. By trading spot income at date 0 the consumers can reallocate spot income at that date and at future dates. In equilibrium the net income vector of each consumer agrees with the consumer's net expenditure vector.

Applying Proposition 2.3.A it is seen that $((\bar{c}^i)_{i\in\mathbb{I}}, (\bar{r}^i)_{i\in\mathbb{I}}, p, \beta)$ is a spot-market equilibrium if and only if $((\bar{c}^i)_{i\in\mathbb{I}}, P)$, with $P(t) = \beta(t)p(t)$, is a Walras equilibrium.

We have just shown that a Walras equilibrium may arise in an economy where consumers have common expectations regarding goods prices on spot-markets and can trade spot income at the current date, to be used in the spot-markets at future dates. The case where $\beta(t) = 1$, for $t \in \mathbb{T}$, corresponds to the case, with spot-markets where the date 0 price of t-crowns is 1, for $t \in \mathbb{T}_1$. Then prices on the forward market, in case each commodity can be traded at the current date, equal the expected spot prices, for the case when there are spot-markets for goods at every date.

In the interpretation with spot-markets a consumer may, disregarding its cost at the current date, acquire any net income vector at future dates. In the sequel we shall see that when this is the case, then there is for a spot-market equilibrium a corresponding Walras equilibrium. Hence Walras equilibria may arise with market institutions different from those discussed in Section 2.2.

2.4 The Spot-Market Demand Function

In Sections 2.2 and 2.3 we gave two examples of institutions both leading to Walras equilibrium allocations. In the second interpretation the possibilities of transferring income between the spot-markets were important.

Since these possibilities were, in a sense, maximal the Walrasian demand was the relevant concept. We will study economies with spot-markets for goods where there are more or less restricted possibilities of transferring income between the spot-markets. With this in mind, it is of interest to try to separate the choice of income transfers between dates from the choice of goods bundles in the spot-markets. To do so we first study the spot-market Consumer Spot-Market Problem with a given net income vector and no possibilities of income transfers. We can then derive the *spot-market demand function.* Using the spot-market demand function we define the *indirect utility function*, which for given prices describes a consumer's preferences over net income vectors. These preferences are generated from the possible uses of income on the spot-markets for goods. It follows that the preferences for net income vectors will depend on the spot prices for goods. Applying the indirect utility function we decompose the general consumer spot-market problem, where the feasible income transfers are assumed to satisfy some arbitrary restrictions, into a first stage of choosing a "best" net income vector and a second stage of finding a solution to the consumer spot-market problem with this net income vector.

Let $p \in (\mathbb{R}^L_{++})^{T+1}$ and let

$$E(p) = \left\{ r \in \mathbb{R}^{T+1} \mid r(t) > -p(t)e(t) \quad \text{for } t \in \mathbb{T} \right\}$$

With a net income vector $r \notin E(p)$, we will have $r(t) + p(t)e(t) \leq 0$ for some $t \in \mathbb{T}$, and there will then be no consumption in C; the consumption set, satisfying the budget restrictions for the consumer.

Let

$$D = \left\{ (p, r) \in (\mathbb{R}^L_{++})^{T+1} \times \mathbb{R}^{T+1} \mid r \in E(p) \right\}$$

Note that $(p, r) \in D$ if and only if there is some $c \in C$ satisfying the consumer's budget restrictions, at prices p and net income vector r.

Consider the problem of choosing a "best" consumption with the net income vector given. We refer to this problem as The Consumer Spot-Market Problem given r.

$$\text{Max}_c \quad u(c(0), c(1), \ldots, c(T)) \quad \text{s. to } c \in C \text{ and}$$
$$\begin{cases} p(0)(c(0) - e(0)) & \leq \; r(0) \\ p(1)(c(1) - e(1)) & \leq \; r(1) \\ & \vdots \\ p(T)(c(T) - e(T)) & \leq \; r(T) \end{cases} \qquad \textbf{2.4.A}$$

The following proposition shows that the Spot-Market Consumer Problem given r determines a differentiable demand function.

Proposition 2.4.A *Let (C, u, e) be a consumer. For $(p, r) \in D$ consider the Consumer Spot-Market Problem given r;* 2.4.A.

(a) *The Consumer Spot-Market Problem given r,* 2.4.A, *has a unique solution $\bar{c}$ and $p(t)(\bar{c}(t) - e(t)) = r(t)$ for $t \in T$.*

(b) *Let $g(p, r)$ be the solution to* 2.4.A. *The function $g : D \longrightarrow (\mathbb{R}^L_{++})^{T+1}$ is a differentiable function.*

(c) *For $\beta = (\beta(0), \beta(1), \ldots, \beta(T)) \in \mathbb{R}^{T+1}_{++}$, let $\hat{p}(t) = \beta(t)p(t)$ and $\hat{r}(t) = \beta(t)r(t)$, for $t \in T$. Then $g(p, r) = g(\hat{p}, \hat{r})$.*

Proof: To prove **(a)** note that for $(p, r) \in D$, the set

$$B(p, r) \;=\; \{c \in C \mid c \text{ satisfies the restrictions in 2.4.A}\}$$

is a non-empty set. Let $\tilde{c} \in B(p, r)$. The set

$$\hat{B}(p, r) \;=\; \{c \in C \mid c \in B(p, r) \text{ and } u(c) \geq u(\tilde{c})\}$$

is a non-empty, convex and compact set. Since u is continuous and strictly quasi-concave there is a unique $\bar{c} \in \hat{B}(p, r)$ such that $u(\bar{c}) \geq u(c)$ for $c \in \hat{B}(p, r)$. It follows that $u(\bar{c}) \geq u(c)$ also for $c \in B(p, r)$ and hence $\bar{c}$ is the unique solution to 2.4.A.

(b) If $\bar{c}$ is a solution to 2.4.A then $\bar{c}$ is a solution to

$$\text{Max}_c \; u(c) \quad \text{s. to } c \in C$$

and the restrictions in 2.4.A, with equalities rather than inequalities.

Thus by Theorem B in Appendix A there exist Lagrange multipliers $\lambda(t)$ such that for $t \in \mathbb{T}$,

$$\begin{cases} \operatorname{grad}_{c(t)} u(\bar{c}) - \lambda(t)p(t) = F_t(c,\lambda,p,r) = 0 \\ p(t)\,[c(t) - e(t)] - r(t) = H_t(c,\lambda,p,r) = 0 \end{cases} \qquad \textbf{2.4.B}$$

where $\operatorname{grad}_{c(t)} u(\bar{c})$ is the vector of partial derivatives with respect to $c(t) = (c_1(t), \ldots, c_L(t))$, evaluated at $\bar{c}$.

On the other hand, if $(\bar{c}, \lambda)$ satisfies 2.4.B then a reasoning similar to the one used in the proof of Proposition 1.1.B shows that $\bar{c}$ is a solution to 2.4.A.

To prove that 2.4.B determines (c,λ) as a differentiable function of (p, r) we apply the Implicit Function Theorem; Theorem C in Appendix A, to the functions $F_0, \ldots, F_T, H_0, \ldots, H_T$, defined by 2.4.B, at the point $(\bar{c}, \lambda, p, r)$. The crucial point is to show that a certain bordered matrix is non-singular. Assume that $T = 1$ and $L = 2$ in order to exhibit this matrix explicitly.

To begin with, calculate the Jacobian for the function defined in 2.4.B to get

$$\hat{E} = \left[\begin{array}{cccc|cc} & & & & p_1(1) & 0 \\ & & & & p_2(1) & 0 \\ & D^2u(\bar{c}) & & & 0 & p_1(2) \\ & & & & 0 & p_2(2) \\ \hline p_1(1) & p_2(1) & 0 & 0 & 0 & 0 \\ 0 & 0 & p_1(2) & p_2(2) & 0 & 0 \end{array}\right]$$

If we can prove that $\det \hat{E} \neq 0$ then we can finish the proof as in Proposition 1.1.B and conclude that (c, λ) is a differentiable function of (p, r).

To prove that $\det \hat{E} \neq 0$ perform the following operations on $\hat{E}$

(i) Multiply row 5 and column 5 by $\lambda(1)$

(ii) Multiply row 6 and column 6 by $\lambda(2)$

(iii) Add row 6 to row 5 and add column 6 to column 5

After performing these operations, column 5 and row 5 of the resulting matrix, denoted E, are both equal to the gradient of u evaluated at $\bar{c}$, as can be seen from 2.4.B. By the maintained assumption (C5) the quadratic form given by the matrix $D^2u(\bar{c})$ is negative definite on the hyperplane with normal $\operatorname{grad} u(\bar{c})$. A fortiori, the quadratic form is negative definite on the subspace orthogonal to the row vectors of the left-hand lower 2×4 matrix of E. We can now apply Theorem A in Appendix B to conclude that $\det E$ is not 0. The operations in (iii) does not affect the determinant. Hence $\det \hat{E} = \lambda(1)\lambda(2)\det E$ and it follows that $\det \hat{E} \neq 0$.

(c) Since

$$B(p, r) = B(\beta(0)p(0), \ldots, \beta(T)p(T), \ \beta(0)r(0), \ldots, \beta(T)r(T))$$

the definition of g implies $g(p, r) = g(\hat{p}, \hat{r})$.

Definition of the spot-market demand function

The consumer's *spot-market demand function* is the function $g : D \longrightarrow (\mathbb{R}^L_{++})^{T+1}$ whose value at (p, r),

$$g(p, r) = (g_0(p, r), g_1(p, r), \ldots, g_T(p, r))$$

is the unique solution to the Consumer Spot-Market Problem given r. The Walrasian demand function is derived assuming maximal possibilities of income transfers. The spot-market demand function is at the other extreme and is derived assuming a given net income vector and no possibilities of income transfers. The maintained assumption (C3) ensures that the consumer chooses a consumption using all her income at each date. Hence the consumer's *net expenditure vector* will agree with her net income vector.

2.5 The Indirect Utility Function

Using the spot demand function we can define the indirect utility function.

Definition 2.5.A *Let* $v : D \longrightarrow \mathbb{R}$ *be the function defined by* $v(p, r) = u(g(p, r))$. *The function* v *is the* **indirect utility function.**

Thus v assigns to the price system $p = (p(0), p(1), \ldots, p(T))$ and the net income vector $r = (r(0), r(1), \ldots, r(T))$ the utility of the "best" consumption at these prices and with this net income vector. For given spot prices, the indirect utility function depicts the consumer's preferences regarding net income vectors. We will see below that the derived preferences on net income vectors are well behaved but first we give two examples related to the indirect utility function.

Example 2.5.A (Case of $L = 1$).

Let $L = 1$. Since the solution to the Consumer Spot-Market Problem given r satisfies the budget restrictions with equality we get

$$g(p, r) = \left(e(0) + \frac{r(0)}{p(0)},\ e(1) + \frac{r(1)}{p(1)}, \ldots,\ e(T) + \frac{r(T)}{p(T)}\right) \quad \text{and}$$

$$v(p, r) = u(g(p, r)) = u\left(e(0) + \frac{r(0)}{p(0)},\ e(1) + \frac{r(1)}{p(1)}, \ldots,\ e(T) + \frac{r(T)}{p(T)}\right)$$

For $L = 1$ and all spot prices equal to 1 the indirect utility function, over net income vectors, is equal to the (direct) utility function over net trades in the goods. For $L = 1$ the Consumer Spot-Market Problem given r reduces to the choice of a time profile of consumption. The problem of choosing a suitable goods bundle at different dates is in this case trivial.

□

Example 2.5.B (Case of time separable utility function).

If the consumer has a utility function

$$u(c) = u_0(c(0)) + u_1(c(1)) + \ldots + u_T(c(T))$$

which is separable over time then the Consumer Spot-Market Problem given r, 2.4.A, is solved by solving, for $t \in \mathbb{T}$,

$$\begin{aligned} \text{Max}_{c(t)} \quad u_t(c(t)) \quad \text{s. to } c(t) \in \mathbb{R}^L_{++} \text{ and} \\ p(t)(c(t) - e(t)) \leq r(t) \end{aligned}$$

In this case, for $t \in \mathbb{T}$, the function g_t depends only on $p(t)$ and $r(t)$. Hence there are functions $\hat{g}_t$ defined on subsets of $\mathbb{R}^L \times \mathbb{R}$ such that, for $t \in \mathbb{T}$,

$$\hat{g}_t(p(t), r(t)) = g_t(p, r) \quad \text{for } (p, r) \in D, t \in \mathbb{T}$$

For $t \in \mathbb{T}$, define v_t by $v_t(p(t), r(t)) = u_t(\hat{g}_t(p(t), r(t)))$. v_t and $\hat{g}_t$ are defined for $(p(t), r(t)) \in \mathbb{R}^L_{++} \times \mathbb{R}$ such that $p(t)e(t) + r(t) > 0$. Since the indirect utility function is

$$\begin{aligned} v(p, r) &= u(g(p, r)) \\ &= u_0(\hat{g}_0(p(0), r(0))) + u_1(\hat{g}_1(p(1), r(1))) + \ldots + u_T(\hat{g}_T(p(T), r(T))) \\ &= v_0(p(0), r(0)) + v_1(p(1), r(1)) + \ldots + v_T(p(T), r(T)) \end{aligned}$$

it follows that also the indirect utility function is separable over time.

Properties of the Indirect Utility Function

The indirect utility function would be of little use unless it inherited some of the pleasant properties of the (direct) utility function. The following proposition shows that the indirect utility function, given spot prices for the goods, indeed inherits the properties (C3), (C4) and (C5) (with the consumption set $C = E(p)$). Thus the implications of the Maintained Assumptions given in Lemma 1.1.A apply also to the derived preferences for net income vectors.

The proof below points to an interesting relation between the Lagrange multipliers from the Spot-Market Consumer Problem and the partial derivatives of the indirect utility function. Unfortunately the proof that $v(p, \cdot)$ satisfies (C5) is rather lengthy so we have relegated the proof for the case $L > 1$ to a starred section.

Proposition 2.5.B *Let (C, u, e) be a consumer and let $v : D \longrightarrow \mathbb{R}$ be the consumer's indirect utility function. The function v is a differentiable function of (p, r) and*

(a) *for $(p, r) \in D$ and $\beta = (\beta(0), \beta(1), \ldots, \beta(T)) \in \mathbb{R}^{T+1}_{++}$, let $\hat{p}(t) = \beta(t)p(t)$ and $\hat{r}(t) = \beta(t)r(t)$, for $t \in T$. Then $v(p, r) = v(\hat{p}, \hat{r})$.*

(b) *$v(p, \cdot) : E(p) \longrightarrow \mathbb{R}$ is twice differentiable and satisfies* (C3)-(C5) *of the Maintained Assumptions.*

Proof: We have

$$(p, r) \longrightarrow g(p, r) \longrightarrow u(g(p, r))$$

and since g, by Proposition 2.4.A, is a differentiable function we get that $u \circ g$ is a differentiable function, being the composition of differentiable functions. Since $v = u \circ g$ it follows that v is a differentiable function of (p, r)

(a) follows from the corresponding property for the pure spot-market demand function; (c) of Proposition 2.4.A. Hence $\hat{p}(t) = \beta(t)p(t)$ and $\hat{r}(t) = \beta(t)r(t)$ implies

$$v(p, r) \; = \; u(g(p, r)) \; = \; u(g(\hat{p}, \hat{r})) \; = \; v(\hat{p}, \hat{r})$$

(b) To prove that $v(p, \cdot)$ is twice differentiable and satisfies (C4), note that the pure spot-market demand function $(g_t)_{t \in \mathbb{T}}$ satisfies, for $t \in \mathbb{T}$,

$$p(t)g_t(p, r) \; = \; p(t)e(t) \; + \; r(t) \quad \text{for } (p, r) \in D$$

Choose a fixed $t \in \mathbb{T}$ and differentiate both the left hand and right hand side of this relation with respect to $r(0)$

$$p_1(t)\frac{\partial g_{t1}(p, r)}{\partial r(0)} + p_2(t)\frac{\partial g_{t2}(p, r)}{\partial r(0)} + \ldots + p_L(t)\frac{\partial g_{tL}(p, r)}{\partial r(0)} \; = \; \begin{cases} 1 & \text{if } t = 0 \\ 0 & \text{if } t \neq 0 \end{cases} \qquad \textbf{2.5.A}$$

Let $(g_t(p, r))_{t \in \mathbb{T}} = \bar{c}$. The proof of Proposition 2.4.A shows that $\bar{c}$ also satisfies relation 2.4.B. Since $v(p, r) = u(g_0(p, r), g_1(p, r), \ldots, g_T(p, r))$ we get using relations 2.4.B and 2.5.A, for $t = 0$

$$\begin{aligned}\frac{\partial v(p,r)}{\partial r(0)} &= \sum_{t\in\mathbb{T}}\sum_{l\in\mathbb{L}}\frac{\partial u(\bar{c})}{\partial c_l(t)}\frac{\partial g_{tl}(p,r)}{\partial r(0)} \\ &= \sum_{t\in\mathbb{T}}\sum_{l\in\mathbb{L}}\lambda(t)p_l(t)\frac{\partial g_{tl}(p,r)}{\partial r(0)} \\ &= \sum_{t\in\mathbb{T}}\lambda(t)\sum_{l\in\mathbb{L}}p_l(t)\frac{\partial g_{tl}(p,r)}{\partial r(0)} = \lambda(0)\end{aligned}$$

where $\lambda(t)$ are the Lagrange multipliers from relation 2.4.B. In the same way it is seen that for $t \in \mathbb{T}_1$

$$\frac{\partial v(p,\bar{r})}{\partial r(t)} = \lambda(t) \qquad \textbf{2.5.B}$$

so that the Lagrange multipliers of the Consumer Spot-Market Problem given r are equal to the partial derivatives of the indirect utility function but the Lagrange multipliers are positive numbers. Hence $\text{grad}_r\, v(p,r) \in \mathbb{R}^{T+1}_{++}$. From the proof of Proposition 2.4.A also follows that the Lagrange multipliers are differentiable functions of r. Thus the second derivatives of $v(p,\cdot)$ exist and are continuous functions and hence $v(p,\cdot)$ is a twice differentiable function.

To prove that $v(p,\cdot)$ satisfies (C3) of the Maintained Assumptions recall that

$$E(p) = \{r \in \mathbb{R}^{T+1} \mid r(t) > -p(t)e(t) \quad \text{for } t \in \mathbb{T}\}$$

and choose a fixed $p \in \mathbb{R}^{(T+1)L}_{++}$.

Let $(r^n)_{n\in\mathbb{N}}$ be a sequence with $r^n \in E(p)$, such that $v(p,r^n) = \bar{v} = u(g(p,r^n)) = \bar{u}$ and such that $(p,r^n) \to (p,\bar{r})$. We will show that $\bar{r} \in E(p)$, that is, $0 < p(t)e(t) + \bar{r}(t)$ for $t \in \mathbb{T}$.

Put $g(p,r^n) = c^n$. Then $p(t)\,[c^n(t) - e(t)]) - r^n(t) = 0$ for $t \in \mathbb{T}$. The terms of the sequence $(c^n)_{n\in\mathbb{N}}$ all belong to the compact set

$$\left\{c \in \mathbb{R}^{(T+1)L} \mid p(t)c^n(t) \leq p(t)e(t) + \sup_{n\in\mathbb{N}} r^n(t) \text{ and } u(c) \geq \bar{u}\right\}$$

Hence $(c^n)_{n\in\mathbb{N}}$ has a convergent subsequence, $(c^n)_{n\in\mathbb{N}_1}$ with $\mathbb{N}_1 \subset \mathbb{N}$. Let $c^n \to \bar{c}$ as $n \in \mathbb{N}_1$. Then $(c^n, r^n) \to (\bar{c},\bar{r})$ as $n \in \mathbb{N}_1$ and by the Maintained Assumption (C3) the limit point $\bar{c}$ belongs to $\mathbb{R}^{(T+1)L}_{++}$. Since $p(t) \in \mathbb{R}^L_{++}$

we get $0 < p(t)\bar{c}(t)$ and since the function

$$(c, r) \longrightarrow p(t)\left[c(t) - e(t)\right] - r(t)$$

is continuous we get $p(t)\left[\bar{c}(t) - e(t)\right] - \bar{r}(t) = 0$. Hence $0 < p(t)\bar{c}(t) = p(t)e(t) + \bar{r}(t)$ as was to be proved.

To show that $v(p, \cdot)$ satisfies the Maintained Assumption (C5) consider first the case where $T = 3$ and $L = 1$. Then the indirect utility function is

$$v(p, r) = u\left(e(0) + \frac{r(0)}{p(0)}, e(1) + \frac{r(1)}{p(1)}, e(2) + \frac{r(2)}{p(2)}, e(3) + \frac{r(3)}{p(3)}\right) \qquad \textbf{2.5.C}$$

and a calculation shows that, with p fixed, the matrix of second derivatives with respect to r is

$$D^2 v(p, r) = \left[\frac{D^2_{ij} u}{p(i)p(j)}\right]$$

where the second derivatives of u are evaluated at the point given in relation 2.5.C. Let $h \neq 0$ be orthogonal to $Dv(p, r)$. Then with $D_i v(p, r)$ denoting the derivative with respect to $r(i)$, for $i = 0, \ldots, 3$

$$\begin{aligned} Dv(p, r) \cdot h &= D_0 v(p, r) h_0 + D_1 v(p, r) h_1 + D_2 v(p, r) h_2 + D_3 v(p, r) h_3 \\ &= D_0 u \frac{h_0}{p(0)} + D_1 u \frac{h_1}{p(1)} + D_2 u \frac{h_2}{p(2)} + D_3 u \frac{h_3}{p(3)} \\ &= Du \cdot \left(\frac{h_0}{p(0)}, \frac{h_1}{p(1)}, \frac{h_2}{p(2)}, \frac{h_3}{p(3)}\right) = Du \cdot \hat{h} = 0 \end{aligned}$$

showing that the the vector with components $\hat{h}(t) = h(t)/p(t), t \in \mathbb{T}$ is orthogonal to Du. Since u satisfies the Maintained Assumption (C5) we get

$$0 > \hat{h}^t D^2 u \hat{h} = h^t \left[\frac{D^2_{ij} u}{p(i)p(j)}\right] h = h^t D^2 v(p, r) h$$

Hence if $L = 1$ then the indirect utility function satisfies the Maintained Assumption (C5).

Proof that $v(p,\cdot)$ satisfies (C5)

The purpose of this section is to give the fairly lengthy prove that $v(p,\cdot)$ satisfies (C5).

Proof of the remaining part of Proposition 2.5.B: Rather than doing the proof for the general case we will restrict ourselves to the case $T = 1$ and $L = 2$, since this will allow us to be fairly explicit in the notation and still retain the main ideas of the proof and it is evident how the reasoning generalizes.

Consider the function S with values given by

$$S(q_{01}, q_{02}, q_{11}, q_{12}) = u\left(e_1(0) + \frac{q_{01}}{p_1(0)},\ e_2(0) + \frac{q_{02}}{p_2(1)},\ e_1(1) + \frac{q_{11}}{p_1(1)},\ e_2(1) + \frac{q_{12}}{p(1)}\right)$$

and defined for $(q_{01}, q_{02}, q_{11}, q_{12})$ which make the arguments in u positive. Since S is, given p, an indirect utility function with $L = 1$ the function S satisfies the Maintained Assumption (C5).

The value of the function $v(p,\cdot)$ at r is the the value of the function S at the solution to

$$\text{Max}_q \quad S(q_{01}, q_{02}, q_{11}, q_{12}) \quad \text{s. to} \quad \begin{cases} q_{01} + q_{02} &= r(0) \\ q_{11} + q_{12} &= r(1) \end{cases} \qquad \textbf{2.5.D}$$

It is solved by q such that

$$\begin{aligned} q_{01} &= p_1(0)(g_{01}(p,r) - e_1(0)) & q_{02} &= p_2(0)(g_{02}(p,r) - e_2(0)) \\ q_{11} &= p_1(1)(g_{11}(p,r) - e_1(1)) & q_{12} &= p_2(1)(g_{12}(p,r) - e_2(1)) \end{aligned}$$

which shows that the solution q varies differentiably with r.

Choose (h_0, h_1) orthogonal to $(D_1 v(p,\bar{r}), D_2(p,\bar{r}))$ and such that $\bar{r} + \alpha h \in E(p)$ for $\alpha \in [0,1]$. Consider the function

$$v(p, (\bar{r}(0), \bar{r}(1)) + \alpha(h_0, h_1)) = v(p, \bar{r} + \alpha h)$$

Let $\bar{q}$ be the solution to the problem in 2.5.D for $\bar{r} = (\bar{r}(1), \bar{r}(0))$ and $q(\bar{r} + \alpha h)$ the solution as $\alpha \in [0,1]$. We have, trivially, for $\alpha \in]0,1]$,

$$S(q(\bar{r} + \alpha h)) = S\left(\bar{q} + \alpha\left[\frac{1}{\alpha}\left[q(\bar{r} + \alpha h) - \bar{q}\right]\right]\right)$$

Put

$$\eta(\alpha) = (\eta_{01}(\alpha), \eta_{02}(\alpha), \eta_{11}(\alpha), \eta_{12}(\alpha)) = \frac{1}{\alpha}[q(\bar{r} + \alpha h) - \bar{q}]$$

and define the functions $\hat{v}, \hat{S} : [0,1] \longrightarrow \mathbb{R}$ by

$$\hat{v}(\alpha) = v(p, \bar{r} + \alpha h) \quad \text{and} \quad \hat{S}(\alpha) = S(\bar{q} + \alpha\eta(\alpha))$$

Then $\hat{v}(\alpha) = \hat{S}(\alpha)$ and we have

$$\begin{aligned} q_{01}(\bar{r} + \alpha h) + q_{02}(\bar{r} + \alpha h) &= \bar{r}(0) + \alpha h_0 \text{ and } q_{01}(\bar{r}) + q_{02}(\bar{r}) = \bar{r}(0) \\ q_{11}(\bar{r} + \alpha h) + q_{12}(\bar{r} + \alpha h) &= \bar{r}(1) + \alpha h_1 \text{ and } q_{11}(\bar{r}) + q_{12}(\bar{r}) = \bar{r}(1) \end{aligned}$$

and thus for $\alpha \in [0,1]$

$$\begin{aligned} \eta_{01}(\alpha) + \eta_{02}(\alpha) &= h_0 \\ \eta_{11}(\alpha) + \eta_{12}(\alpha) &= h_1 \end{aligned} \qquad \textbf{2.5.E}$$

which implies

$$D_\alpha\eta_{01}(\alpha) + D_\alpha\eta_{02}(\alpha) = 0 \quad \text{and} \quad D_\alpha\eta_{11}(\alpha) + D_\alpha\eta_{12}(\alpha) = 0 \qquad \textbf{2.5.F}$$

We can now calculate the first derivatives of $\hat{v}(\cdot)$ and $\hat{S}(\cdot)$,

$$\begin{aligned} D_\alpha\hat{v}(\alpha) &= D_0 v(p, \bar{r} + \alpha h)h_0 + D_1 v(p, \bar{r} + \alpha h)h_1 \\ D_\alpha\hat{S}(\alpha) &= D_{01}S(\bar{q} + \alpha\eta(\alpha))\,[\eta_{01} + \alpha D_\alpha\eta_{01}] \\ &\quad\vdots \\ &+ D_{12}S(\bar{q} + \alpha\eta(\alpha))\,[\eta_{12} + \alpha D_\alpha\eta_{12}] \end{aligned}$$

Here, for example, D_{01} is the partial derivative with respect to the variable 01 (which is the first argument of S).

Since $\bar{q} + \alpha\eta(\alpha)$ is the solution to the problem in 2.5.D we have $D_{01}S(\bar{q} + \alpha\eta(\alpha)) = D_{02}S(\bar{q} + \alpha\eta(\alpha))$ and $D_{11}S(\bar{q} + \alpha\eta(\alpha)) = D_{12}S(\bar{q} + \alpha\eta(\alpha))$ and we get using relation 2.5.F

$$\begin{aligned} D_\alpha\hat{S}(\alpha) &= D_{01}S(\bar{q} + \alpha\eta(\alpha))\eta_{01} + D_{02}S(\bar{q} + \alpha\eta(\alpha))\eta_{02} \\ &+ D_{11}S(\bar{q} + \alpha\eta(\alpha))\eta_{11} + D_{12}S(\bar{q} + \alpha\eta(\alpha))\eta_{12} = 0 \end{aligned}$$

Hence for $\alpha = 0$ we get $D_\alpha \hat{v}(0) = D_\alpha \hat{S}(0) = 0$ and $\operatorname{grad} S(\bar{q}) \cdot \eta(0) = D_\alpha \hat{S}(0) = 0$ showing that $\eta(0)$ is orthogonal to $\operatorname{grad} S(\bar{q})$.

We can now calculate the second derivatives of $\hat{v}$ and $\hat{S}$. It is easy to see that $D^2_\alpha \hat{v}(\alpha) = h^t D^2 v(p, \bar{r} + \alpha h)h$ and, with the derivatives of S evaluated at $\bar{q} + \alpha\eta(\alpha)$,

$$\begin{aligned} D^2_\alpha \hat{S}(\alpha) &= D^2_{01,01} S \left[\eta_{01} + \alpha D_\alpha \eta_{01}\right] \eta_{01} + D_{01} S \cdot D_a \eta_{01} \\ &+ D^2_{01,02} S \left[\eta_{02} + \alpha D_\alpha \eta_{02}\right] \eta_{01} \\ &+ D^2_{01,11} S \left[\eta_{11} + \alpha D_\alpha \eta_{11}\right] \eta_{01} \\ &+ D^2_{01,12} S \left[\eta_{12} + \alpha D_\alpha \eta_{12}\right] \eta_{01} \\ &\qquad \vdots \\ &+ D^2_{12,11} S \left[\eta_{11} + \alpha D_\alpha \eta_{11}\right] \eta_{12} \\ &+ D^2_{12,12} S \left[\eta_{12} + \alpha D_\alpha \eta_{12}\right] \eta_{12} + D_{12} S \cdot D_a \eta_{12} \end{aligned}$$

or

$$\begin{aligned} D^2_\alpha \hat{S}(\alpha) &= \sum_{i,j \in \mathbb{T} \times \mathbb{L}} D^2_{i,,j} S(\bar{q} + \alpha\eta(\alpha))\eta_i \eta_j \\ &\quad + \sum_{i \in \mathbb{T} \times \mathbb{L}} D_i S(\bar{q} + \alpha\eta(\alpha)) D_a \eta_i \\ &\qquad + \alpha \sum_{i,j \in \mathbb{T} \times \mathbb{L}} D^2_{i,,j} S(\bar{q} + \alpha\eta(\alpha))\eta_i D_\alpha \eta_j \end{aligned}$$

Since $D_{01} S(\bar{q} + \alpha\eta(\alpha)) = D_{02} S(\bar{q} + \alpha\eta(\alpha))$ and $D_{11} S(\bar{q} + \alpha\eta(\alpha)) = D_{12} S(\bar{q} + \alpha\eta(\alpha))$ we get using 2.5.F that the middle sum is 0. Let α tend to 0. Then the last sum tends to 0 and thus

$$D^2_\alpha \hat{S}(0) = \textstyle\sum_{i,j \in \mathbb{T} \times \mathbb{L}} D^2_{i,,j} S(\bar{q})\eta_i \eta_j$$

and

$$D^2_\alpha \hat{v}(0) - D^2_\alpha \hat{S}(0) = h^t D^2 v(p, \bar{r})h - \textstyle\sum_{i,j \in \mathbb{T} \times \mathbb{L}} D^2_{i,,j} S(\bar{q})\eta_i(0)\eta_j(0) = 0$$

The vector $\eta(0)$ is orthogonal to $\operatorname{grad} S$ and from relation 2.5.E we have $\eta \neq 0$. Hence

$$\textstyle\sum_{i,j \in \mathbb{T} \times \mathbb{L}} D^2_{i,,j} S(\bar{q})\eta_i(0)\eta_j(0)$$

is a negative number and it follows that $h^t D^2 v(p, \bar{r})h < 0$.

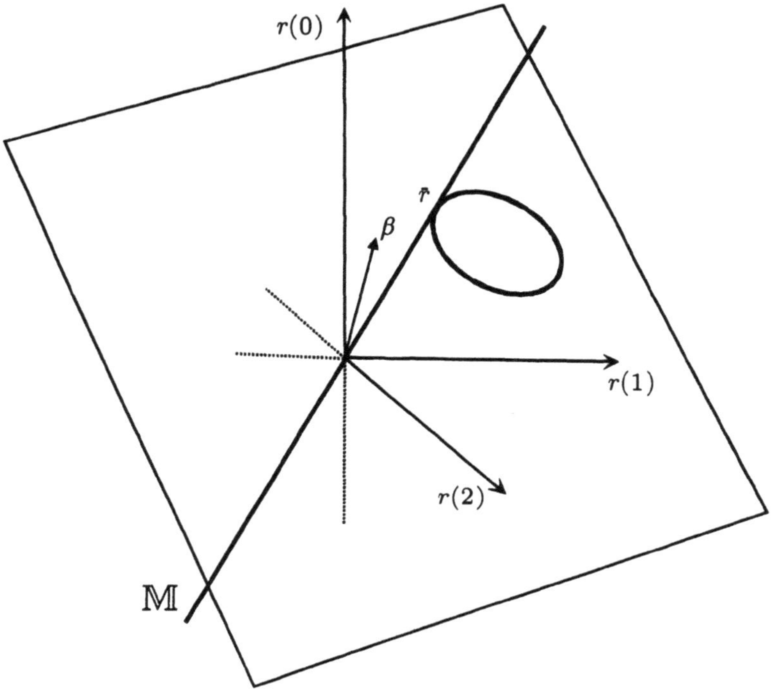

Figure 2.5.A: *A one-dimensional subspace of feasible income transfers;* $\mathbb{M}$

2.6 Decomposition of the Consumer Problem

Choice of a net income vector and goods bundles

The indirect utility function suggests that it is possible to decompose the Consumer Problem into a problem of choosing, at given spot prices for goods, a net income vector and the problem of making a "best" choice in the spot-markets given the spot prices and a net income vector. In Theorem 2.6.A we show that this is indeed the case also when the net income vector is restricted to belong to some arbitrary subset, $\mathbb{M}$, of $\mathbb{R}^{T+1}$. In the Consumer Problem from the interpretation with spot-markets for goods and a market for income at date 0 this subset was taken to be a homogenous hyperplane in $\mathbb{R}^{T+1}$ with a normal which was positive in each component. When defining the spot-market demand function the subset $\mathbb{M}$ was the singleton $\{r\}$. In the sequel we will see other specifications of the set $\mathbb{M}$. In particular, $\mathbb{M}$ might be generated by the assets of the economy.

We will refer to the consumer problem of choosing a "best" consumption, when the net income vector is required to belong to a subset $\mathbb{M}$ of $\mathbb{R}^{T+1}$, as the Consumer Spot-Market Problem relative to $\mathbb{M}$.

$$\text{Max}_{(c,r)} \quad u(c(0), c(1), \ldots, c(T)) \quad \text{s. to } c \in C, r \in E(p) \text{ and}$$

$$\left\{ \begin{array}{rcl} p(0)(c(0)-e(0)) & \leq & r(0) \\ p(1)(c(1)-e(1)) & \leq & r(1) \\ \vdots & & \vdots \\ p(T)(c(T)-e(T)) & \leq & r(T) \\ & \text{and} & \\ (r(0), r(1), \ldots, r(T)) & \in & \mathbb{M} \end{array} \right. \qquad \textbf{2.6.A}$$

Of course, we would not expect this problem to have a solution unless the set of feasible net income vectors satisfies some reasonable conditions.

Using the indirect utility function we can also study the problem of choosing a "best" net income vector, for given spot prices. We have the Consumer Net Income Vector Problem.

$$\begin{array}{l} \text{Max}_r \quad v(p, r(0), r(1), \ldots, r(0)) \quad \text{s. to } r \in E(p) \text{ and} \\ \qquad (r(0), r(1), \ldots, r(T)) \in \mathbb{M} \end{array} \qquad \textbf{2.6.B}$$

We have illustrated the problem for a simple case in Figure 2.5.A which exhibits the vector β as a normal to a hyperplane containing the one-dimensional subspace of feasible income transfers; $\mathbb{M}$. The intersection of the contour set of the indirect utility function with the hyperplane is shown. The net income vector $\bar{r}$ is a best choice in $\mathbb{M}$ for the consumer.

The spot-market problem and the net income vector problem

Theorem 2.6.A below relates the solutions of the Consumer Spot-Market Problem given r, the Consumer Net Income Vector Problem and the Consumer Spot-Market Problem relative to $\mathbb{M}$.

Theorem 2.6.A (Decomposition of the Consumer Spot-Market Problem) *Let (C, u, e) be a consumer. Let $p \in (\mathbb{R}^L_{++})^{T+1}$ and let $\mathbb{M}$ be a subset of $\mathbb{R}^{T+1}$. Then*

$(\bar{c}, \bar{r})$ is a solution the Consumer Spot-Market Problem rel. to $\mathbb{M}$

if and only if

$$\begin{cases} \textbf{(i) } \bar{r} \text{ is a solution to the Consumer Net Income Vector Problem} \\ \textbf{(ii) } \bar{c} \text{ is a solution to the Consumer Spot-Market Problem given } \bar{r}. \end{cases}$$

Proof: Note that in 2.6.A the maximization is over (c, r) but in 2.6.B only over c a given r.

"**only if**". Let $(\bar{c}, \bar{r})$ be a solution to 2.6.A. Then $\bar{c} = g(p, \bar{r})$ and $u(\bar{c}) \geq u(g(p, r))$ for $r \in \mathbb{M}$. In order to show that $\bar{r}$ is a solution to 2.6.B, we show that $v(p, \bar{r}) \geq v(p, r)$ for $r \in \mathbb{M}$. By the definition of the indirect utility function and the pure spot-market demand function g

$$v(p, \bar{r}) = u(g(p, \bar{r})) = u(\bar{c}) \geq u(g(p, r)) = v(p, r) \quad \text{for } r \in \mathbb{M}$$

Hence $\bar{r}$ is a solution to 2.6.B.

Next we show that $\bar{c}$ is a solution to 2.4.A with $r = \bar{r}$. Let c be any point satisfying the restrictions in 2.4.A with $r = \bar{r}$. By the definition of g we get

$$u(g(p, \bar{r})) \geq u(c)$$

Hence $\bar{c} = g(p, \bar{r})$ is a solution to 2.4.A.

"**if**". Since $\bar{r}$ is a solution to 2.6.B we have $v(p, \bar{r}) \geq v(p, r)$ for $r \in \mathbb{M}$. Since $g(p, \bar{r})$ is the unique solution to 2.4.A and $\bar{c}$ also solves 2.4.A we have $g(p, \bar{r}) = \bar{c}$. Let (c, r) satisfy the restrictions of 2.6.A at prices p. Then, by the definition of g and v,

$$\begin{aligned} u(g(p, r)) &\geq u(c) \\ u(g(p, \bar{r})) &= v(p, \bar{r}) \geq v(p, r) = u(g(p, r)) \quad \text{for } r \in \mathbb{M} \end{aligned}$$

It follows that $u(g(p, \bar{r})) = u(\bar{c}) \geq u(c)$ and since $(g(p, \bar{r}), \bar{r}) = (\bar{c}, \bar{r})$ satisfies the restrictions in 2.6.A $(\bar{c}, \bar{r})$ is a solution to 2.6.A.

Example 2.6.A (The indirect utility function and demand for a net income vector with a Cobb-Douglas utility function)

In this example we calculate the indirect utility function of a consumer with (a logarithmic transformation of) a Cobb-Douglas utility function. Note that such a utility function is separable over time.
Let $L = 2$ and $T = 2$

$$u(c(0), c(1), c(2)) = \sum_{t\in\mathbb{T}}[\alpha_1(t)\ln c_1(t) + \alpha_2(t)\ln c_2(t)]$$

where $\alpha_l(t) > 0$ and $\alpha_1(t) + \alpha_2(t) = 1$, for $t \in \mathbb{T}$, $l \in \mathbb{L}$.
The spot-market demand function is, for $t \in \mathbb{T}$,

$$\begin{cases} g_{t1}(p,r) = \alpha_1(t)\dfrac{p(t)e(t)+r(t)}{p_1(t)} \\ g_{t2}(p,r) = \alpha_2(t)\dfrac{p(t)e(t)+r(t)}{p_2(t)} \end{cases} \qquad \textbf{2.6.C}$$

Using $v(p,r) = u(g_0(p,r), g_1(p,r), \ldots, g_T(p,r))$ and $\alpha_1(t) + \alpha_2(t) = 1$ we get

$$\begin{aligned} v(p,r) &= \sum_{t\in\mathbb{T}}\left\{\ln[p(t)e(t)+r(t)] + \alpha_1(t)\ln\frac{\alpha_1(t)}{p_1(t)} + \alpha_2(t)\ln\frac{\alpha_2(t)}{p_2(t)}\right\} \\ &= \sum_{t\in\mathbb{T}}\ln[p(t)e(t)+r(t)] + \sum_{t\in\mathbb{T}}K_t(\alpha_1(t),\alpha_2(t),p_1(t),p_2(t)) \end{aligned}$$

where K_t, $t \in \mathbb{T}$, are functions not depending on r.
Assume that the consumer may choose the net income vector in the set

$$\mathbb{M} = \left\{r \in \mathbb{R}^3 \mid \beta(0)r(0) + \beta(1)r(1) + \beta(2)r(2) = 0\right\}$$

where $\beta(0) = 1$ and $\beta(1), \beta(2)$ are positive.

The Consumer Net Income Vector Problem, 2.6.B, is

$$\begin{aligned} \text{Max}_r \quad & v(p,r) \quad \text{s. to } r \in E(p) \text{ and} \\ & r(0) + \beta(1)r(1) + \beta(2)r(2) = 0 \end{aligned} \qquad \textbf{2.6.D}$$

Calculating the gradient of v with respect to r we get

$$\operatorname{grad}_r v(p,r) = \begin{bmatrix} \dfrac{1}{p(0)e(0)+r(0)} \\ \dfrac{1}{p(1)e(1)+r(1)} \\ \dfrac{1}{p(2)e(2)+r(2)} \end{bmatrix}$$

Put $p(t)e(t) = w(t)$, $t = 0,1,2$. At a solution to 2.6.D the following conditions are satisfied for some $\lambda \in \mathbb{R}_{++}$

$$\begin{bmatrix} \dfrac{1}{w(0)+r(0)} \\ \dfrac{1}{w(1)+r(1)} \\ \dfrac{1}{w(2)+r(2)} \end{bmatrix} - \lambda \begin{bmatrix} 1 \\ \beta(1) \\ \beta(2) \end{bmatrix} = 0 \qquad \textbf{2.6.E}$$

Since $\beta(0)r(0) + \beta(1)r(1) + \beta(2)r(2) = 0$, we get

$$\beta(0)r(0) + \beta(1)r(1) + \beta(2)r(2) = 0$$

if and only if

$$\Sigma_{t\in\mathbb{T}}\beta(t)[w(t)+r(t)] = \Sigma_{t\in\mathbb{T}}\beta(t)w(t)$$

Put $\hat{w} = \Sigma_{t\in\mathbb{T}}\beta(t)w(t)$. Then $\hat{w}$ is the discounted, to date 0, value of the consumer's initial endowment and using 2.6.E we get $\lambda = 3/\hat{w}$. The solution to 2.6.E is

$$\begin{bmatrix} r(0) \\ r(1) \\ r(2) \end{bmatrix} = \begin{bmatrix} \dfrac{1}{3}\dfrac{\hat{w}}{\beta(0)} - w(0) \\ \dfrac{1}{3}\dfrac{\hat{w}}{\beta(1)} - w(1) \\ \dfrac{1}{3}\dfrac{\hat{w}}{\beta(2)} - w(2) \end{bmatrix}$$

Hence the consumer chooses r so that her discounted gross income $\beta(t)$ $[w(t) + r(t)] = \hat{w}/3$, for $t \in \mathbb{T}$. This equality of the discounted values depends on the fact that the sum of coefficients, for each date, in the utility function are equal.

Now we can use the pure spot-market demand function 2.6.C to calculate, say, the demand for good 1 at date 0

$$\begin{aligned} g_{01}(p,r) &= \alpha_1(0)\frac{p(0)e(0)+r(0)}{p_1(0)} \\ &= \alpha_1(0)\frac{p(0)e(0)+\frac{1}{3}\frac{\hat{w}}{\beta(0)}-p(0)e(0)}{p_1(0)} \\ &= \frac{\alpha_1(0)}{\sum_{t\in\mathbb{T}}[\alpha_1(t)+\alpha_2(t)]}\frac{\hat{w}}{\beta(0)p_1(0)} \end{aligned}$$

which is equal to the Walrasian demand at the prices $\beta(t)p(t)$, for $t \in \mathbb{T}$. This was to be expected since the feasible set of income transfers is maximal.

2.7 Spot-Market Equilibria Relative to a Set of Income Transfers

Definition of a spot-market equilibrium relative to $\mathbb{M}$

The results of the preceding section show that we can separate the choice of a net income vector from the choice of goods in the spot-markets also when the net income vector is restricted to belong to some subset $\mathbb{M} \subset \mathbb{R}^{T+1}$.

All the equilibria which we will study in Chapter 2 to 6, are particular instances of spot-market equilibria relative to a set of income transfers. The set of net income vectors available to the consumer is a subset, $\mathbb{M}$, of $\mathbb{R}^{T+1}$ or a corresponding set of feasible net income vectors in problems with time and uncertainty. The set $\mathbb{M}$ will always be independent of the consumer so that each consumer has the same set of net income vectors available.

Although the set $\mathbb{M}$ will, in the applications, often be a linear subspace we state the definition of a spot-market equilibrium relative a set of income transfers for the general case.

Definition 2.7.A *Let* $\mathcal{E} = (C^i, u^i, e^i)_{i\in\mathbb{I}}$ *be an economy and* $\mathbb{M}$ *a subset of* $\mathbb{R}^{T+1}$ *The pair* $((\bar{c}^i)_{i\in\mathbb{I}}, p)$ *is a* **spot-market equilibrium for** $\mathcal{E}$ **relative to** $\mathbb{M}$ *if, with* $\bar{r}^i(t) = p(t)(\bar{c}^i(t) - e^i(t))$, *for* $t \in \mathbb{T}$:

(a) *for* $i \in \mathbb{I}$, $\bar{r}^i$ *is a solution to*

$$\begin{aligned} Max_r \quad & v^i(p, r^i(0), r^i(1), \ldots, r^i(T)) \quad s.\ to\ r \in E(p)\ and \\ & (r^i(0), r^i(1), \ldots, r^i(T)) \in \mathbb{M} \end{aligned}$$

(b) *for* $i \in \mathbb{I}$, $\bar{c}^i$ *is a solution to:*

$$\text{Max}_{\,c} \quad u^i(c^i(0), c^i(1), \ldots, c^i(T)) \quad s.\ to\ c \in C\ and$$

$$\begin{cases} p(0)(c^i(0) - e^i(0)) \leq \bar{r}^i(0) \\ p(1)(c^i(1) - e^i(1)) \leq \bar{r}^i(1) \\ \qquad\vdots \\ p(T)(c^i(T) - e^i(T)) \leq \bar{r}^i(T) \end{cases}$$

(c) *at each date goods markets balance,*

$$\begin{aligned} \textstyle\sum_{i\in\mathbb{I}} \bar{c}^i(0) &= \textstyle\sum_{i\in\mathbb{I}} e^i(0) \\ \textstyle\sum_{i\in\mathbb{I}} \bar{c}^i(1) &= \textstyle\sum_{i\in\mathbb{I}} e^i(1) \\ &\vdots \\ \textstyle\sum_{i\in\mathbb{I}} \bar{c}^i(T) &= \textstyle\sum_{i\in\mathbb{I}} e^i(T) \end{aligned}$$

(d) *the date* 0 *market for spot income balances;* $\sum_{i\in\mathbb{I}} \bar{r}^i = 0$

In a spot-market equilibrium the net income vectors are implicitly defined by the consumptions and the prices and the net expenditure vector agrees with the net income vector. Note that by Theorem 2.6.A, (a) and (b) in the definition, implies that, for $i \in \mathbb{I}$, the pair $(\bar{c}^i, \bar{r}^i)$ is a solution to the Consumer Spot-Market Problem relative to $\mathbb{M}$. Independent of the number of goods we have by condition (d), which follows from the market balance conditions for the goods markets, (c), that the net income vectors sum to 0 in a spot-market equilibrium relative to $\mathbb{M}$. Thus an individual consumer may transfer purchasing power between dates but the aggregate "savings" and "dissavings" equal 0, at each date.

In case $L = 1$ the consumer choice of goods in each of the spot-markets is trivial and we then have the standard model of the theory of finance which focuses on the possibilities of income transfers over dates or, as will be seen in Chapter 4, between dates-events. In case $L = 1$ the market balance conditions for the goods markets, (c), follows from the market balance conditions for the date 0 market for spot income.

The definition allows for the case that $\mathbb{M}$ depends on the spot prices and $\mathbb{M}$ in the definition must then be understood to denote the subset of net income vectors induced by the equilibrium prices. As noted above, in most applications $\mathbb{M}$ will be a linear subspace. We now give some examples of how $\mathbb{M}$, might be specified.

The set of feasible net income vectors

In Section 2.3 we studied the case where date t income, could be bought and sold at date 0. The date 0 price of t-crowns in 0-crowns was some positive number, $\beta(t), t \in \mathbb{T}$, where $\beta(0) = 1$. Then

$$\mathbb{M} = \left\{r \in \mathbb{R}^{T+1} \mid \sum\nolimits_{t \in \mathbb{T}} \beta(t) r(t) = 0\right\} = \left\{r \in \mathbb{R}^{T+1} \mid \beta r = 0\right\}$$

In this case $\mathbb{M}$ is a homogenous hyperplane in $\mathbb{R}^{T+1}$ and thus has dimension T. As we shall see below if $\mathbb{M}$ is a linear subspace then for an equilibrium to exist it is necessary that the dimension of $\mathbb{M}$, the subset in the definition of a spot-market equilibrium, is at most T.

At the other extreme we have $\mathbb{M} = \{0\}$. Then there are no possibilities of transferring purchasing power between dates, and a spot-market equilibrium relative to $\mathbb{M}$, is a *pure spot-market equilibrium.*

In general $\mathbb{M}$ can be a linear subspace of $\mathbb{R}^{T+1}$ of dimension J, where $J \leq T$. Such a subspace can always be described as the orthogonal subspace to (the span of) $(T+1) - J$ linearly independent vectors. For example, let $T = 3$ and

$$\mathbb{M} = \left\{r \in \mathbb{R}^4 \mid r(0) + r(1) = 0 \text{ and } r(2) + r(3) = 0\right\}$$

$\mathbb{M}$ is a linear subspace. The vectors $a = (1, -1, 0, 0)$ and $b = (0, 0, 1, -1)$ form a basis for $\mathbb{M}$. The vectors $\beta^1 = (1, 1, 0, 0)$ and $\beta^2 = (0, 0, 1, 1)$

are linearly independent and, both of them, orthogonal to both a and b. They form a basis for the linear subspace orthogonal to $\mathbb{M}$. Hence

$$\begin{aligned}\mathbb{M} = \Big\{ r \in \mathbb{R}^{T+1} \quad &\Big| \quad \textstyle\sum_{t\in\mathbb{T}} \beta^1(t)r(t) = 0 \text{ and } \sum_{t\in\mathbb{T}} \beta^2(t)r(t) = 0 \Big\} \\ &= \{ r \in \mathbb{R}^4 \mid r = \alpha a + \gamma b \quad \text{for some } \alpha, \gamma \in \mathbb{R} \}\end{aligned}$$

$\mathbb{M}$ is a set of net income vectors which allows the consumer to transfer (net) income between date 0 and date 1. $\mathbb{M}$ also allows the consumer to transfer income between date 2 and date 3. There are, however, no means of transferring income between the first two dates and the last two dates.

Since β^1 and β^2 are linearly dependent we lose one dimension for each of the restrictions. Hence $\dim \mathbb{M} = (T+1) - J = (3+1) - 2 = 2$.

Properties of the set of feasible income transfers

In the definition of a spot-market equilibrium relative to a set of net income vectors there are no restrictions on $\mathbb{M}$. When $\mathbb{M}$ is a linear subspace, it is a necessary condition for the existence of a spot-market equilibrium relative to $\mathbb{M}$, as the proposition below shows, that $\mathbb{M}$ is not completely arbitrary.

Proposition 2.7.B *Let $\mathcal{E} = (C^i, u^i, e^i)_{i\in\mathbb{I}}$ be an economy and $((\vec{c}^i)_{i\in\mathbb{I}}, p)$ be a spot-market equilibrium relative to $\mathbb{M}$. If $\mathbb{M}$ is a linear subspace then*

(a) $\mathbb{M} \cap \mathbb{R}^{T+1}_{+} = \{0\}$

(b) $\dim \mathbb{M} \leq T$

(c) *there is $\beta \in \mathbb{R}^{T+1}_{++}$ with $\beta(0) = 1$ such that $\mathbb{M} \subset \{ r \in \mathbb{R}^{T+1} \mid \beta r = 0 \}$.*

Proof: To prove **(a)** assume, in order to arrive at a contradiction, that $r \in \mathbb{M} \cap \mathbb{R}^{T+1}_{+}$ and $r \neq 0$. Then λr, $\lambda > 0$, also belongs to $\mathbb{M}$. In this case there can be no solution to, say, consumer 1's problem of choosing a net income vector since her indirect utility function is increasing in income for a given p. From this contradiction follows that there is no such r. Hence $\mathbb{M} \cap \mathbb{R}^{T+1}_{+} = \{0\}$ as asserted.

(b) $\mathbb{M}$ is a linear subspace and $\mathbb{M} \cap \mathbb{R}_+^{T+1} = \{0\}$ from (a). Hence $\mathbb{M}$ is not equal to $\mathbb{R}^{T+1}$ and it follows that $\dim \mathbb{M} < \dim \mathbb{R}^{T+1}$. Since $\dim \mathbb{R}^{T+1} = T+1$ we have $\dim \mathbb{M} \leq T$.

(c) Since $\mathbb{M}$ is a linear subspace which does not intersect $\mathbb{R}_+^{T+1} \backslash \{0\}$, Farkas' lemma, Lemma F in Appendix C, implies the existence of a vector $\beta \in \mathbb{R}_{++}^{T+1}$ with $\beta(0) = 1$, such that $\mathbb{M} \subset \left\{ r \in \mathbb{R}^{T+1} \mid \beta r = 0 \right\}$.

2.8 Marginal Analysis

In this section we use the indirect utility function to show that for a small enough variation in the net income vector a consumer could use the gradient of the indirect utility function to determine whether the variation is desirable or not.

The First Theorem of Welfare Economics shows that when commodities can be exchanged freely, then the exchange is characterized by the equalization of the subjective (relative) evaluations, for each of the consumers. Here we show that a similar result is true for the exchange of net income between dates. However, we extend the characterization to cases where there are restrictions on the exchange of net income. An example shows that when exchange possibilities are restricted then the subjective evaluations of different consumers' are, in general, not equal in equilibrium.

Evaluation of a small variation in the net income vector

Recall that by the definition of the indirect utility function, v, we have for $(p, r) \in D$,

$$v(p, r(0), r(1), \ldots, r(T)) = u(g_0(p, r), g_1(p, r), \ldots, g_T(p, r))$$

where $(g_t(p, r))_{t \in \mathbb{T}}$ is the solution to

$$\begin{aligned} \text{Max}_c \quad & u(c(0), \ldots, c(T)) \quad \text{s. to } c \in C \text{ and} \\ & p(t)\,[c(t) - e(t)] \leq r(t) \quad \text{for } t \in \mathbb{T} \end{aligned}$$

The gradient of v with respect to r

$$\operatorname{grad}_r v(p,r) \;=\; \begin{bmatrix} \dfrac{\partial v(p,r)}{\partial r(0)} \\ \dfrac{\partial v(p,r)}{\partial r(1)} \\ \vdots \\ \dfrac{\partial v(p,r)}{\partial r(T)} \end{bmatrix}$$

is the *vector of marginal utilities of income.* By Proposition 2.5.B the vector $\operatorname{grad}_r v(p,r)$ has all components positive. Transforming u, by composing it with a differentiable real valued function with positive derivative, to $\hat{u}$, results in a new indirect utility function, say $\hat{v}$, such that $\operatorname{grad}_r v(p,r)$ is proportional to $\operatorname{grad}_r v(p,r)$. Clearly the gradient of the indirect utility function gives only the *relative* evaluation of (net) spot income at different dates.

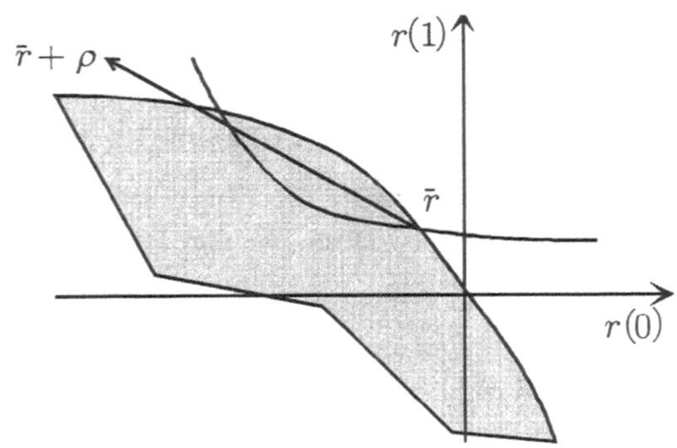

Figure 2.8.A: $\mathbb{M}$ *is the shaded set. A contour set for the indirect utility function is shown. The consumer prefers* $\bar{r}+\tau\rho$ *to* $\bar{r}$ *for* $\tau > 0$ *and small enough*

To see that the gradient will determine which (small) variations in the net income vector that are favorable to the consumer let p be given and let $\bar{r}$ be a net income vector. Consider a variation in the direction of ρ where $\operatorname{grad}_r v(p,\bar{r})\rho > 0$. Let τ be the multiple of ρ chosen so that the consumer gets the net income vector $\bar{r}+\tau\rho$. See Figure 2.8.A.

Taylor expanding $v(p,\cdot)$ around the point $\bar{r}$ we get

$$v(p,\bar{r}+\tau\rho) \;=\; v(p,\bar{r}) + \tau \operatorname{grad}_r v(p,\bar{r})\cdot\rho + h(\tau) \quad \text{where } \lim_{|\tau|\to 0}\frac{h(\tau)}{|\,\tau\,|} = 0$$

Since $\text{grad}_r\, v(p,\bar{r})\cdot\rho > 0$ we get $v(p,\bar{r}+\tau\rho) > v(p,\bar{r})$ for $\tau > 0$ and small enough. An analogous argument shows that if $\text{grad}_r\, v(p,\bar{r})\cdot\rho < 0$, then $v(p,\bar{r}+\tau\rho) > v(p,\bar{r})$ for $\tau < 0$ and τ small enough.

Consider a consumer whose best choice of net income vector at prices p in a linear subspace $\mathbb{M}$ is $\bar{r}$. See Figure 2.8.A. For any vector $\rho \in \mathbb{M}$ and $\tau \in \mathbb{R}$, the vector $\bar{r}+\tau\rho$ belongs to $\mathbb{M}$. Since for τ small enough (positive or negative) the vector $\bar{r}+\tau\rho$ is possible for the consumer but does not give higher utility than $\bar{r}$ we get by the reasoning above

$$\text{grad}_r\, v(p,\bar{r})\cdot\rho = 0 \quad \text{for } \rho\in\mathbb{M}$$

Hence $\text{grad}_r\, v(p,\bar{r})$ is orthogonal to ρ. Since this is true for each $\rho \in \mathbb{M}$, $\text{grad}_r\, v(p,\bar{r})$ belongs to the subspace orthogonal to $\mathbb{M}$. This subspace will be denoted $\mathbb{M}^{\perp}$.

Equalization of the relative evaluations of spot income

Since the non-equalization of evaluations of spot income is a major theme in the theory of incomplete of markets, we begin by giving a slightly different proof the fact that the gradient of the indirect utility function, evaluated at the equilibrium net income vector, belongs to the orthogonal subspace of $\mathbb{M}$.

Proposition 2.8.A *Let $\mathcal{E} = (C^i, u^i, e^i)_{i\in\mathbb{I}}$ be an economy and let $((\bar{c}^i)_{i\in\mathbb{I}}, p)$ be a spot-market equilibrium for $\mathcal{E}$ relative to $\mathbb{M}$ with $\bar{r}^i(t) = p(t)(\bar{c}^i(t) - e^i(t))$ for $i \in \mathbb{I}$ and $t \in \mathbb{T}$. If $\mathbb{M}$ is a linear subspace then $\text{grad}_r\, v^i(p,\bar{r}^i) \in \mathbb{M}^{\perp}$, for $i\in\mathbb{I}$.*

Proof: Let $\mathbb{M}$ have dimension J, where $0 \le J \le T$ and let A be a $(T+1)\times J$ matrix whose column vectors form a basis for $\mathbb{M}$. Let $\bar{r}$ be the income vector chosen by a consumer in the equilibrium. Then $\bar{r}$ is a solution to: $\text{Max}_r\, v(p,r)$ s. to $r\in\mathbb{M}$. Let $\bar{r} = A\bar{\theta}$. Then $\bar{\theta}$ is a solution to

$$\text{Max}_\theta\, v(p,r) \quad \text{s. to } r = A\theta$$

Substituting $A\theta$ into $v(p,r)$ we get from the first order conditions that each of the partial derivatives of v, with respect to $\theta_1,\ldots,\theta_J$ evaluated at $\bar{r} = A\bar{\theta}$, are equal to 0. However, the partial derivative with respect to, say, θ_1, is

$$D_{r(0)}v(p, A\bar{\theta})a_{01} \;+\; D_{r(1)}v(p, A\bar{\theta})a_{11} \;+\; \ldots \;+\; D_{r(T)}v(p, A\bar{\theta})a_{T1}$$

Since this is just the scalar product of $\operatorname{grad}_r v(p, r)$, evaluated at $r = A\bar{\theta}$, and the first column of A it follows that $\operatorname{grad}_r v(p, A\bar{\theta})$ is orthogonal to the first column of A. Considering the partial derivatives with respect to $\theta_2, \ldots, \theta_J$, it is seen that $\operatorname{grad}_r v(p, A\bar{\theta})$ is orthogonal to each of the column vectors of A. Hence $\operatorname{grad}_r v(p, \bar{r})$ is orthogonal to each vector in $\mathbb{M}$ which implies that $\operatorname{grad}_r v(p, \bar{r}) \in \mathbb{M}^\perp$.

□

By Proposition 2.7.B there can not exist a spot-market equilibrium relative to $\mathbb{M}$ unless $\dim \mathbb{M} \leq T$. If possibilities of income transfers are maximal, that is, $\dim \mathbb{M} = T$ then $\dim \mathbb{M}^\perp = 1$. Since the consumers' gradients all belong to $\mathbb{M}^\perp$ they are in this case proportional and the subjective evaluations of spot income are equalized in equilibrium.

In general, however, when $\dim \mathbb{M} < T$ one can only conclude that the relative evaluations of net spot income belong to the orthogonal subspace to $\mathbb{M}$. There is in a spot-market equilibrium relative to $\mathbb{M}$ complete agreement about the relative evaluation of net income vectors in $\mathbb{M}$. In case the dimension of $\mathbb{M}$ is small then this agreement puts few restrictions on relative evaluations; one can only conclude that the relative evaluations, as given by the gradients, belong to the orthogonal subspace, $\mathbb{M}^\perp$.

It might occur that the subjective evaluations of net spot income are equalized also in case $\dim \mathbb{M} < T$. To construct an example, take an economy with two consumers, a and b, $L = 1$, $T = 3$ and a spot-market equilibrium relative to $\mathbb{M} = \{r \in \mathbb{R}^3 \mid \mathbf{1}r{=}0\}$ where $\mathbf{1} = (1, 1, 1)$. $\mathbb{M}$ has dimension 2 . Let $\bar{r}^a$ and $\bar{r}^b$ be the net income vectors in the equilibrium. Assume that $\bar{r}^a \neq 0$. Since we have $\bar{r}^a + \bar{r}^b = 0$, both $\bar{r}^a$ and $\bar{r}^b$ belong to the 1-dimensional subspace spanned by $\bar{r}^a$ and the spot-market equilibrium is a spot-market equilibrium also relative to the smaller subspace spanned by $\bar{r}^a$. Still, of course, the gradients of the indirect utility functions are proportional, in this equilibrium.

Example 2.8.A below illustrates that one can not, in general, expect the subjective evaluations of net income to be equalized, unless the linear subspace of net income vectors has $\dim \mathbb{M} = T$.

Example 2.8.A (When $\dim \mathbb{M} < T$ it is, in general, not true that $\text{grad}_r\, v^a(p, r^a)$ is proportional to $\text{grad}_r\, v^b(p, r^b)$ in a spot-market equilibrium relative to $\mathbb{M}$).
Consider an economy with two consumers, a and b, where $T = 3$ and $L = 1$. The consumers have utility functions

$$u^i(c^i) = \sum_{t \in \mathbb{T}} \alpha^i(t) \ln c^i(t) \quad \text{for } i \in \{a, b\}$$

with parameters given in Table 2.8.A along with two vectors, β^1 and β^2, defining the set $\mathbb{M}$.

Table 2.8.A: *Parameters for the example*

Date	$\beta^1(\cdot)$	$\beta^2(\cdot)$	$\alpha^a(\cdot)$	$\alpha^b(\cdot)$	$e^a(\cdot)$	$e^b(\cdot)$
0	1	0	1/6	2/6	2	1
1	1	0	2/6	1/6	2	1
2	0	1	1/6	2/6	1	2
3	0	1	2/6	1/6	1	2

Thus

$$\mathbb{M} = \{r \in \mathbb{R}^4 \mid \beta^1 r = r(0) + r(1) = 0 \text{ and } \beta^2 r = r(2) + r(3) = 0\}$$

and since β^1 and β^2 are linearly independent we get $\dim \mathbb{M} = 2$. It is not possible to transfer income between the first two dates and the last two dates.

We want to find a spot-market equilibrium relative to $\mathbb{M}$. We begin by finding prices and net income vectors for the consumers.

For $i \in \{a, b\}$ the net income vectors chosen in an equilibrium $\hat{r}^a$ and $\hat{r}^b$ are solutions to

$$\text{Max}_r \quad v^i(p, r^i) \quad \text{s. to } (p, r^i) \in D^i \text{ and}$$
$$\begin{cases} \beta^1(0) r^i(0) + \beta^1(1) r^i(1) & = 0 \\ \beta^2(2) r^i(2) + \beta^2(3) r^i(3) & = 0 \end{cases}$$

There exist $(\lambda_1^a, \lambda_2^a)$ and $(\lambda_1^b, \lambda_2^b)$ such that $\hat{r}^a$ and $\hat{r}^b$ satisfy

$$\begin{cases} \operatorname{grad}_r v^a(p, \hat{r}^a) \; - \; \lambda_1^a \beta^1 \; - \; \lambda_2^a \beta^2 = 0 \\ \operatorname{grad}_r v^b(p, \hat{r}^b) \; - \; \lambda_1^b \beta^1 \; - \; \lambda_2^b \beta^2 = 0 \end{cases}$$

By Example 2.6.A on page 54, the indirect utility functions are

$$v^a(p, \hat{r}^a) = \sum_{t \in \mathbb{T}} \alpha^a(t)[\ln(p(t)e^a(t) + r^a(t)] \; + \; K^a$$
$$v^b(p, \hat{r}^b) = \sum_{t \in \mathbb{T}} \alpha^b(t)[\ln(p(t)e^b(t) + r^b(t)] \; + \; K^b$$

where K^a and K^b are functions that do not depend on r^a or r^b. Thus we get

$$\frac{1}{6} \begin{bmatrix} \dfrac{1}{2p(0) + \hat{r}^a(0)} \\ \dfrac{2}{2p(1) + \hat{r}^a(1)} \\ \dfrac{1}{p(2) + \hat{r}^a(2)} \\ \dfrac{2}{p(3) + \hat{r}^a(3)} \end{bmatrix} - \lambda_1^a \begin{bmatrix} 1 \\ 1 \\ 0 \\ 0 \end{bmatrix} - \lambda_2^a \begin{bmatrix} 0 \\ 0 \\ 1 \\ 1 \end{bmatrix} = 0$$

and

$$\frac{1}{6} \begin{bmatrix} \dfrac{2}{p(0) + \hat{r}^b(0)} \\ \dfrac{1}{p(1) + \hat{r}^b(1)} \\ \dfrac{2}{2p(2) + \hat{r}^b(2)} \\ \dfrac{1}{2p(3) + \hat{r}^b(3)} \end{bmatrix} - \lambda_1^b \begin{bmatrix} 1 \\ 1 \\ 0 \\ 0 \end{bmatrix} - \lambda_2^b \begin{bmatrix} 0 \\ 0 \\ 1 \\ 1 \end{bmatrix} = 0$$

Using that

$$\hat{r}^a(t) \; + \; \hat{r}^b(t) \; = \; 0 \quad \text{for } t \in \mathbb{T}$$

and that $\hat{r}^a$ and $\hat{r}^b$ both belong to $\mathbb{M}$ we get

$$\hat{r}^a(1) \; = \; -\hat{r}^a(0) = \hat{r}^b(0) = -\hat{r}^b(1)$$
$$\hat{r}^a(3) \; = \; -\hat{r}^a(2) = \hat{r}^b(2) = -\hat{r}^b(3)$$

Table 2.8.B: *A spot-market equilibrium*

Date	$p(\cdot)$	$\hat{r}^a(\cdot)$	$\hat{r}^b(\cdot)$	$c^a(\cdot)$	$c^b(\cdot)$
0	1	−1/2	1/2	3/2	3/2
1	5/4	1/2	−1/2	12/5	3/5
2	1	−1/2	1/2	3/2	3/2
3	5/4	1/2	−1/2	3/5	12/5

and some further calculations show that the values in Table 2.8.B give a spot-market equilibrium relative to $\mathbb{M}$.

Calculating the gradients at the equilibrium we get

$$\operatorname{grad}_r v^a(p, \hat{r}^a) = \frac{1}{18}\begin{bmatrix} 2 \\ 2 \\ 4 \\ 4 \end{bmatrix}$$

$$\operatorname{grad}_r v^b(p, \hat{r}^b) = \frac{1}{18}\begin{bmatrix} 4 \\ 4 \\ 2 \\ 2 \end{bmatrix}$$

Hence the gradients are not proportional and the consumers' subjective evaluations of net income are not equalized in the equilibrium. However, as could be expected, the possibility of transferring income between date 0 and date 1, implies that the relative evaluations are equalized for these dates and similarly for dates 3 and 4.

Pareto Optimal Spot-Market Equilibrium Allocations

In a spot-market equilibrium there are no restrictions on the exchange of goods in the spot-markets but possibly on the exchange of income over dates. Theorem 2.8.B below shows that it is precisely the limitations in the exchange of income over dates that results in spot-market equilibrium allocations which are not Pareto optimal allocations. Note that there are no assumptions on $\mathbb{M}$ in the theorem but the subjective evaluations of spot income are assumed to be equalized in the equilibrium. This will always be the case if $\mathbb{M}$ is a linear subspace and $\dim \mathbb{M} = T$.

Theorem 2.8.B (Income Allocations and Pareto Optimal Allocations) *Let $\mathcal{E} = (C^i, u^i, e^i)_{i \in \mathbb{I}}$ be an economy and $((\bar{c}^i)_{i \in \mathbb{I}}, p)$ a spot-market equilibrium for $\mathcal{E}$ relative to $\mathbb{M} \subset \mathbb{R}^{T+1}$ with $\bar{r}^i(t) = p(t)(\bar{c}^i(t) - e^i(t))$ the net income vector in the equilibrium, for $i \in \mathbb{I}$ and $t \in \mathbb{T}$. The equilibrium allocation, $(\bar{c}^i)_{i \in \mathbb{I}}$, is a Pareto optimal allocation if and only if there is a vector $\beta \in \mathbb{R}^{T+1}_{++}$ and $\alpha^i \in \mathbb{R}$, $i \in \mathbb{I}$, such that*

$$grad_r\, v^i(p, \bar{r}^i) = \alpha^i \beta \quad for\ i \in \mathbb{I}$$

Proof: From Corollary 1.2.H we know that an allocation, $(c^i)_{i \in \mathbb{I}}$, is a Pareto optimal allocation if and only if there is a vector $\hat{p} \in \mathbb{R}^{L(T+1)}$ and $\gamma^i \in \mathbb{R}$, $i \in \mathbb{I}$, such that

$$\operatorname{grad} u^i(c^i) = \gamma^i \hat{p} \qquad \textbf{2.8.A}$$

Consider any consumer $i \in \mathbb{I}$. $\bar{c}^i(t)$ is a solution to the Consumer Spot-Market Problem given $r^i = \bar{r}^i$ and satisfies 2.4.B which implies

$$\operatorname{grad}_{c(t)} u^i(\bar{c}) - \lambda^i(t) p(t) = 0 \quad \text{for } t \in \mathbb{T}$$

From the proof of Proposition 2.5.B it is seen that

$$D_{r(t)} v^i(p, \bar{r}^i) = \lambda^i(t) \quad \text{for } t \in \mathbb{T} \qquad \textbf{2.8.B}$$

so that the Lagrange multipliers of the Consumer Spot-Market Problem given r are equal to the partial derivatives of the indirect utility function.

Now consider two different consumers, a and b. We have, using 2.4.B and 2.8.B,

$$\operatorname{grad}_c u^a(\bar{c}^a) = \begin{bmatrix} \frac{\partial v^a(p, \bar{r}^a)}{\partial r(0)} p(0) \\ \frac{\partial v^a(p, \bar{r}^a)}{\partial r(1)} p(1) \\ \vdots \\ \frac{\partial v^a(p, \bar{r}^a)}{\partial r(T)} p(T) \end{bmatrix} \qquad \operatorname{grad}_c u^b(\bar{c}^b) = \begin{bmatrix} \frac{\partial v^b(p, \bar{r}^b)}{\partial r(0)} p(0) \\ \frac{\partial v^b(p, \bar{r}^b)}{\partial r(1)} p(1) \\ \vdots \\ \frac{\partial v^b(p, \bar{r}^b)}{\partial r(T)} p(T) \end{bmatrix}$$

Clearly $\operatorname{grad}_c u^a(\bar{c}^a)$ is proportional to $\operatorname{grad}_c u^b(\bar{c}^b)$ if and only if the vector $\operatorname{grad}_r v^a(p, \bar{r}^a)$ is proportional to $\operatorname{grad}_r v^a(p, \bar{r}^b)$. It follows that the allocation is a Pareto optimal allocation if and only if the gradients of the indirect utility functions are all proportional.

Summary

In this chapter we have studied economies over time. We have seen that two different set of institutions; a complete market for all commodities at the current date *or* a market for future income at date 0 together with spot-markets for the goods at each date both resulted in a Walras equilibrium.

Using the interpretation over time we introduced the spot-market demand function which we then applied in the definition of the indirect utility function. The indirect utility function allowed the consumer to evaluate different net income vectors and made possible a decomposition of the consumer problem into a choice of a net income vector and a choice of commodity bundles in the spot-markets at different dates.

The market institutions with maximal possibilities to transfer income and with no such possibilities were seen to be particular instances of spot-market equilibria relative a set of income transfers.

Finally we considered how the possibilities of equalizing the relative evaluation of spot income between the consumers were related to the set of feasible net income vectors. We showed that it was precisely the restrictions on net income transfers which opened up for the possibilities of spot-market equilibrium allocations which are not Pareto optimal allocations.

Exercises

In the Exercises 2.A to 2.C below we will consider a family of economies $\mathcal{E} = (C^i, u^i, e^i)_{i \in \mathbb{I}}$ with $\mathbb{I} = \{a, b\}$, $\mathbb{T} = \{0, 1, 2\}$ and $L = 2$ where initial endowments vary. The utility functions are

$$u^a(c) = \sum_{t=0}^{2} \alpha_1(t) \ln c_1(t) + \alpha_2(t) \ln c_2(t) \quad \text{and}$$
$$u^b(c) = \sum_{t=0}^{2} \gamma_1(t) \ln c_1(t) + \gamma_2(t) \ln c_2(t)$$

where

$$\begin{array}{llll} \alpha_1(0) = \frac{1}{4} & \alpha_2(0) = \frac{1}{4} & \gamma_1(0) = \frac{3}{8} & \gamma_2(0) = \frac{1}{8} \\ \alpha_1(1) = \frac{1}{12} & \alpha_2(1) = \frac{1}{12} & \gamma_1(1) = \frac{3}{12} & \gamma_2(1) = \frac{1}{12} \\ \alpha_1(2) = \frac{3}{12} & \alpha_2(2) = \frac{1}{12} & \gamma_1(2) = \frac{1}{12} & \gamma_2(2) = \frac{1}{12} \end{array}$$

and with initial endowments to be specified.

Exercise 2.A Let $\sum_{t\in\mathbb{T}}[e_1^a(t) + e_2^a(t)] = 48$.

(a) Find initial endowments where $e^a = e^b$ such that the economy has a Walras equilibrium with $(p(0), p(1), p(2)) = ((1,1),(1,1),(1,1))$ as equilibrium price system. Find a Walras equilibrium $(p, \bar{c}^a, \bar{c}^b)$ for the economy.

(b) Show that the Walras equilibrium found in (a) can be interpreted as an equilibrium with spot-markets for the goods and a market for income at date 0. What are the spot-prices for the goods and the date 0 prices of date 1 and date 2 income ? What are the associated net income vectors?

(c) Let $\beta = (\beta(0), \beta(1), \beta(2))$, where $\beta(0) = 1$ and $\beta(1), \beta(2) \in \mathbb{R}_{++}$ are the date 0 prices of income for date 1 and date 2. Find a spot-market equilibrium with the date 0 prices for date 1 and date 2 income equal to given $\beta(1)$ and $\beta(2)$ respectively. Show that consumer a has the same consumptions available in the spot-market equilibrium as in the Walras equilibrium. Find the consumers' actions in the market for income and show that this market balances.

(d) Find endowments e^a and e^b with

$$\sum_{t\in\mathbb{T}}[e_1^a(t) + e_2^a(t)] = \sum_{t\in\mathbb{T}}[e_1^b(t) + e_2^b(t)] = 48$$

such that the economy has a no-trade equilibrium (in which the consumers demand their initial endowment) with $p = ((1,1),(1,1),(1,1))$ as an equilibrium price system.

Exercise 2.B Let the economy be defined as in Exercise 2.A but with the initial endowments $e^a = (12,12,4,4,12,4)$ and $e^b = (18,6,12,4,4,4)$.

(a) Find the spot-market demand functions.

(b) Find the indirect utility function.

(c) Solve the Consumer Net Income Vector Problem at date 0 prices for income $\beta = (\beta(0),\beta(1),\beta(2))$, where $\beta(0) = 1$ and $\beta \in \mathbb{R}^3_{++}$ and spot-prices $p \in \mathbb{R}^6_{++}$. State the equilibrium conditions for the market for spot income at date 0.

(d) For given spot-prices, $(p(0),p(1),p(2)) = ((1,1),(2,2),(4,4))$, find date 0 prices for income, $\beta = (\beta(0),\beta(1),\beta(2))$, where $\beta(0) = 1$, such that β gives market balance in the date 0 market for income. Find the net income vectors demanded by consumer a and b and show that in the equilibrium the gradients of the indirect utility functions are proportional. Is the the equilibrium allocation a Pareto optimal allocation?

(e) Use your result from (a) to show that $p = ((1,1),(2,2),(4,4))$ gives market balance for the goods market at each date.

Exercise 2.C Assume that the initial endowments are

$$
\begin{aligned}
e^a(0) &= \begin{bmatrix} 4 \\ 2 \end{bmatrix} & e^b(0) &= \begin{bmatrix} 1 \\ 1 \end{bmatrix} \\
e^a(1) &= \begin{bmatrix} 1 \\ 1 \end{bmatrix} & e^b(1) &= \begin{bmatrix} 4 \\ 2 \end{bmatrix} \\
e^a(2) &= \begin{bmatrix} 1 \\ 1 \end{bmatrix} & e^b(2) &= \begin{bmatrix} 4 \\ 2 \end{bmatrix}
\end{aligned}
$$

and the set of net income vectors is

$$\mathbb{M} = \{r \in \mathbb{R}^3 \mid r = \theta(-1,1,1), \theta \in \mathbb{R}\}$$

(a) Formulate the Consumer Net Income Vector Problem for consumer a and b as a choice of θ^a and θ^b. What are the marginal conditions that a solution must satisfy?

(b) Show that for spot-prices $(p(0), p(1), p(2)) = ((1,1),(1,1),(1,1))$, $\theta^a = -\theta^b = 2$ induce solutions to the Consumer Net Income Vector Problem for consumer a and b.

(c) Check that $(p(0), p(1), p(2)) = ((1,1),(1,1),(1,1))$ gives market balance on the spot-markets for the goods. Find a spot-market equilibrium relative to $\mathbb{M}$.

(d) Evaluate the gradients of the indirect utility functions at the equilibrium. Is the equilibrium allocation a Pareto optimal allocation?

Exercise 2.D Let $T = 1$ and L arbitrary and let the consumer (C, u, e) satisfy the Maintained Assumptions. Consider the indirect utility function $v : D \longrightarrow \mathbb{R}$.

(a) Show that the function $p(1) \longrightarrow v(p(0), \cdot, r(0), r(1))$ is a decreasing function of $p(1)$, that is, $\bar{p}(1) > \hat{p}(1)$ implies

$$v(p(0), \bar{p}(1), r(0), r(1)) < v(p(0), \hat{p}(1), r(0), r(1)$$

(b) Show that the indirect utility function with prices given, that is the function, $v(p, \cdot) : E(p) \longrightarrow \mathbb{R}$ is a strictly quasi-concave function using that the (direct) utility function u is increasing in each argument and a strictly quasi-concave function.

(c) Show that, for $\lambda > 0$,

$$D_{r(1)}v(p(0), \lambda p(1), 0, 0) = \frac{1}{\lambda} D_{r(1)}v(p(0), p(1), 0, 0) \quad \text{for } \lambda > 0$$

and interpret.

Exercise 2.E Let $((\bar{c}^i)_{i\in\mathbb{I}}, p)$ be a spot-market equilibrium relative to $\mathbb{M}$, with $\bar{r}^i(t) = p(t)(\bar{c}^i(t) - e^i(t))$, for $t \in \mathbb{T}, i \in \mathbb{I}$ and let

$$Z = \left\{ r \in \mathbb{R}^{T+1} \mid r = \textstyle\sum_{i\in\mathbb{I}_1} \bar{r}^i \text{ for some subset } \mathbb{I}_1 \subset \mathbb{I} \right\}$$

(a) Show that $0 \in Z$ and that $r \in Z$ implies $-r \in Z$.

(b) Assume that $T = 1, I = 4$ and $\bar{r}^1 = (-7,6), \bar{r}^2 = (-1,2), \bar{r}^3 = (6,-7)$ and $\bar{r}^4 = (2,-1)$. Check that this the net income vectors balance. Let $\mathbb{M} = \{\bar{r}^1, \bar{r}^2, \bar{r}^3, \bar{r}^4\}$. Is there an economy with a spot-market equilibrium with these net income vectors and with $\mathbb{M}$ so defined ? Find the set Z and illustrate in a diagram. Is it true that $Z \cap \mathbb{R}^{T+1} = \{0\}$?

Assume in part (c) and (d) that $\mathbb{M}$ satisfies $\mathbb{M} + \mathbb{M} = \mathbb{M}$

(c) Show that $Z \subset \mathbb{M}$ and hence $v^i(p, \bar{r}^i) \geq v^i(p, r)$ for $r \in Z$

(d) Show that $Z \cap \mathbb{R}_+^{T+1} = \{0\}$.

Exercise 2.F Let $T = 2$ and $\mathbb{I} = \{a, b\}$ and let $((\bar{c}^i)_{i \in \mathbb{I}}, p)$ be a spot-market equilibrium relative to$\mathbb{M}$, with $\bar{r}^i(t) = p(t)(\bar{c}^i(t) - e^i(t))$, for $t \in \mathbb{T}$,$i \in \mathbb{I}$. Let the set of net income vectors be

$$\mathbb{M} = \left\{r \in \mathbb{R}^2 \mid [(r(0)+1)^2 + (r(1)+1)^2 + (r(2)+1)^2]^{1/2} \leq 3^{1/2}\right\}$$

(a) Let $\bar{r}^a$ and $\bar{r}^b$ be solutions to consumer a's and b's Net Income Vector Problem at spot-prices $p \in \mathbb{R}^6_{++}$. The solutions belong to the boundary of $\mathbb{M}$ and $\bar{r}^i \geq (-1,-1,-1)$, $i = a, b$. Why?

(b) Assume that there is a spot-market equilibrium for $\mathcal{E}$ relative to $\mathbb{M}$. Show that $\bar{r}^a = \bar{r}^b = 0$ in the equilibrium and that the equilibrium is also a spot-market equilibrium relative to

$$\mathbb{H} = \left\{r \in \mathbb{R}^3 \mid r(0) + r(1) + r(2) = 0\right\}$$

Chapter 3

ASSETS AND INCOME TRANSFERS

Introduction

This chapter will be concerned with equilibrium over time but as in Chapter 2 all the definitions, concepts and results carry over to economies with uncertainty and two dates.

In Chapter 2 we saw that it is possible to decompose the Consumer Problem into a problem of choosing a net income vector and given the net income vector a choice of a goods bundle at each date. We defined an equilibrium relative to a set of net income vectors and gave examples where the possibilities of transferring income between dates were more or less restricted.

Here we specify in more detail the institutions used to allow the agents to make income transfers between dates. We will see how the set of feasible net income vectors may arise from the exchange of *assets*. The net income vector is in this case the dividend vector resulting from a choice of a *portfolio*, that is, a bundle of assets. The assets may be *nominal assets* paying dividends in the units of account used at the different dates or *real assets* promising the holder goods bundles as dividends or giving him an obligation to deliver some goods bundles at future dates. These are studied in Section 3.1 and 3.3. When the asset structure is rich enough to allow the agents to freely transfer income between dates we have *complete (asset) markets.* The distinction between complete and incomplete asset markets is the subject of Section 3.1.

The introduction of assets in the economy leads to the problem of *pricing assets,* usually at date 0, in order to achieve market balance in the asset market. The consumers are assumed to be interested in getting as much of the goods as possible. This leads to a desire for large incomes to be used in the spot-markets. If there are *arbitrage* possibilities then there is a kind of inconsistency in asset prices. This inconsistency makes it possible for a consumer to acquire an arbitrarily large net income to be used at some date without giving up income at any other date. (This is often described as the possibility of a "free lunch"). If there are arbitrage possibilities then the commodity markets can not balance. Hence it is a necessary condition for the existence of an equilibrium that assets are priced so as not to allow arbitrage. The notion of arbitrage is closely related to the existence of *discount factors* defining the date 0 prices of the assets. This is the subject of Section 3.2.

For an economy with assets we define a *Radner equilibrium* in Section 3.5. This is a particular case of an equilibrium relative to a set of income transfers. In a Radner equilibrium the set of income transfers is generated by the assets of the economy. In the equilibrium we obtain spot-prices of the goods at each date and date 0 prices for all the assets. Each consumer chooses a portfolio and (plans) a sequence of consumptions in the spot markets. In the equilibrium prices of commodities and assets are adjusted so that the choices of the agents are consistent in the sense that commodity and asset markets balance.

With *complete asset* markets there is underlying a Radner equilibrium an implicit Walras equilibrium and the Radner equilibrium is simply a more detailed specification of how the equilibrium is attained. This means that many of the results which are true for Walras equilibria also hold for Radner equilibria when asset markets are complete.

We apply the concepts introduced to an economy with production in Section 3.6. Apart from the traditional role of the producer of transforming inputs into outputs; an activity typically carried out over time, the producer performs a second role of creating assets through the choice of a production plan. Independently of whether markets are complete or not the consumers-owners are not concerned about the acts of the producer in his second role in case the producer is merely able to replicate existing assets. This is the Modigliani-Miller theorem [1958]. Another theme is the objective function of the producer and its acceptance by the owners. With complete asset markets maximization of profits is a well-defined objective upon which all owners will agree. With incomplete asset markets profit maximization is not well-defined and there are in general many production plans with the property that there is no other unanimously preferred production plan by the owners.

In the sequel it is of interest to distinguish carefully between assets and fiat money. The distinction is brought out by the study an *economy with (fiat) money* and a *monetary equilibrium* in Section 3.7.

3.1 Nominal Assets

In Chapter 2 we studied the case where consumers could trade spot income to be delivered at future dates at date 0. With asset trade at the current date consumers are not necessarily able to trade spot income for each date by itself. An asset is defined by a dividend vector and these dividend vectors are the objects of trade. This feature of trade in assets can be illustrated by the following example which relates the exchange of assets to the exchange of baskets of goods.

Example 3.1.A Trade in baskets of goods
Consider an economy with $T = 0$ and $L = 3$ so that there is only one date and three goods. Assume that the consumers are allowed to trade only some specified *baskets* of the goods. A three course meal in a restaurant gives an example of this. The appetizer, main course and dessert can not be bought separately but only as a basket. Another example is a subscription to American Economic Review which comes together with subscriptions to Journal of Economic Literature and Economic Perspectives. We disregard the cost of the consumptions or baskets and thus concentrate on what can be achieved regardless of cost. We start by considering the case usually studied where each good may be traded by itself.

Case 1. Simple Baskets

$$\begin{bmatrix} 1 \\ 0 \\ 0 \end{bmatrix}, \begin{bmatrix} 0 \\ 1 \\ 0 \end{bmatrix} \quad \text{and} \quad \begin{bmatrix} 0 \\ 0 \\ 1 \end{bmatrix}$$

It corresponds to the case where the baskets are such that the j'th basket contains only a single unit of good j, $j = 1, 2, 3$.

Case 2a. Baskets not Restricting Consumer Choice

Assume that the consumers are restricted to exchange baskets of goods defined by

$$\begin{bmatrix} 1 \\ 2 \\ 2 \end{bmatrix}, \begin{bmatrix} 3 \\ 2 \\ 1 \end{bmatrix} \quad \text{and} \quad \begin{bmatrix} 1 \\ 1 \\ 0 \end{bmatrix}$$

Basket 1, for example, contains 1 unit of good 1, 2 units of good 2 and 2 units of good 3. A consumer who obtains $\theta = (\theta_1, \theta_2, \theta_3) = (5, -1, 3)$ units of the baskets gets the consumption c where

$$c = \begin{bmatrix} 1 \\ 2 \\ 2 \end{bmatrix} 5 + \begin{bmatrix} 3 \\ 2 \\ 1 \end{bmatrix} (-1) + \begin{bmatrix} 1 \\ 1 \\ 0 \end{bmatrix} 3 = \begin{bmatrix} 5 \\ 11 \\ 9 \end{bmatrix}$$

Let M denote the matrix with the vectors defining the baskets as columns. It is easy to check that M is invertible. Hence a consumer who desires the consumption c may buy (or sell) $\theta = (\theta_1, \theta_2, \theta_3) \in \mathbb{R}^3$ of the baskets where

$$M\theta = c \quad \text{or} \quad \theta = M^{-1}c$$

to obtain the consumption c. Since M is invertible the equations have a solution in θ for any $c \in \mathbb{R}^3$ so the exchange of goods is not hampered by the restriction that exchange may occur only in the baskets.

Case 2b. Baskets Restricting Consumer Choice

On the other hand if only the first two baskets are available and the consumer desires the consumption $c = (1, 1, 1)$ this will not be possible since

$$\begin{bmatrix} 1 \\ 2 \\ 2 \end{bmatrix} \theta_1 + \begin{bmatrix} 3 \\ 2 \\ 1 \end{bmatrix} \theta_2 = \begin{bmatrix} 1 \\ 1 \\ 1 \end{bmatrix}$$

does not have a solution. In this case the restriction to exchange of baskets rather than exchange of individual goods affects the choices open to the to consumer.

□

An asset is, loosely speaking, a contract which promises the holder income at different points in time or obliges the holder to make some particular payments. In relation to Example 3.1.A the assets define the "baskets of spot income" which the consumers may trade. In Chapter 2 the agents could trade date t income directly corresponding to Case 1 in Example 3.1.A. Here we will study more general "baskets of spot income"; assets.

Short and long positions

In real economies it is usually the case that the consumers can not buy and sell unlimited quantities of an asset. One reason for this is the possibility of default. There are also many contracts which the consumer may buy but which are not possible for the consumer to sell. For example, a consumer may borrow from a bank at some interest rate but it is, in general, not possible for the consumer to lend to the bank at the same interest rate.

The assets in the economy is a way of defining which contracts are allowed. Any agent may buy or sell any asset but agents may not enter into contracts of exchange of spot incomes between them which are not among the assets. Thus the assets specify the objects of trade.

It is assumed that agents have no initial endowment of assets which implies that the asset trade results in promises of delivery or claims to delivery of units of account at different dates which sum to 0. An agent who holds a positive amount of an asset is said to have a *long position* in the asset. An agent who has sold, or issued, the asset and thus has a negative amount of the asset is said to have a *short position* in the asset. Since assets are in 0 net supply the sum of what the agents hold who are in long positions and what those who are in short positions have issued is 0. When the agents may buy or sell the assets whose prices are given in 0−crowns it is an equilibrium condition that demand and supply of any asset equals 0. This in turn ensures that the claims to spot income and obligations to deliver spot income are equal at each future date. The exchange of assets thus results in a redistribution of purchasing power over dates for the consumers.

An example of complete asset markets

Let us first give an example where the assets clearly allow the consumers to obtain any net income vector at future dates.

Example 3.1.B Arrow-Debreu assets
Assume that there are $J = T$ assets. For a future date, $\tau \in \mathbb{T}_1$, asset τ is a contract to deliver 1 τ-crown at date τ and nothing at the other dates. Hence the dividend vectors of the assets are

	Asset					
Date	1	2	...	τ	...	T
1	1	0	...	0	...	0
2	0	1	...	0	...	0
$\vdots$	$\vdots$					
τ	0	0	...	1	...	0
$\vdots$	$\vdots$	$\vdots$		$\vdots$		$\vdots$
T	0	0	...	0	...	1

Consider a consumer holding a portfolio $\theta \in \mathbb{R}^T$ where $\theta = (\theta_1, \theta_2, ..., \theta_T)$. If $\theta_1 > 0$ so that the consumer has a long position in asset 1 then the consumer will *receive* the amount θ_1 of 1−crowns at date 1.

If $\theta_1 < 0$ so that the consumer has a short position in asset 1 then she is obliged to *deliver* the amount $|\theta_1|$ of 1−crowns at date 1. Since assets are in 0 net supply the sum of (planned) deliveries to and from the consumers at date t, for $t \in \mathbb{T}_1$, in the economy is 0.

A consumer holding the portfolio θ obtains the net income vector $(r(1), r(2), ..., r(T)) = (\theta_1, \theta_2, ..., \theta_T)$ at future dates.

Asset t is simply spot-income to be delivered at date t. This is precisely the situation that we encountered in Chapter 2 and we will see that the date 0 prices for income (discount factors) will agree with the date 0 prices of the Arrow-Debreu assets.

Nominal assets, dividends and portfolios

We now proceed to the general case. A *nominal asset* is defined by its *dividend vector* for future dates $v = (v(1), v(2), ..., v(T)) \in \mathbb{R}^T$. The holder of 1 unit of the asset is entitled to receive $v(t)$ t-crowns at date t if $v(t) > 0$ and is obliged to deliver the amount $|v(t)|$ of t-crowns at date t if $v(t) < 0$.

Let $\mathbb{J} = \{1, 2, ..., J\}$ be the set of assets making up the *asset structure*. It is convenient to define a matrix with the dividend vectors as columns to describe the asset structure. We then get a $T \times J$ matrix.

	Asset			
Date	1	2	...	J
1	$v^1(1)$	$v^2(1)$	...	$v^J(1)$
2	$v^1(2)$	$v^2(2)$	...	$v^J(2)$
$\vdots$	$\vdots$	$\vdots$		$\vdots$
t	$v^1(t)$	$v^2(t)$	...	$v^J(t)$
$\vdots$	$\vdots$	$\vdots$		$\vdots$
T	$v^1(T)$	$v^2(T)$	...	$v^J(T)$

This matrix is the *dividend matrix* of the asset structure and will be denoted by V. Assume that the assets are traded only at date 0 but for the moment disregard the date 0 cost of the assets. A *portfolio* is a vector $\theta \in \mathbb{R}^J$, where θ_j is a positive or negative number, showing the amount of asset j in the portfolio. The owner, or holder, of a portfolio θ is entitled to receive or has an obligation to deliver t-crowns for $t \in \mathbb{T}$. The payments made to, or by, the holder of a portfolio θ as a consequence of his holding of asset j are:

- if $\theta_j > 0$ then the owner, or holder, of the portfolio *receives* the amount $\theta_j v^j(t)$ of t-crowns at date t if $v^j(t) > 0$ and *delivers* $|\theta_j v^j(t)|$ at date t if $v^j(t) < 0$.

- if $\theta_j < 0$ then the owner, or holder, of the portfolio *delivers* the amount $|\theta_j v^j(t)|$ of t-crowns at date t if $v^j(t) > 0$ and *receives* $|\theta_j v^j(t)|$ at date t if $v^j(t) < 0$.

Hence no matter what the signs of θ_j and $v^j(t)$, if $\theta_j v^j(t) > 0$ then there is a *delivery to* the portfolio holder of t-crowns at date t and if $\theta_j v^j(t) < 0$ then there is a *delivery from* the agent at date t.

The total payment received at date t from holding the portfolio is simply the sum of payments induced by the different assets. Thus an agent holding a portfolio $\theta \in \mathbb{R}^J$ receives (if positive) or delivers (if negative) at date t the following amount of t-crowns

$$v^1(t)\theta_1 + v^2(t)\theta_2 + \dots + v^J(t)\theta_J$$

The amount of t-crowns received or delivered at date t is thus the sum of what is received from holding asset j, for $j \in \mathbb{J}$. Hence a consumer holding a portfolio θ gets a net income vector, r, for future dates which is a linear combination of the columns of V.

$$r = \begin{bmatrix} r(1) \\ \vdots \\ r(t) \\ \vdots \\ r(T) \end{bmatrix} = \begin{bmatrix} v^1(1) \\ \vdots \\ v^1(t) \\ \vdots \\ v^1(T) \end{bmatrix} \theta_1 + \begin{bmatrix} v^2(1) \\ \vdots \\ v^2(t) \\ \vdots \\ v^2(T) \end{bmatrix} \theta_2 + \dots + \begin{bmatrix} v^J(1) \\ \vdots \\ v^J(t) \\ \vdots \\ v^J(T) \end{bmatrix} \theta_J = V\theta$$

Complete and Incomplete Asset Markets

The net income vectors, for $t \in \mathbb{T}_1$, which can be achieved by holding some portfolio is the linear subspace spanned by the column vectors of V. This subspace will be denoted by $\langle V \rangle$. It is the *subspace of (net) income transfers*. If the vectors $v^1, v^2 \dots, v^J$ are linearly independent then $\dim \langle V \rangle = J$. Since the row rank and column rank of a matrix agrees and V has T rows, it will always be the case that $\dim \langle V \rangle \leq T$, independent of the number of assets.

It turns out that in the definition of complete and incomplete asset markets it is convenient to disregard the date 0 cost of the assets to begin with.

Definition 3.1.A *If* $\dim \langle V \rangle = T$ *then we have* **complete asset markets**. *If* $\dim \langle V \rangle < T$ *we have* **incomplete asset markets**.

If dim $\langle V \rangle = T$ then a consumer can, disregarding the date 0 cost, obtain any net income vector at future dates. If dim $\langle V \rangle < T$ then there will always be some net income vector which is not available to the consumer at *any* date 0 cost.

We now turn to the cost of the assets at date 0. Let $q \in \mathbb{R}^J$ be the date 0 prices of the assets in the unit 0-crowns. Since an asset may be an obligation to deliver t-crowns at date t for one or more dates, one can not expect asset prices necessarily to be positive. If the price of asset j is negative, $q_j < 0$, this means that the seller of asset j has to deliver q_j 0-crowns for each unit of asset j, that is delivered to the buyer at date 0. Note that q_j is the date 0 cost of a portfolio containing a single unit of asset j and nothing of the other assets. This suggests that it is not the prices of individual assets that are of importance but the date 0 cost of portfolios.

The extended dividend matrix

By enlarging the dividend matrix, V, with a first row containing the date 0 prices of the assets (corrected for sign) we get the *extended dividend matrix*.

	Asset			
Date	1	2	...	J
0	$-q_1$	$-q_2$	...	$-q_J$

Thus the extended dividend matrix is

$$W = \begin{bmatrix} -q \\ V \end{bmatrix}$$

In order to acquire the portfolio θ an agent must deliver 0-crowns in the amount of

$$(-q_1)\theta_1 + (-q_2)\theta_2 + \ldots + (-q_T)\theta_J$$

which is (the negative of) the date 0 cost of the portfolio.

The net income vector for dates $t \in \mathbb{T}$ accruing to a consumer holding a portfolio θ, is $W\theta$ which is a linear combination of the columns of the extended dividend matrix. $W\theta$ is a vector in $\mathbb{R}^{T+1}$. If the t'th component

is positive the consumer *receives* t-crowns and if the t'th component is negative the consumer *delivers* t-crowns at date t.
The linear subspace spanned by the columns of W is the *extended subspace of income transfers*. Hence in an economy with assets having an extended dividend matrix W the consumer has access to the following net income vectors

$$\mathbb{M} = \{r \in \mathbb{R}^{T+1} \mid r = W\theta \quad \text{for some } \theta \in \mathbb{R}^J\} = \langle W \rangle$$

Here $\langle W \rangle$ denotes the linear subspace spanned by the column vectors of W. Note that a consumer is not restricted in any way in his choice of a portfolio since the matrix W incorporates the date 0 cost of the assets in the first row. (However some choices of θ may result in net income vectors which does not allow the consumer to choose a consumption in the consumption set.)

Example 3.1.C Prices of spot income and Arrow-Debreu assets
For the Arrow-Debreu assets considered in Example 3.1.B the date 0 prices ought to be simply the date 0 prices of spot-income at different dates. These prices will be determined by the equilibrium conditions. The extended dividend matrix is given in Table 3.1.A

Table 3.1.A: *The Arrow-Debreu assets*

	Asset					
Date	1	2	$\dots$	τ	$\dots$	T
0	$-\beta(1)$	$-\beta(2)$	$\dots$	$-\beta(\tau)$	$\dots$	$-\beta(T)$
1	1	0	$\dots$	0	$\dots$	0
2	0	1	$\dots$	0	$\dots$	0
$\vdots$	$\vdots$	$\vdots$		$\vdots$		$\vdots$
τ	0	0	$\dots$	1	$\dots$	0
$\vdots$	$\vdots$	$\vdots$		$\vdots$	$\ddots$	
T	0	0	$\dots$	0	$\dots$	1

With the portfolio choice $\theta \in \mathbb{R}^T$ the consumer obtains the net income vector

$$\begin{bmatrix} r(0) \\ r(1) \\ r(2) \\ \vdots \\ r(t) \\ \vdots \\ r(T) \end{bmatrix} = \begin{bmatrix} -\Sigma_{t\in\mathbb{T}_1}\beta(t)\theta_t \\ \theta_1 \\ \theta_2 \\ \vdots \\ \theta_t \\ \vdots \\ \theta_T \end{bmatrix} = W\theta$$

□

The example above indicates that it is not the assets themselves that are of interest to the agents but the net income vectors that can be achieved by holding a portfolio of the assets.

We may think of the assets as defining some net income vectors which the consumers can exchange between them. When the asset market is complete it is of no consequence to the consumers that they can not at date 0 directly exchange income to be used in the spot-markets. They can still achieve any desired net income vector by trading in the assets. When the asset market is incomplete then the restriction of exchanges of income to those that can be achieved through asset trade is usually important to the consumers. Cf. Example 3.1.A.

3.2 Arbitrage and Date 0 Prices of Income

Replicating portfolios

Let $(v^1, v^2, ..., v^T)$ be the Arrow-Debreu assets. Thus

$$v^t(\tau) = \begin{cases} 1 & \text{if } \tau = t \\ 0 & \text{if } \tau \neq t \end{cases}$$

Let $q \in \mathbb{R}^T$ be the date 0 asset prices of the Arrow-Debreu assets.

Consider now an asset with dividend vector $(\hat{v}(t))_{t\in\mathbb{T}_1}$. Then $\hat{v}$ is the unique linear combination $\theta = (\hat{v}(t))_{t\in\mathbb{T}_1} \in \mathbb{R}^T$ of the Arrow-Debreu assets. The portfolio θ *replicates* $\hat{v}$ in the sense that θ promises the same future dividends as $\hat{v}$. Since $\hat{v}$ gives the holder the same future net income vector as the portfolio θ its date 0 cost ought to be the same as the date 0 cost of the portfolio θ. Denoting by $\hat{q}$ the date 0 price of $\hat{v}$ we should have

$$\hat{q} = q_1\theta_1 + q_2\theta_2 + \ldots + q_T\theta_T = c(\theta)$$

where $c(\theta)$ is the date 0 cost of the portfolio θ.

Assume that $\hat{q} - c(\theta) \neq 0$. Multiply by a real number z and consider $z(\hat{q} - c(\theta))$. If $\hat{q} - c(\theta) > 0$ then the agent can sell z units of the asset $\hat{v}$ and buy the portfolio $z\theta$ of the Arrow-Debreu assets. For this she receives net the positive amount $z\hat{q} - zc(\theta)$ 0-crowns at date 0. On the other hand the dividends from the Arrow-Debreu assets suffice to cover the deliveries of t-crowns she is obliged to make for $t \in \mathbb{T}_1$ from the sale of z units of $\hat{v}$. If $\hat{q} - c(\theta) < 0$ she should reverse the trade buying $\hat{v}$ and selling a corresponding amount of the portfolio θ. In either case the agent makes a profit in 0-crowns and gets "something for nothing" by the arbitrage operation.

Assets can not in equilibrium be priced so that there is a possibility of arbitrage. If there is a possibility of arbitrage then, for example, consumer 1's problem will not have a solution (provided she has monotone preferences). The absence of arbitrage possibilities is thus a necessary condition for the existence of an equilibrium.

We will now show that the constraint that assets are priced so as not to allow arbitrage implies the existence of date 0 prices for income (discount factors) with the property that for each asset its price at date 0 equals its discounted value. This is true both for the case of complete asset markets and for incomplete asset markets.

Absence of arbitrage and existence of discount factors

Let V be an asset structure with J assets. Hence V is a $T \times J$ matrix. Let $q \in \mathbb{R}^J$ be the prices of the assets and

$$W = \begin{bmatrix} -q \\ V \end{bmatrix}$$

the extended dividend matrix. Recall that $\langle W \rangle$ denotes the linear subspace spanned by the column vectors of W.

Definition 3.2.A *Let* (V, q) *be an asset structure and asset prices.* (V, q) *is* **arbitrage free** *if* $\langle W \rangle \cap \mathbb{R}_{+}^{T+1} = \{0\}$.

This definition makes sense since $\langle W \rangle$ is the set of net income vectors available to a consumer. If the condition in the definition is not satisfied then there is some portfolio $\theta \in \mathbb{R}^J$ such that $W\theta = r$ and $r > 0$ (so that $r(t) \geq 0$ for $t \in \mathbb{T}$ with some strict inequality). Such a portfolio will be referred to as an *arbitrage portfolio.*

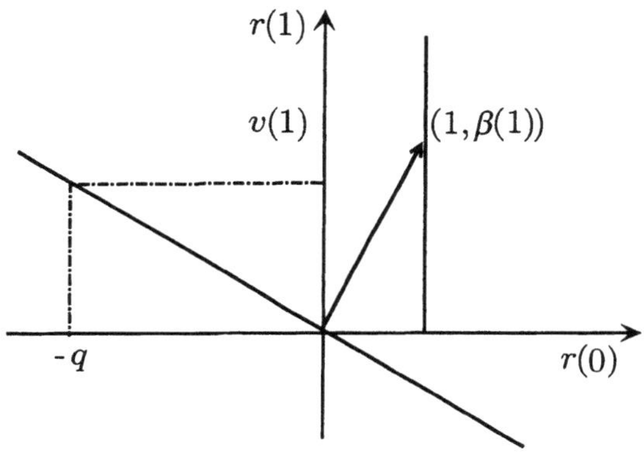

Figure 3.2.A: *The single asset with extended dividend vector* $(-q, v(1))$ *does not allow for arbitrage. Hence there is* $(1, \beta(1))$ *such that* $1 \cdot -q + \beta(1)v(1) = 0$

Given V there always *exists* asset prices q such that (V, q) is arbitrage free. To see this choose any vector $\beta \in \mathbb{R}_{++}^{T+1}$ with $\beta(0) = 1$. For each $j \in J$, let $q_j = -\Sigma_{t \in \mathbb{T}_1} \beta(t) v^j(t)$, or equivalently, choose q_j to satisfy

$$\beta(0)q_j \; + \; \Sigma_{t \in \mathbb{T}} \beta(t) v^j(t) \; = \; 0$$

Let W be the extended dividend matrix and assume, in order to get a contradiction, that θ is a portfolio such that $W\theta > 0$, that is the holder of the portfolio θ gets a positive net income at some date and a nonnegative income at each date (including date 0). Then $\beta W \theta > 0$ since $\beta \in \mathbb{R}_{++}^{T+1}$. Now change the order of multiplication; βW is a row vector whose components are the scalar products of β with the columns of W. By definition of the vector q, $\beta W = 0$. Hence $\beta W \theta = 0$; a contradiction. Hence there can not be a portfolio θ such that $W\theta > 0$ and it follows that the prices q are such that (V, q) is arbitrage free. Thus any $\beta \in \mathbb{R}^{T+1}_{++}$ defines asset prices, q, so that (V, q) is arbitrage free. The vector β is a vector of *discount factors.* The theorem below states that a converse is true.

Theorem 3.2.B Existence of Discount Factors *Let V be a dividend matrix, $q \in \mathbb{R}^J$ the asset prices and W the associated extended dividend matrix. Then*

(V, q) is arbitrage free

if and only if

there exists a vector $\beta \in R^{T+1}_{++}$, with $\beta(0) = 1$ such that $\beta W = 0$.

Proof: Let w^j denote the j'th column vector of W.

To prove the "if" part assume that there exists a vector $\beta = (\beta(0), \beta(1), ..., \beta(T))$ in $\mathbb{R}^{T+1}_{++}$ such that $\beta W = 0$. $\beta W = 0$ is equivalent to $\beta w^j = 0$ for $j \in \mathbb{J}$, that is β is orthogonal to each column vector of W.

Assume, in order to obtain a contradiction, that there is a vector $w \in \langle W \rangle \cap (\mathbb{R}^{T+1}_{+} \setminus \{0\})$. Then $w > 0$. Since $w \in \langle W \rangle$ there is a portfolio $\theta \in \mathbb{R}^J$ such that $w = w^1\theta_1 + w^2\theta_2 + ... + w^J\theta_J$. Since β is a positive vector we get

$$\beta w = \beta w^1\theta_1 + \beta w^2\theta_2 + ... + \beta w^J\theta_J > 0$$

which contradicts $\beta w^j = 0$ for each $j \in \mathbb{J}$. Hence $\langle W \rangle$ does not intersect $\mathbb{R}^{T+1} \setminus \{0\}$.

"only if". Assume that $\langle W \rangle$ does not intersect $\mathbb{R}^{T+1}_{+} \setminus \{0\}$. By Farkas' lemma (see Appendix C) there is a vector $\beta \in \mathbb{R}^{T+1}_{++}$ such that $\beta w = 0$ for each $w \in \langle W \rangle$. Since the column vectors of W, the vectors w^j for $j \in \mathbb{J}$, belongs to $\langle W \rangle$ we get $\beta W = 0$. Since β is a positive vector we can normalize so that $\beta(0) = 1$. Cf. Figure 3.2.A.

Date 0 asset prices determined by discount factors

Let us expand a bit on the calculations and interpretation. $\beta W = 0$ implies that, for $j \in \mathbb{J}$,

$$\beta w^j = 1(-q_j) + \beta(1)v^j(1) + \beta(2)v^j(2) + \ldots + \beta(T)v^j(T) = 0$$

or equivalently

$$q_j = \beta(1)v^j(1) + \beta(2)v^j(2) + \ldots + \beta(T)v^j(T)$$

This shows that the date 0 price of asset j, for $j \in \mathbb{J}$, is the discounted value of the dividends of the asset at future dates. The numbers $\beta(t)$, for $t \in \mathbb{T}_1$, are the discount factors or date 0 prices for date t spot-income.

Hence

$$W = \begin{bmatrix} -q_1 & -q_2 & \ldots & -q_J \\ v^1(1) & v^2(1) & \ldots & v^J(1) \\ v^1(2) & v^2(2) & \ldots & v^J(2) \\ \vdots & \vdots & & \vdots \\ v^1(t) & v^2(t) & \ldots & v^J(t) \\ \vdots & \vdots & & \vdots \\ v^1(T) & v^2(T) & \ldots & v^J(T) \end{bmatrix}$$

$$\text{with } q_j = -\Sigma_{t \in \mathbb{T}_1} \beta(t) v^j(t) \quad \text{for } j \in \mathbb{J}$$

Clearly the first row of W is a linear combination of the last T rows; the rows of V. Hence it will *always* be the case if assets are priced so that (V, q) is arbitrage free that $\operatorname{rank} W = \operatorname{rank} V$. Since V has only T rows it follows that $\operatorname{rank} V \leq T$ and $\operatorname{rank} W \leq T$.

The corollary below shows that when asset markets are complete then the net income vectors available to a consumer are precisely those whose value is 0 for the appropriate prices of future income. With complete markets the implicit prices of future spot income are uniquely determined.

Corollary 3.2.C *Let (V, q) be arbitrage free. If asset markets are complete so that* $\operatorname{rank} W = T$ *then*

(a) *there is a unique $\beta \in \mathbb{R}^{T+1}_{++}$, with $\beta(0) = 1$ such that $\beta W = 0$*

(b) $\langle W \rangle = \{r \in \mathbb{R}^{T+1} \mid \beta r = 0\}$.

Proof: To prove **(a)** note that since W has rank T the linear subspace $\langle W \rangle$ has dimension T. By Theorem 3.2.B there exist a $\beta \in \mathbb{R}^{T+1}_{++}$ such that $\beta W = 0$. This shows that β belongs to the orthogonal subspace to $\langle W \rangle$. Assume that $\hat{\beta}$ is another vector with $\hat{\beta}(0) = 1$ in the linear subspace orthogonal to $\langle W \rangle$. Since the linear subspace orthogonal to $\langle W \rangle$ has dimension 1, β and $\hat{\beta}$ are linearly dependent. Hence there are α and $\hat{\alpha}$ not both 0, such that $\alpha\beta + \hat{\alpha}\hat{\beta} = 0$ but then $\beta(0) = \hat{\beta}(0) = 1$ implies $\alpha = -\hat{\alpha}$. Thus $\beta = -(\hat{\alpha}/\alpha)\hat{\beta} = \hat{\beta}$.

(**b**) From (**a**) we have $\langle W \rangle \subset \{r \in \mathbb{R}^{T+1} \mid \beta r = 0\}$. Since the rank of W is T the dimension of $\langle W \rangle$ is T which is also the dimension of the homogenous hyperplane $\{r \in \mathbb{R}^{T+1} \mid \beta r = 0\}$. Hence $\langle W \rangle = \{r \in \mathbb{R}^{T+1} \mid \beta r = 0\}$.

□

In Chapter 4 we will show that the unique prices for spot income resulting from complete asset markets can be used to price any new but redundant asset. The reason is clear. With complete markets we can always construct a replicating portfolio giving the same future dividends as the new asset. The no-arbitrage condition implies that the price of the new asset must be the date 0 cost of the replicating portfolio.

Example 3.2.A Incomplete markets and non-uniqueness of discount rates

Let $T = 2$ and $J = 1$ so that there is a single asset which pays -1 at date 1 and 1 at date 2. Hence

$$V = \begin{bmatrix} -1 \\ 1 \end{bmatrix}$$

Clearly any date 0 price $q \in \mathbb{R}$ makes (V, q) arbitrage free. We do not have complete asset markets since $\dim \langle V \rangle = 1 < 2 = T$. We get

$$W = \begin{bmatrix} -q \\ -1 \\ 1 \end{bmatrix}$$

Let $q \geq 0$. It is easy to see that

$$\begin{aligned} \beta_1 &= (1, 1, 1+q) \\ \beta_2 &= (1, 0, q) \end{aligned}$$

are linearly independent and $\beta_1 W = \beta_2 W = 0$. The vectors $\beta_1$and β_2 span a linear subspace of dimension 2 which equals the orthogonal subspace of W, which we will denote $W^\perp$. Hence for each $\beta \in W^\perp$ we have $\beta W = 0$. Thus the price $q = 2$ is in this case consistent with many different "prices" for future income. Although it is tempting to think of the, not uniquely determined, discount factors as defining several budget restrictions which the consumer must meet it might be slightly misleading. With V as above the set of net income vectors available to the consumer is

$$\{r \in \mathbb{R}^3 \mid r = W\theta \text{ for some } \theta \in \mathbb{R}\} = \{r \in \mathbb{R}^3 \mid \beta_1 r = 0 \text{ and } \beta_2 r = 0\}$$

and this set is not the same as $\{r \in \mathbb{R}^3 \mid \beta_1 r \leq 0 \text{ and } \beta_2 r \leq 0\}$. Cf. Figure 3.2.B.

□

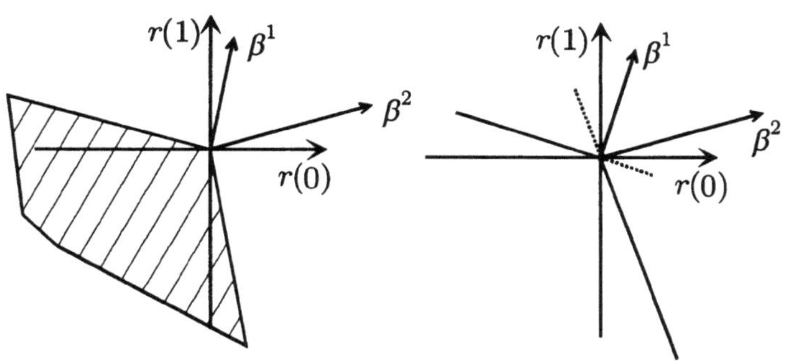

Figure 3.2.B: *The set of net income vectors satisfying $\beta^1 r \leq 0$ and $\beta^2 r \leq 0$ differs from the set of income vectors satisfying $\beta^1 r = 0$ and $\beta^2 r = 0$. Here the last set is $\{0\}$*

Budget restrictions with incomplete asset markets

Let (V, q) be an asset structure and asset prices such that there is no arbitrage. Consider the Consumer Problem for the case where the consumer is restricted to choose a consumption giving equality in each of the budget restrictions and a portfolio to obtain the desired net income vector.

$$r(0) = -\sum_{j\in\mathbb{J}} q^j\theta^j \quad \text{and} \quad p(0)(c(0)-e(0)) = r(0)$$

$$\begin{cases} r(1) = -\sum_{j\in\mathbb{J}} v^j(1)\theta^j \\ r(2) = -\sum_{j\in\mathbb{J}} v^j(2)\theta^j \\ \quad\vdots \\ r(T) = -\sum_{j\in\mathbb{J}} v^j(T)\theta^j \end{cases} \quad \text{and} \quad \begin{cases} p(1)(c(1)-e(1)) = r(1) \\ p(2)(c(2)-e(2)) = r(2) \\ \quad\vdots \\ p(T)(c(T)-e(T)) = r(T) \end{cases}$$

By Theorem 3.2.B there is a $\beta = (1, \beta(1), ..., \beta(T)) \in \mathbb{R}^{T+1}_{++}$ such that $q^j = \sum_{t\in\mathbb{T}_1} \beta(t)v^j(t)$. We can then rewrite the date 0 budget restriction

$$\begin{aligned} p(0)(c(0)-e(0)) &= r(0) = -\sum_{j\in\mathbb{J}} q^j\theta^j \\ &= -\sum_{j\in\mathbb{J}} \left[\sum_{t\in\mathbb{T}_1} \beta(t)v^j(t)\right] \theta^j \\ &= -\sum_{t\in\mathbb{T}_1} \sum_{j\in\mathbb{J}} \beta(t)v^j(t)\theta^j \\ &= -\sum_{t\in\mathbb{T}_1} \beta(t)r(t) \\ &= -\sum_{t\in\mathbb{T}_1} \beta(t)\left[p(t)(c(t)-e(t))\right] \end{aligned}$$

which is equivalent to

$$\sum_{t\in\mathbb{T}} \beta(t)[p(t)(c(t)-e(t))] = 0 \qquad \textbf{3.2.A}$$

If asset markets are complete 3.2.A is the only restriction which has to be satisfied since then the requirement $(r(1), ..., r(T)) \in V$ will be satisfied trivially.

If asset markets are incomplete the further requirement $(r(1), ..., r(T)) \in V$ has to be added to the budget restriction 3.2.A. It is a condition on the future net income vector which in this case coincides with the future net expenditure vector, $(p(t)[c(t)-e(t))]_{t\in\mathbb{T}_1}$.

Determinacy of Discount Factors*

The results of the previous section show that when markets are complete then there are uniquely (given that $\beta(0) = 1$) defined discount factors. When markets are not complete then the span of the extended dividend matrix has a dimension smaller than T and the set of prices for future income (discount factors) consistent with no arbitrage prices of the assets is correspondingly larger.

Proposition 3.2.D concerns "prices for future income" that are consistent with a given set of asset prices not allowing arbitrage.

Proposition 3.2.D *Let (V, q) be arbitrage free and let W be the associated extended dividend matrix. If* $\operatorname{rank} W = J$ *and* $J \leq T$ *then there are* $K = (T+1) - J$ *linearly independent vectors* $\beta^1, \beta^2, ..., \beta^K \in \mathbb{R}^{T+1}_{++}$, *such that* $\beta^k(0) = 1$ *and* $\beta^k W = 0$ *for* $k = 1, ..., K$.

Proof: By assumption $\dim \langle W \rangle = J$. Hence $\dim \langle W \rangle^{\perp} = K = (T+1) - J$. By Theorem 3.2.B there exists $\hat{\beta}^1 \in \mathbb{R}^{T+1}_{++}$ which belongs to $\langle W \rangle^{\perp}$. We can extend with $\hat{\beta}^2, ..., \hat{\beta}^K$ so that $\hat{\beta}^1, \hat{\beta}^2, ..., \hat{\beta}^K$ form a basis for $\langle W \rangle^{\perp}$. Choose $\delta \in]0, 1[$ and consider the vectors

$$\hat{\beta}^1,\ \delta\hat{\beta}^2 + (1-\delta)\hat{\beta}^1, ...,\ \delta\hat{\beta}^K + (1-\delta)\hat{\beta}^1$$

Choose a fixed δ small enough to get $\hat{\beta}^1 \in \mathbb{R}^{T+1}_{++}$ and $\delta\hat{\beta}^k + (1-\delta)\hat{\beta}^1 \in \mathbb{R}^{T+1}_{++}$ for $k = 2, ..., K$.

We now show that $\hat{\beta}^1$ and the vectors $\delta\hat{\beta}^k + (1-\delta)\hat{\beta}^1$ for $k = 2, ..., K$ are linearly independent. Assume, in order to get a contradiction, that this is not the case. Then there are $\alpha_1, ..., \alpha_K$, not all 0, such that

$$\alpha_1\hat{\beta}^1 + \alpha_2[\delta\hat{\beta}^2 + (1-\delta)\hat{\beta}^1] + ... + \alpha_K[\delta\hat{\beta}^K + (1-\delta)\hat{\beta}^1] = 0$$

or equivalently

$$[\alpha_1\frac{1}{\delta} + \Sigma_{k=2}^K\alpha_k\frac{1-\delta}{\delta}]\hat{\beta}^1 + \alpha_2\hat{\beta}^2 + ... + \alpha_K\hat{\beta}^K = 0$$

Since $\hat{\beta}^1, \hat{\beta}^2, ..., \hat{\beta}^K$ are linearly independent the last sum is 0 only if

$$\alpha_1\frac{1}{\delta} + \Sigma_{k=2}^K\alpha_k\frac{1-\delta}{\delta} = \alpha_2 = ... = \alpha_K = 0$$

which implies $\alpha_1 = \alpha_2 = ... = \alpha_K = 0$. From this follows that $\hat{\beta}^1$ and $\delta\hat{\beta}^k + (1-\delta)\hat{\beta}^1$, for $k = 2, ..., K$, are linearly independent, as asserted. Put

$$\beta_1 = \frac{1}{\hat{\beta}^1(0)}\beta^1 \quad \text{and}$$

$$\beta_k = \frac{1}{\delta\hat{\beta}^k(0) + (1-\delta)\hat{\beta}^1(0)}[\delta\hat{\beta}^k + (1-\delta)\hat{\beta}^1] \quad \text{for } k = 2,3, ..., K$$

Then $\beta^1, \beta^2, ..., \beta^K \in \mathbb{R}^K_{++}$ are linearly independent and each of them has a first component equal to 1.

□

Proposition 3.2.D suggests a convenient way to describe the subspace of income transfers when there is no arbitrage but asset markets may be incomplete. Let (V, q) be arbitrage free and let W be the corresponding extended dividend matrix, with $\operatorname{rank} W = J$. By Proposition 3.2.D , there are linearly independent vectors such that $\langle W \rangle$ is the orthogonal subspace to the span of $\beta^1, \beta^2, ..., \beta^K$. Hence

$$\langle W \rangle = \{r \in \mathbb{R}^{T+1} \mid \beta^1 r = \beta^2 r = \ldots = \beta^K r = 0\}$$

3.3 Real Assets

In this section we consider real assets where the dividends are goods bundles. To introduce the idea of a real asset we consider an example.

Example 3.3.A Future contracts for the delivery of good L.

Assume $T = 2$ and $L = 2$. There are $J = 2$ real assets. The assets pay dividends in commodities. The first asset yields 1 unit of good 2, at date 1, the second asset yields 1 unit of good 2, at date 2. This is described by the real dividend matrix A.

	Asset	
Date	1	2
1	$\begin{bmatrix} 0 \\ 1 \end{bmatrix}$	$\begin{bmatrix} 0 \\ 0 \end{bmatrix}$
2	$\begin{bmatrix} 0 \\ 0 \end{bmatrix}$	$\begin{bmatrix} 0 \\ 1 \end{bmatrix}$

Assume that spot prices $(p(t))_{t=0}^2$ are given. We then get the following value dividend matrix, V,

Date	Asset 1	Asset 2
1	$p(1)\begin{bmatrix}0\\1\end{bmatrix}$	$p(1)\begin{bmatrix}0\\0\end{bmatrix}$
2	$p(2)\begin{bmatrix}0\\0\end{bmatrix}$	$p(2)\begin{bmatrix}0\\1\end{bmatrix}$

or

Date	Asset 1	Asset 2
1	$p_2(1)$	0
2	0	$p_2(2)$

If $p_2(1)$ and $p_2(2)$ are both positive, the value dividend matrix, $V(p)$, will have rank 2 so that we have complete markets. In this case, the rank is independent of the (positive) commodity prices.

An agent may, if we disregard the date 0 cost, acquire any net income vector $r = (r(1), r(2)) \in \mathbb{R}^2$, for future dates, by choosing a portfolio θ such that

$$r = V\theta = \begin{bmatrix} p_2(1)\theta_1 \\ p_2(2)\theta_2 \end{bmatrix} \quad \text{or} \quad \begin{bmatrix} \theta_1 \\ \theta_2 \end{bmatrix} = \begin{bmatrix} \frac{r(1)}{p_2(1)} \\ \frac{r(2)}{p_2(2)} \end{bmatrix}$$

□

With spot-prices given real assets are not much different from nominal assets. The important novel feature is, that as spot-prices vary asset markets may be more or less complete since the rank of the value dividend matrix is a function of the spot-prices.

A *real asset* is characterized by its *(real) dividend vector*, $a \in (\mathbb{R}^L)^T$. This is shown in Table 3.3.A. A holder of 1 unit of asset a has a claim to the goods vector $a(t)$ at date t. If $a_l(t) > 0$ this means that the amount $a_l(t)$ of good l is *delivered to* him at date t; if $a_l(t) < 0$ then there is a *delivery from* him of the amount $|a_l(t)|$ of good l at date t.

Assume that we have J real assets, $a^1, a^2, ..., a^J$. These assets make up the *asset structure*, which can be summarized in the *real dividend matrix* of dimension $TL \times J$. This matrix is perhaps more conveniently thought of as a matrix of order $T \times J$, whose elements are

Table 3.3.A: *Dividend vector of a real asset*

$$a = \begin{bmatrix} a(1) \\ \vdots \\ a(t) \\ \vdots \\ a(T) \end{bmatrix} = \begin{bmatrix} \begin{bmatrix} a_1(1) \\ \vdots \\ a_L(1) \end{bmatrix} \\ \vdots \\ \begin{bmatrix} a_1(t) \\ \vdots \\ a_L(t) \end{bmatrix} \\ \vdots \\ \begin{bmatrix} a_1(T) \\ \vdots \\ a_L(T) \end{bmatrix} \end{bmatrix}$$

column vectors in $\mathbb{R}^L$. Each column represents the deliveries *to* (positive) or *from* (negative) a consumer holding 1 unit of the asset. This is shown in Table 3.3.B

Table 3.3.B: *Real dividend matrix*

	Asset			
Date	1	2	...	J
1	$a^1(1)$	$a^2(1)$	...	$a^J(1)$
2	$a^1(2)$	$a^2(2)$	...	$a^J(2)$
$\vdots$	$\vdots$	$\vdots$		$\vdots$
t	$a^1(t)$	$a^2(t)$	...	$a^J(t)$
$\vdots$	$\vdots$	$\vdots$		$\vdots$
T	$a^1(T)$	$a^2(T)$	...	$a^J(T)$

The agents trade in the assets to achieve income transfers between dates. What matters to the agent is the value in the spot-market of the goods bundles she receives at date $t \in \mathbb{T}$. Correctly foreseeing the spot prices that will prevail the agent calculates the associated value dividend matrix, $V(p)$, which is a function of the spot-prices. (For the moment disregard the row corresponding to date 0, in Table 3.3.C.)

Table 3.3.C: *The value dividend matrix*

	Asset			
Date	1	2	...	J
0	$-q_1$	$-q_2$	...	$-q_J$
1	$p(1)a^1(1$	$p(1)a^2(1)$	...	$p(1)a^J(1)$
2	$p(2)a^1(2)$	$p(2)a^2(2)$	...	$p(2)a^J(2)$
$\vdots$	$\vdots$	$\vdots$		$\vdots$
t	$p(t)a^1(t)$	$p(t)a^2(t)$	...	$p(t)a^J(t)$
$\vdots$	$\vdots$	$\vdots$		$\vdots$
T	$p(T)a^1(T)$	$p(T)a^2(T)$	...	$p(T)a^J(T)$

The *value dividend matrix* is the (sub)matrix in Table 3.3.C above for dates $1, ..., T$. The row corresponding to date 0, contains the prices of the assets (in 0-crowns).

Let $W(p, q)$ denote the extended dividend matrix, which is derived by enlarging V with a new first row, containing the negative of the asset prices. Thus the matrix above for dates $t = 0, 1, 2, ..., T$ is $W(p, q)$.

The fundamental difference between nominal assets and real assets is that with real assets, the value dividend matrix depends on the prices. In particular rank $W(p, q)$ may depend on the spot-prices, as the following example shows.

Example 3.3.B Drop in rank of the value dividend matrix

Consider an economy with $T = 2$ and $L = 2$. Assume that there are at date 0 only two assets so that $J = 2$. The first asset gives the holder 1 unit of good 1, at dates 1 and 2, and the second asset gives the holder 1 unit of good 2, at dates 1 and 2. Thus the real dividend matrix is

$$A = \begin{bmatrix} \begin{bmatrix} 1 \\ 0 \end{bmatrix} & \begin{bmatrix} 0 \\ 1 \end{bmatrix} \\ \begin{bmatrix} 1 \\ 0 \end{bmatrix} & \begin{bmatrix} 0 \\ 1 \end{bmatrix} \end{bmatrix}$$

The value dividend matrix depends on the spot-prices for goods. It is

$$V(p) = \begin{bmatrix} p_1(1) & p_2(1) \\ p_1(2) & p_2(2) \end{bmatrix}$$

The rank of the value dividend matrix is (since we are assuming positive commodity prices) 1 or 2. The rank is 2 unless the spot-prices at date 1 are proportional to the spot-prices, at date 2. If the rank is 2 then the asset markets are complete.

Except for a very small set of commodity prices, namely those where $p(1) = \alpha p(2)$, for some $\alpha > 0$, we have complete asset markets. The new feature is that the spot-prices determine whether asset markets are complete or not.

We will consider a similar example, due to Hart [1975], and study the consequences for the existence of equilibrium.

3.4 Assets Traded at Future Dates

So far we have assumed that assets can be traded only at date 0 but many assets that are available in actual economies, like bonds, are traded at many dates. Although an asset may pay a dividend only, at the last date, frequent trading of the asset, gives rise to new possibilities of income transfers. We will illustrate this in an example and show how it can be accommodated with the concepts already introduced. We will see in Chapter 4 that also with nominal assets that are traded at several dates the rank of the dividend matrix may vary as in the case of real assets and two dates.

Example 3.4.A Spanning by a frequently traded asset

Let $T = 4$ and $L = 1$. There is to start with, a single real asset which delivers 1 unit of the good, at date 3. The extended value dividend matrix is given in Table 3.4.A and the date 0 price of the asset is denoted by $q(0)$. Let $q(t)$ denote the price of the asset at date t, for $t \in \mathbb{T}$. We have indicated in Table 3.4.B how our single asset generates new assets. For

Table 3.4.A: *Value dividend vector of the given asset*

Date	Given Asset 1
0	$-q(0)$
1	0
2	0
3	$p(3)$

example, asset 3 is derived from buying the asset at date 1 and selling it at date 2. Asset 4 is derived from buying the asset at date 2 and receiving the dividend, $p(3)$, at date 3. There are other derived assets; for example, to buy the asset at date 1 and hold onto it until date 3 but we will see that assets 2, 3 and 4 are enough to give us complete markets.

Table 3.4.B: *Value dividend vectors of the induced assets*

	Induced Assets		
Date	2	3	4
0	$-q(0)$	0	0
1	$q(1)$	$-q(1)$	0
2	0	$q(2)$	$-q(2)$
3	0	0	$p(3)$

Since $p(3)$ is positive we have if there is to be no arbitrage, $q(t) > 0$, for $t = 0, 1, 2, 3$. The submatrix from Table 3.4.B, corresponding to dates 1, 2 and 3, is the value dividend matrix which will be denoted $V(p, q)$. This matrix will now depend on p and the spot-prices, $q = (q(1), q(2))$ of the assets. The row corresponding to date 0 gives the date 0 prices of the induced assets. We can imagine that, for example, asset 3 as being traded at date 0 with a date 0 price of 0. The holder of 1 unit of asset 3, is obliged to deliver the amount $q(1)$ of 1-crowns, at date 1, and receives the amount $q(2)$ of 2−crowns at date 2.

The determinant of $V(p, q)$ is the product of its diagonal entries. Since the determinant is a positive number we have rank $V(p, q) = 3$, which implies that we have complete markets. Thus the frequent trading of the single given asset, generates new assets and these new assets give us complete asset markets. In Chapter 4 we will see that this idea carries over to the case of uncertainty and several dates.

3.5 Radner Equilibrium

In this section we will introduce the notion of a Radner equilibrium and relate it to equilibrium relative to a subspace of net income vectors.

Portfolios and feasible net income vectors

In Chapter 2 we studied the case where the agents at date 0 traded in spot income, to be delivered at dates 1,2,..,T. Here we will see that an asset structure induces a set of feasible net income vectors for the agents to choose from. We can then define a Radner equilibrium, where the agents choose a consumption together with a portfolio which finances the consumption.

Definition 3.5.A *An* **economy with an asset structure** *is an economy and a dividend matrix. It will be denoted by* $\mathcal{E} = \{(C^i, u^i, e^i)_{i\in\mathrm{I}}, V\}$.

In general, we allow the matrix V to depend on the commodity prices. This will be the case if V is the value dividend matrix induced by a real dividend matrix and commodity prices. It will be clear from the context whether nominal or real assets are considered. In the case of nominal assets the value dividend matrix is constant with respect to p.

In Proposition 3.5.B below we consider an economy where the set of feasible net income vectors, is generated by an asset structure. We may then determine the portfolio choices necessary for the consumers to carry out their consumption plans.

Proposition 3.5.B *Let* $\mathcal{E} = \{(C^i, u^i, e^i)_{i\in\mathrm{I}}, V\}$ *be an economy with an asset structure and* $((\bar{c}^i)_{i\in\mathrm{I}}, p)$ *a spot market equilibrium for* $\mathcal{E}$ *relative to* $\mathbb{M}$, *where*

$$\mathbb{M} = \left\langle \begin{bmatrix} -q \\ V(p) \end{bmatrix} \right\rangle = \langle W(p,q) \rangle$$

for some $q \in \mathbb{R}^{T+1}$. *Let* $\bar{r}^i(t) = p(t)[\bar{c}^i(t) - e^i(t)]$, *for* $i \in \mathbb{I}$, $t \in \mathbb{T}$.

Then there are $\bar{\theta}^i, i \in \mathbb{I}$, *such that*

(a) $\Sigma_{i\in\mathbb{I}}\bar{\theta}^i = 0$ *and for* $i \in I$, $\bar{r}^i = W(p,q)\bar{\theta}^i$ *and* $\bar{\theta}^i$ *is a solution to*

$$\text{Max}_{\theta^i}\ v^i(p, r^i(0), r^i(1), ..., r^i(T)) \quad s.\ to\ r^i = W(p,q)\theta^i \qquad \textbf{3.5.A}$$

(b) *for* $i \in I$, *with* $\bar{r}^i = W(p,q)\bar{\theta}^i$, *the consumption* $\bar{c}^i$ *is a solution to*

$$\text{Max}_{c^i}\ u^i(c^i(0), c^i(1), ..., c^i(T)) \quad s.\ to\ c^i \in C^i \text{ and } p(t)(c^i(t) - e^i(t)) \leq \bar{r}^i(t) \quad for\ t \in \mathbb{T} \qquad \textbf{3.5.B}$$

Proof: Let $\bar{r}^i, i \in \mathbb{I}$, be the net income vectors in the equilibrium relative to $\mathbb{M}$. Then $\Sigma_{i\in\mathbb{I}}\bar{r}^i = 0$ and, for $i \in \mathbb{I}$, $\bar{r}^i$ solves

$$\text{Max}_{r^i}\ v^i(p, r^i(0), r^i(1), ..., r^i(T)) \quad \text{s. to } r^i \in \langle W(p,q)\rangle \qquad \textbf{3.5.C}$$

Choose $i \in \{2, 3, ..., I\}$. Since $\bar{r}^i \in \langle W(p,q)\rangle$ there is $\bar{\theta}^i$ such that $\bar{r}^i = W(p,q)\bar{\theta}^i$. Furthermore, it is easy to see that

$\bar{r}^i$ is a solution to 3.5.C

if and only if

$\bar{\theta}^i$ is a solution to 3.5.A

This shows that, for $i \in \{2, 3, ..., I\}$, the portfolio $\bar{\theta}^i$ is a solution to 3.5.A in (**a**). From the definition of a spot-market equilibrium relative to $\mathbb{M}$ follows that for $i \in \mathbb{I}$, the consumption $\bar{c}^i$ is a solution to 3.5.B. This proves (**b**).

Put $\bar{\theta}^1 = -\Sigma_{i=2}^I\bar{\theta}^i$, to get

$$\bar{r}^1 = -\Sigma_{i=2}^I\bar{r}^i = W(p,q)[-\Sigma_{i=2}^I\bar{\theta}] = W(p,q)\bar{\theta}^1$$

and since the net income vector $\bar{r}^1$ is a solution to 3.5.C, for $i = 1$, the portfolio $\bar{\theta}^1$ is a solution to 3.5.A, for $i = 1$. Clearly, this definition of $\bar{\theta}^1$, implies that $\Sigma_{i\in\mathbb{I}}\bar{\theta}^i = 0$, as asserted in (**a**).

□

In Chapter 2 we were able to decompose the Consumer Problem, into a problem of choosing a net income vector and the choice of a consumption, for a given net income vector.

It is clear from the first part of the proof of Proposition 3.5.B above, that a similar decomposition holds for the choice of portfolio and the choice of a consumption given a portfolio..

Hence the conditions of Proposition 3.5.B imply that, in an equilibrium, where the set of net income vectors, is generated by assets, the consumption-portfolio pair $(\bar{c}^i, \bar{\theta}^i)$ is a solution to

$$\begin{aligned} \text{Max}_{(c^i,\theta^i)} \quad & u^i(c^i(0), c^i(1), ..., c^i(T)) \quad \text{s. to } c^i \in C^i \text{ and} \\ & p(t)[c^i(t) - e^i(t)] \leq r^i(t) \quad \text{for } t \in \mathbb{T} \text{ and } r^i = W(p,q)\theta^i \end{aligned}$$

From Proposition 2.8.A, we know that $\text{grad}_r\, v^i(p, \bar{r}^i)$ belongs to the orthogonal subspace to $\mathbb{M}$. Since $\mathbb{M}$, in Proposition 3.5.B equals $\langle W(p,q)\rangle$ we get

$$\text{grad}_r\, v^i(p, \bar{r}^i) W(p,q) = 0 \quad \text{for } i \in \mathbb{I}$$

so that each consumer's subjective evaluation of the net income vectors spanned by the assets is 0.

Relating equilibrium relative to a subspace to a Radner equilibrium

We have seen above how the set of net income vectors, available to the consumers, is generated by the assets in the economy. The definition of a Radner equilibrium is similar to the definition of a spot-market equilibrium relative to a subspace. The only new feature is that the asset choices needed to achieve the net income vector are explicitly considered. Let $\mathcal{E} = \{(C^i, u^i, e^i)_{i\in\mathbb{I}}, V\}$ be an economy with an asset structure. A *Radner equilibrium* for $\mathcal{E}$ is a tuple $((\bar{c}^i, \bar{\theta}^i)_{i\in\mathbb{I}}, p, q)$ with

- portfolio-consumption choices, $(\bar{c}^i, \bar{\theta}^i)_{i\in\mathbb{I}}$,
- commodity prices, p, and date 0 asset prices, q

such that $((\bar{c}^i)_{i\in\mathbb{I}}, p)$ is an equilibrium relative to the subspace $\mathbb{M} = \langle W(p,q)\rangle$ and, for $i \in \mathbb{I}$, $p(t)[\bar{c}^i(t) - e^i(t)]_{t\in\mathbb{T}} = W(p,q)\bar{\theta}^i$.

These conditions imply that the asset markets balance.

A detailed definition of a Radner equilibrium*

For easy reference we spell out the conditions for a Radner equilibrium in more detail below.

Definition 3.5.C *A* **Radner equilibrium** *for an economy with an asset structure,* $\mathcal{E} = \{(C^i, u^i, e^i)_{i \in \mathbb{I}}, V\}$, *is a tuple*

$$((\bar{c}^i, \bar{\theta}^i)_{i \in \mathbb{I}}, p, q)$$

where

- $p \in (\mathbb{R}^L_{++})^{T+1}$ *are the commodity prices,*
- $q \in \mathbb{R}^J$ *are the asset prices and,*
- *for* $i \in I$, $\bar{c}^i \in C^i$ *is a consumption plan and* $\bar{\theta}^i \in \mathbb{R}^J$ *is a portfolio,*

and where $((\bar{c}^i, \bar{\theta}^i)_{i \in \mathbb{I}}, p, q)$ *satisfies:*

(a) *for* $i \in I$, $(\bar{c}^i, \bar{\theta}^i)$ *solves*

$$\text{Max}_{(c^i,\theta^i)} \quad u^i(c^i(0), c^i(1), ..., c^i(T)) \quad \text{s. to } c^i \in C^i$$
$$\begin{cases} p(t)(c^i(t) - e^i(t)) \leq r^i(t) & \text{for } t \in T \\ r^i = W(p,q)\theta^i \end{cases}$$

(b) *markets balance,*

$$\begin{cases} \sum_{i \in \mathbb{I}} \bar{c}^i(0) = \sum_{i \in \mathbb{I}} e^i(0) \\ \sum_{i \in \mathbb{I}} \bar{c}^i(1) = \sum_{i \in \mathbb{I}} e^i(1) \\ \vdots \\ \sum_{i \in \mathbb{I}} \bar{c}^i(T) = \sum_{i \in \mathbb{I}} e^i(T) \end{cases} \qquad \textit{(Market Balance for Goods at Each Date)}$$

$$\sum_{i \in \mathbb{I}} \bar{\theta}^i = 0 \qquad \textit{(Market Balance for Assets)}$$

The **rank** *of a Radner equilibrium is the rank of the dividend matrix* $W(p,q)$.

Note that (**a**) holds, for $i \in \mathbb{I}$, if and only if $\bar{\theta}^i$solves

$$\mathrm{Max}_{\theta^i}\ v^i(p, r^i(0), r^i(1), ..., r^i(T)) \quad \text{s. to} \quad r^i \;=\; W(p,q)\theta^i$$

and $\bar{c}^i$ solves

$$\mathrm{Max}_{c^i}\quad u^i(c^i(0), c^i(1), ..., c^i(T)) \quad \text{s. to } c^i \in C^i \text{ and}$$
$$p(t)(c^i(t) - e^i(t)) \;\leq\; \bar{r}^i(t) \quad \text{for } t \in \mathbb{T}$$

with $\bar{r}^i = W(p,q)\bar{\theta}^i$.

3.6 Applications to Economies with Production

The introduction of producers, or firms, in the economy has two kinds of consequences. Firstly, the choice of a production plan affects the market balance conditions for commodities since production typically involves the transformation of inputs at one date to outputs at another date. This kind of activity will most likely influence the consumers' demand for transferring purchasing power. To finance a production plan a producer may have to engage in asset transactions, for example, issuing bonds at the current date which are to be repaid in the future. Thus, secondly, the choice of a production plan and action in the asset market determine a dividend plan for the producer. As the ownership of the stock is a claim to this dividend stream the producer creates a "new" asset, which the consumers can exchange.

The Modigliani-Miller theorem [1958], gives conditions under which the producer's actions in the asset market is inessential and the role of the producer is essentially his choices in the commodity market. These conditions also imply that the trade in the stock is superfluous; consumers can attain the same net income vector by trading in the (other) given assets. The Modigliani-Miller theorem was originally stated for economies with complete asset markets and the results were extended to incomplete asset markets by De Marzo [1988].

The producer's role in creating a new asset implies that the owners might disagree as to which production plan should be chosen. Consumers judge different production plans according to the net income vectors that will be available to them and when asset markets are incomplete it is likely that opinions will differ as to which asset the producer should create.

We will, to ease on the notation, introduce only a single producer.

Economy with Assets and Production

Let us define an *economy with an asset structure and production*, by taking an economy with an asset structure and adjoining a single producer. It is a tuple

$$\mathcal{E} = \{(C^i, u^i, e^i, \alpha^i)_{i \in \mathbf{I}}, V, Y\}$$

where the new elements in the description are the initial shares of the consumers, $(\alpha^i)_{i \in \mathbf{I}}$, and the production set, Y. The shares $(\alpha^i)_{i \in \mathbf{I}}$ are assumed to be nonnegative and sum to 1.

Assumptions on the producer

A production plan

$$y = (y(0), y(1), ..., y(T)) \quad \text{where} \quad y(t) \in \mathbb{R}^L \text{ for } t \in \mathbb{T}$$

is a vector of deliveries of inputs *to* (negative sign) and of outputs *from* (positive sign) the producer. In the sequel we will assume that the producer satisfies the following Assumption (Y1), unless otherwise stated.

(Y1) *The producer has a (non-empty) production set $Y \subset \mathbb{R}^{L(T+1)}$, consisting of the feasible production plans given by*

$$Y = \{y \in \mathbb{R}^{L(T+1)} \mid F(y) \leq 0\}$$

where $F : \mathbb{R}^{L(T+1)} \longrightarrow \mathbb{R}$ is a strictly convex, differentiable function with $\operatorname{grad} F(y) \in \mathbb{R}^{L(T+1)}_{++}$ *and* $F(0) = 0$.

Since F is increasing in each variable this assumption entails that if $\bar{y} \in Y$ then each point of $\{\bar{y}\} - \mathbb{R}^{L(T+1)}_{+}$, except possibly $\bar{y}$, is an interior point of Y.

Spot net revenue vector

Let the spot commodity prices be given by $p \in (\mathbb{R}^L)^{T+1}_{++}$ and let y be a production plan. Because of the sign conventions the vector

$$(p(0)y(0),\, p(1)y(1), \ldots,\, p(T)y(T))$$

is the *spot net revenue vector*, that is, the spot revenue less spot cost for each date.

The producer's portfolio choice

With asset prices given by $q \in \mathbb{R}^J$ let $W(p,q)$ be the extended dividend matrix of the asset structure for the economy (not including the asset created by the producer; the stock). The producer chooses a portfolio of the assets along with his production plan. Denote this portfolio by $\theta^0 \in \mathbb{R}^J$. We thus exclude the possibility that the producer buys his own stock.

Table 3.6.A: *Notation for the economy with production*

Date	Transformed init. end.	Spot Net Revenue Vector	The Stock (Dividend Vector)
0	$e^i(0) + \alpha^i y(0)$	$p(0)y(0)$	$-q_0$
1	$e^i(1) + \alpha^i y(1)$	$p(1)y(1)$	$p(1)y(1) + V_1\theta^0$
2	$e^i(2) + \alpha^i y(2)$	$p(2)y(2)$	$p(2)y(2) + V_2\theta^0$
$\vdots$	$\vdots$	$\vdots$	$\vdots$
T	$e^i(T) + \alpha^i y(T)$	$p(T)y(T)$	$p(T)y(T) + V_T\theta^0$

Ownership; initial shares

The description of the economy has to specify the ownership of the producer. Let $\alpha^i, i \in \mathbb{I}$, be the *initial share* owned by consumer i, where $\Sigma_{i\in\mathbb{I}}\alpha^i = 1$ and $0 \leq \alpha^i \leq 1$, for $i \in \mathbb{I}$. We interpret the ownership as an obligation to deliver and receive goods at each date according to the production plan chosen by the producer. It simplifies the exposition if we assume that this obligation, given by the initial share, can not be avoided by selling the stock which we will achieve by measuring the trade in the stock *net* of the initial share. Then a given production plan y transforms the initial endowment of consumer i to $(e^i(t) + \alpha^i y(t))_{t\in\mathbb{T}}$. See Table 3.6.A. Ownership also obliges the consumer to participate in the deliveries of the producer due to the producer's action in the asset market.

The stock

At the current date a consumer with initial share α^i is obliged to deliver the amount $\alpha^i p(0)y(0)$. At this date the consumers may also buy or sell the stock. Denote by γ^i consumer i's action in the stock market. The consumer ends up holding $\alpha^i + \gamma^i$ of the stock, which may be smaller or larger than α^i since shortselling is allowed. Given the producer's choice of production plan and portfolio we take a unit of the stock to be the claim to the *future* dividend vector. Thus the date 0 price of the stock, q_0, is net of the deliveries required by ownership at date 0, that is, q_0 is the date 0 postdividend price. The *dividend vector* induced by the stock is given in Table 3.6.A where V_t denotes row t, for $t \in \mathbb{T}_1$, of the matrix V. Note that a consumer can annihilate the *future* deliveries which her initial share obliges her to make by choosing $\gamma^i = -\alpha^i$.

The consumer problem

We can now state the Consumer Problem for a consumer acting in an economy with assets and production. The consumers gross income vector will be determined by his initial endowment and initial share of the producer. At the current date the consumer chooses a consumption, a portfolio of the J assets and an amount of the stock. This action at date 0 determines his future net income vector. Thus

$$\mathrm{Max}_{(c^i,\gamma^i,\theta^i)} u^i(c) \quad \text{s. to}$$

$$\begin{aligned}
p(0)\left[c^i(0) - e^i(0) - \alpha^i y(0)\right] &\le \alpha^i \cdot -q\theta^0 + \gamma^i \cdot -q_0 - q\theta^i \\
p(t)\left[c^i(t) - e^i(t) - \alpha^i y(t)\right] &\le \alpha^i V_t\theta^0 + \gamma^i[p(t)y(t) + V_t\theta^0] + V_t\theta^i
\end{aligned}$$

3.6.A

for $t \in \mathbb{T}_1$. Here $\gamma^i \in \mathbb{R}$ denotes the net amount of the stock chosen so that $\alpha^i + \gamma^i$ is the gross amount and V_t, $t \in \mathbb{T}$, denotes the t'th row of the dividend matrix V. Choosing $\gamma^i = 0$ and $\theta^i = 0$ the consumer would still be obliged to make the deliveries arising from his initial share.

Radner Equilibrium with a Given Production Plan

We can now define a Radner equilibrium with a given production plan. Note that in the definition there is no restriction on the production plan except that it is required to belong to the production set. The market balance condition for the stock shows that the sum of the *net trades* in the stock are 0 in equilibrium.

Definition 3.6.A *Let* $\mathcal{E} = \{(C^i, u^i, e^i, \alpha^i)_{i \in \mathbf{I}}, V, Y\}$ *be an economy with assets and production. A* **Radner equilibrium with a given production plan** *is a tuple*

$$[(\bar{c}^i, \bar{\theta}^i, \bar{\gamma}^i)_{i \in \mathbf{I}}, (\bar{y}, \bar{\theta}^0), p, q, q_0)$$

where $\bar{y} \in Y$ *and*

(a) *for* $i \in I$, $(\bar{c}^i, \bar{\theta}^i, \bar{\gamma}^i)$ *is a solution to* 3.6.A *given the gross income induced by* $\bar{y}, \bar{\theta}^0, e^i$ *and* α^i *at prices* p, q *and* q_0

(b) *for* $t \in \mathbb{T}$, $\sum_{i \in \mathbf{I}} \bar{c}^i(t) = \sum_{i \in \mathbf{I}} e^i(t) + \bar{y}(t)$ *(Market Balance for Goods)*

(c) $\bar{\theta}^0 + \sum_{i \in \mathbf{I}} \bar{\theta}^i = 0$ *(Market Balance for Assets)*

(d) $\sum_{i \in \mathbf{I}} \bar{\gamma}^i = 0$ *(Market Balance) for the Stock)*

The Modigliani-Miller Theorem

The Modigliani-Miller Theorem concerns a Radner equilibrium with a given production plan where the future net spot revenue vector does not change the subspace of feasible income transfers. In this case the producer's action in the asset market and the possibilities for the consumers to trade in the stock are irrelevant. A consumer can meet a variation in the portfolio or the production plan of the producer by a varying her own portfolio in order to obtain the same net income vector.

Theorem 3.6.B The Modigliani-Miller Theorem *Let $\mathcal{E}$ be an economy with assets and production and*

$$((\bar{c}^i, \bar{\theta}^i, \bar{\gamma}^i)_{i\in\mathbb{I}}, (\bar{y}, \bar{\theta}^0), \bar{p}, \bar{q}, \bar{q}_0)$$

a Radner equilibrium with a given production plan where $(\bar{p}(t)\bar{y}(t))_{t\in\mathbb{T}_1} \in \langle V(\bar{p})\rangle$.

Then there is a Radner equilibrium, with a given production plan,

$$((\hat{c}^i, \hat{\theta}^i, \hat{\gamma}^i)_{i\in\mathbb{I}}, (\hat{y}, \hat{\theta}^0), \hat{p}, \hat{q}, \hat{q}_0)$$

where

(a) $\hat{y} = \bar{y}$ *and, for* $i \in \mathbb{I}$, $\hat{c}^i = \bar{c}^i$ *(Same real actions)*

(b) $\hat{p} = \bar{p}$, $\hat{q} = \bar{q}$, *and* $\hat{q}_0 = \bar{q}_0 - \sum_{j\in\mathbb{J}} q_j \bar{\theta}^0_j$ *(Same prices; after adjusting for the producer's portfolio choice)*

(c) $\hat{\theta}^0 = 0$ *and, for* $i \in \mathbb{I}$, $\hat{\gamma}^i = 0$ *(No trade in the stock; trivial producer portfolio)*

Proof: The idea of the proof is to find a portfolio for consumer i, for $i \in \mathbb{I}$, of the assets $1, \ldots, J$ giving consumer i the same net income vector as in the equilibrium.

Consumer i's net income vector in the equilibrium is

$$\begin{bmatrix} \bar{p}(0)\left[\bar{c}^i(0) - e^i(0) - \alpha^i \bar{y}(0)\right] \\ \vdots \\ \bar{p}(t)\left[\bar{c}^i(t) - e^i(t) - \alpha^i \bar{y}(t)\right] \\ \vdots \\ \bar{p}(T)\left[\bar{c}^i(T) - e^i(T) - \alpha^i \bar{y}(T)\right] \end{bmatrix} = \begin{bmatrix} \bar{r}^i(0) \\ \vdots \\ \bar{r}^i(t) \\ \vdots \\ \bar{r}^i(T) \end{bmatrix} = W_y \begin{bmatrix} \bar{\gamma}^i \\ \bar{\theta}^i + \alpha^i \bar{\theta}^0 \end{bmatrix}$$

where W_y is the extended dividend matrix enlarged with the dividend plan for the producer in the first column.

$$W_y = \begin{bmatrix} -\bar{q}_0 & -\bar{q}_1 & \dots & -\bar{q}_J \\ \bar{p}(1)\bar{y}(1) + V_1\bar{\theta}^0 & v^1(1) & \dots & v^J(1) \\ \vdots & \vdots & & \vdots \\ \bar{p}(t)\bar{y}(t) + V_t\bar{\theta}^0 & v^1(t) & \dots & v^J(t) \\ \vdots & \vdots & & \vdots \\ \bar{p}(T)\bar{y}(T) + V_T\bar{\theta}^0 & v^1(T) & \dots & v^J(T) \end{bmatrix}$$

Since the future spot net revenue vector $(\bar{p}(t)\bar{y}(t))_{t\in\mathbb{T}_1} \in \langle V(\bar{p})\rangle$ there is a portfolio $\bar{\theta}^y \in \mathbb{R}^J$ such that

$$\begin{bmatrix} -\sum_{j\in\mathbb{J}} \bar{\theta}^y_j \bar{q}_j \\ \bar{p}(1)\bar{y}(1) \\ \vdots \\ \bar{p}(t)\bar{y}(t) \\ \vdots \\ \bar{p}(T)\bar{y}(T) \end{bmatrix} = W\bar{\theta}^y$$

The date 0 price of the stock is the sum of the date 0 price of the portfolio replicating future spot net revenue vector $(\bar{p}(t)\bar{y}(t))_{t\in\mathbb{T}_1}$ and the date 0 cost of the producer portfolio $\bar{\theta}^0$ so that

$$\bar{q}_0 = \sum_{j\in\mathbb{J}} \bar{\theta}^y_j \bar{q}_j + \sum_{j\in\mathbb{J}} \bar{q}_j \bar{\theta}^0_j$$

Thus

$$\begin{bmatrix} -\bar{q}_0 \\ \bar{p}(1)\bar{y}(1) + V_1\bar{\theta}^0 \\ \vdots \\ \bar{p}(t)\bar{y}(t) + V_t\bar{\theta}^0 \\ \vdots \\ \bar{p}(T)\bar{y}(T) + V_T\bar{\theta}^0 \end{bmatrix} = \begin{bmatrix} -\sum_{j\in\mathbb{J}} \bar{\theta}^y_j \bar{q}_j \\ \bar{p}(1)\bar{y}(1) \\ \vdots \\ \bar{p}(t)\bar{y}(t) \\ \vdots \\ \bar{p}(T)\bar{y}(T) \end{bmatrix} + \begin{bmatrix} -\sum_{j\in\mathbb{J}} \bar{q}_j \bar{\theta}^0_j \\ V_1\bar{\theta}^0 \\ \vdots \\ V_t\bar{\theta}^0 \\ \vdots \\ V_T\bar{\theta}^0 \end{bmatrix}$$

where the vectors on the right-hand side are equal to $W\bar{\theta}^y$ and $W\bar{\theta}^0$ respectively. Put $\sum_{j\in\mathbb{J}} \bar{\theta}^y_j \bar{q}_j = \hat{q}_0$. Then consumer i's net income vector is

$$\bar{r}^i = W_y \begin{bmatrix} \bar{\gamma} \\ \bar{\theta}^i + \alpha^i \bar{\theta}^0 \end{bmatrix} = \left[W\bar{\theta}^y + W\bar{\theta}^0\right]\bar{\gamma} + W\left[\bar{\theta}^i + \alpha^i\bar{\theta}^0\right]$$
$$= W\left[\bar{\theta}^y\bar{\gamma} + \bar{\theta}^0\bar{\gamma} + \bar{\theta}^i + \alpha^i\bar{\theta}^0\right]$$

For $i \in \mathbb{I}$, let consumer i's portfolio be $\hat{\theta}^i = \bar{\theta}^y\bar{\gamma} + \bar{\theta}^0\bar{\gamma} + \bar{\theta}^i + \alpha^i\bar{\theta}^0$, net trade in the share $\hat{\gamma}^i = 0$ and put $\hat{\theta}^0 = 0$. Then $\bar{r}^i = W\hat{\theta}^i$ so that the consumer gets the same net income vector as in the given equilibrium. The net income vector $\bar{r}^i$ is the solution to the Consumer Net Income Vector Problem as $r \in \langle W_y \rangle$. Since $\langle W_y \rangle = \langle W \rangle$ the assumptions imply that $\bar{c}^i$ is a solution to Consumer i's Spot-Market Problem given $\bar{r}^i$. With these consumption choices goods markets balance at each date.

The stock market and the other asset markets balance since

$$\sum_{i\in\mathbb{I}} \hat{\gamma}^i = 0$$
$$\sum_{i\in\mathbb{I}} \hat{\theta}^i + \hat{\theta}^0 = \sum_{i\in\mathbb{I}} \bar{\theta}_y \bar{\gamma}^i + \sum_{i\in\mathbb{I}} \bar{\theta}^0 \bar{\gamma}^i + \sum_{i\in\mathbb{I}} (\bar{\theta}^i + \alpha^i \bar{\theta}^0)$$
$$= \sum_{i\in\mathbb{I}} \bar{\theta}^i + \bar{\theta}^0 \sum_{i\in\mathbb{I}} \alpha^i = 0$$

Hence

$$((\hat{c}^i, \hat{\theta}^i, \hat{\gamma}^i)_{i\in\mathbb{I}}, (\hat{y}, \hat{\theta}^0), \hat{p}, \hat{q}, \hat{q}_0)$$

is a Radner equilibrium with a given production plan satisfying (a)-(c).

□

If asset markets are complete then the condition $(\bar{p}(t)\bar{y}(t))_{t\in\mathbb{T}_1} \in \langle V(\bar{p}) \rangle$ holds trivially. Once this condition is violated then the producer's asset choice and choice of production plan will influence the "real part" of the economy; the equilibrium allocation. Thus the possibility of separation of the producer's financial decisions and the choice of production plan occurs only rarely when asset markets are incomplete.

Agreement about Choice of Production Plan

Assume that the producer has chosen a production plan $y \in Y$ which at given spot prices generates the spot net revenue vector $(p(t)y(t))_{t\in\mathbb{T}}$. If there is a production plan, $\hat{y} \in Y$, which gives a spot net revenue vector $(p(t)\hat{y}(t))_{t\in\mathbb{T}}$ such that $p(t)\hat{y}(t) \geq p(t)y(t)$, for $t \in \mathbb{T}$, with some strict inequality, then the current owners of the producer will agree that $(p(t)\hat{y}(t))_{t\in\mathbb{T}}$ is to be preferred. What can be said beyond this about the choice of a production plan? We give a partial answer to this question in Theorem 3.6.D below.

A consumer's indirect utility function depicts, for given spot prices, his willingness to substitute income between different dates. For small variations it is enough to know the gradient of the indirect utility function. The lemma below shows that, given spot prices, we can find a vector which describes the producer's ability to, locally, vary the net spot revenue vector. Recall that a vector a belonging to a set A is an efficient point of A if there is no point a' in A such that $a' > a$.

Lemma 3.6.C *Let $(p(t))_{t\in\mathbb{T}} \in \mathbb{R}^{L(T+1)}_{++}$ be a price system and let $\bar{y} \in Y$ be a production plan. If $\bar{z} = (p(t)\bar{y}(t))_{t\in\mathbb{T}}$ is an efficient point of the set*

$$Z = \left\{z \in \mathbb{R}^{T+1} \mid \text{there exists } y \in Y \text{ such that } z = (p(t)y(t))_{t\in\mathbb{T}}\right\}$$

then there exists a unique vector $\lambda = (\lambda_0, \lambda_1, \ldots, \lambda_T) \in \mathbb{R}^{T+1}_{++}$ with $\lambda_0 = 1$ such that $\bar{y}$ is the unique solution to

$$\text{Max}_y \quad \lambda_0 p(0)y(0) + \ldots + \lambda_T p(T)y(T) \quad s. \text{ to } F(y) \leq 0$$

Proof: Consider the linear mapping $h : \mathbb{R}^{L(T+1)} \longrightarrow \mathbb{R}^{T+1}$ which takes the vector $(y(t))_{t\in\mathbb{T}}$ to $(p(t)y(t))_{t\in\mathbb{T}}$. Since h is a linear function the image of the closed, convex set Y is the closed, convex set Z. The point $\bar{z} = (p(t)\bar{y}(t))_{t\in\mathbb{T}}$ belongs to Z. On the other hand, by assumption $\bar{z}$ is an efficient point of Z and thus the convex sets Z and $\{\bar{z}\} + (\mathbb{R}^{T+1}_{+}\backslash\{0\})$ are disjoint. By Theorem B in Appendix A, there is a $\hat{\lambda} \in \mathbb{R}^{T+1}$, $\hat{\lambda} \neq 0$, and $\beta \in \mathbb{R}$ such that

$$\hat{\lambda}(\bar{z} + d) \geq \beta \geq \hat{\lambda}z \quad \text{for } z \in Z \text{ and } d \in \mathbb{R}^{T+1}_{+}$$

Then $\beta = \hat{\lambda}\bar{z}$ and since Y by assumption (Y1) contains $\{\bar{y}\} - (\mathbb{R}^{L(T+1)}_{+}\backslash\{0\})$ in its interior we get, using that h is a linear function, that $\bar{z} - \varepsilon$ belongs to the interior of Z for $\varepsilon \in \mathbb{R}^{T+1}_{+}, \varepsilon \neq 0$. Hence $\hat{\lambda}\bar{z} > \hat{\lambda}(\bar{z} - \varepsilon)$ for $\varepsilon \in \mathbb{R}^{T+1}_{+}$ and it follows that $\hat{\lambda}\varepsilon > 0$ for $\varepsilon \in \mathbb{R}^{T+1}_{+}, \varepsilon \neq 0$ which can be the case only if $\hat{\lambda} \in \mathbb{R}^{T+1}_{++}$. For $t \in \mathbb{T}$, put $\lambda_t = \hat{\lambda}_t / \hat{\lambda}_0$. The vector λ so defined satisfies the conclusion of the lemma.

The vector $\bar{y}$ is the unique solution to the maximum problem since F is a strictly convex function.

□

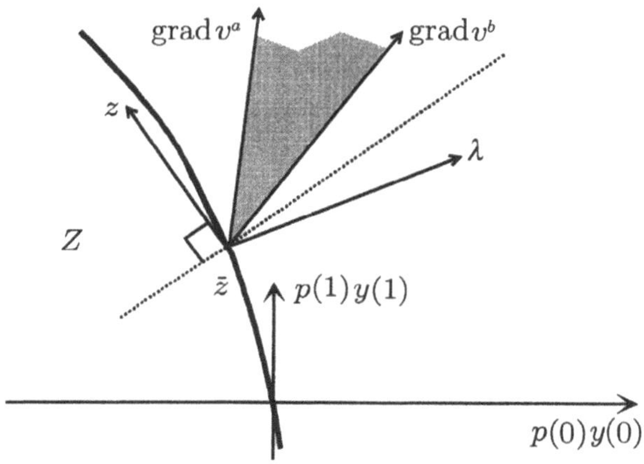

Figure 3.6.A: *The vector λ shows how the producer can, by varying the production plan, transform spot-income between different dates. If λ does not belong to the convex cone generated by the gradients then there exists a spot net revenue vector preferred by the owners*

We can now state the main result of this section. The proof is illustrated in Figure 3.6.A.

Theorem 3.6.D *Let $\mathcal{E}$ be an economy with assets and production and*

$$((\bar{c}^i, \bar{\theta}^i, \bar{\gamma}^i)_{i \in \mathbf{I}}, (\bar{y}, \bar{\theta}^0), \bar{p}, \bar{q}, \bar{q}_0)$$

a Radner equilibrium, with a given production plan, $\bar{y}$. Let

$$\bar{r}^i(t) \;=\; \bar{p}(t)e^i(t) \;+\; \alpha^i \bar{p}(t)\bar{y}(t) \;+\; W\bar{\theta}^i \quad \textit{for } t \in \mathbb{T}$$

and let $\mathbb{I}_0 = \{i \in \mathbb{I} \mid \alpha^i > 0\}$. *Assume that* $\bar{y}$ *generates the efficient spot net revenue vector* $\bar{z} = (\bar{p}(t)\bar{y}(t))_{t\in\mathbb{T}}$ *and let* $\lambda \in \mathbb{R}^{T+1}_{++}$ *be the associated vector from* Lemma 3.6.C. *For a production plan* $y \in Y$ *let*

$$r^i(t) \;=\; \bar{p}(t)e^i(t) \;+\; \alpha^i\bar{p}(t)y(t) \;+\; W\bar{\theta}^i \quad \textit{for } t \in \mathbb{T}$$

Then there is a production plan $y \in Y$ *such that*

$$v^i(p, r^i) \;>\; v^i(p, \bar{r}^i) \textit{ for every consumer } i \in \mathbb{I}_0$$

if and only if

λ *does not belong to the convex cone generated by the vectors* $\mathrm{grad}_r\, v^i(\bar{p}, \bar{r}^i)$ *for* $i \in \mathbb{I}_0$

Proof: Let $\bar{z} = (\bar{p}(t)\bar{y}(t))_{t\in\mathbb{T}}$ and let the set

$$Z \;=\; \left\{z \in \mathbb{R}^{T+1} \mid \text{there exists } y \in Y \text{ such that } z = (\bar{p}(t)y(t))_{t\in\mathbb{T}}\right\}$$

Let λ be the vector associated with $\bar{z}$, in accordance with Lemma 3.6.C.

"only if": Assume that λ belongs to the convex cone generated by the vectors $\mathrm{grad}_r\, v^i(\bar{p}, \bar{r}^i)$ for $i \in \mathbb{I}_0$. If $z \in Z$ is such that

$$v^i(\bar{p}, (\bar{p}(t)e^i(t) \;+\; \alpha^i z(t) \;+\; W\bar{\theta}^i)_{t\in\mathbb{T}}) \;>\; v^i(\bar{p}, \bar{r}^i) \qquad \textbf{3.6.B}$$

then, since $\mathrm{grad}_r\, v^i(\bar{p}, \bar{r}^i)$ defines a supporting hyperplane owning $\bar{r}^i$, for consumer i's upper contour set we have $\mathrm{grad}\, v^i(\bar{p}, \bar{r}^i)\cdot z > \mathrm{grad}_r\, v^i(\bar{p}, \bar{r}^i)\cdot \bar{z}$ for $i \in \mathbb{I}_0$. Since λ is a nonnegative linear combination of the the gradients for the owners we get $\lambda z > \lambda\bar{z}$. By Lemma 3.6.C there can not be such a $z \in Z$. It follows that there can not be a z satisfying 3.6.B and hence not a production plan $y \in Y$ with $z = (\bar{p}(t)y(t))_{t\in\mathbb{T}}$.

"if": Assume that λ does not belong to the convex cone generated by the vectors $\mathrm{grad}_r\, v^i(\bar{p}, \bar{r}^i)$ for $i \in \mathbb{I}_0$.

Let K_1 and K_2 be closed, convex cones with the interior of K_1 owning the vectors $\mathrm{grad}_r\, v^i(\bar{p}, \bar{r}^i)$ for $i \in \mathbb{I}_0$ and the interior of K_2 owning λ where K_1 and K_2 are chosen so that $K_1 \cap K_2 = \{0\}$. By Corollary B in Appendix C there is a vector $z \in \mathbb{R}^{T+1}$, $z \neq 0$, such that the hyperplane with normal

z separates K_1 and K_2. This hyperplane is a homogenous hyperplane and since λ belongs to the interior of K_2 and each of the vectors $\text{grad}_r\, v^i(\bar{p}, \bar{r}^i)$ for $i \in \mathbb{I}_0$ belongs to the interior of K_1 we get

$$\begin{cases} \lambda z < 0 \\ \text{grad}_r\, v^i(\bar{p}, \bar{r}^i) \cdot z > 0 \quad \text{for } i \in \mathbb{I}_0 \end{cases} \qquad \textbf{3.6.C}$$

For $\gamma \in\,]0, 1[$, consider the vectors $\bar{z} + \gamma z$ and, for $i \in \mathbb{I}_0$, $r^i_\gamma = (\bar{p}(t)e^i(t) + \alpha^i[\bar{z}(t) + \gamma z(t)] + W\bar{\theta}^i)_{t\in\mathbb{T}}$. Then, for $i \in \mathbb{I}_0$,

$$v^i(\bar{p}, r^i_\gamma) > v^i(\bar{p}, \bar{r}^i) \qquad \textbf{3.6.D}$$

for γ small enough. We will prove that there is $y \in Y, \gamma \in\,]0, 1[$ such that $(\bar{p}(t)y(t))_{t\in\mathbb{T}} = \bar{z} + \gamma z \in Z$.

By relation 3.6.C we have $\lambda(\bar{z} + \gamma z) < \lambda\bar{z}$ for $\gamma \in]0, 1]$. Assume, in order to obtain a contradiction, that $\bar{z} + \gamma z \notin Z$ for $\gamma \in\,]0, 1[$. Then the closed convex set Z and the line segment $\bar{z} + \gamma z$, $\gamma \in\,]0, 1[$, which is also a convex set can be separated by a hyperplane with normal $\hat{\lambda}$ owning the point $\bar{z}$ so that

$$\hat{\lambda} z \leq \hat{\lambda}\bar{z} \quad \text{for } z \in Z$$

$$\hat{\lambda}(\bar{z} + \gamma z) \geq \hat{\lambda}\bar{z} \quad \text{for } \gamma \in]0, 1]$$

The last inequality gives $\hat{\lambda} z \geq 0$ so that, by relation 3.6.C, $\lambda \neq \hat{\lambda}$ and then the first inequality contradicts that the vector λ, from Lemma 3.6.C, is unique. It follows that $\bar{z} + \gamma z \in Z$ for some $\gamma = \bar{\gamma}$ where $\bar{\gamma} \in\,]0, 1[$, and since Z is a convex set, for each $\gamma \in\,]0, \bar{\gamma}[$. Choose $\hat{\gamma} \in\,]0, \bar{\gamma}[$ so that $\bar{z} + \hat{\gamma} z \in Z$ and so that 3.6.D is satisfied. By the definition of Z there is a $\hat{y} \in Y$ such that $(\bar{p}(t)\hat{y}(t))_{t\in\mathbb{T}} = \bar{z} + \hat{\gamma} z \in Z$.

□

Note that in Theorem 3.6.D we did not allow consumers to trade in the stock or to vary their portfolios along with the suggested variation of the production plan.

With incomplete markets there will, in general, be disagreement among the owners about which production plan to choose, and Theorem 3.6.D shows that it is precisely the lack of agreement among owners about the value of a small variation in the net income vector which is critical for the result. Since the production plan is to be the same for all the owners there is a public goods aspect to the producer's choice problem. Any reasonable solution would also have to take into account differences in the size of ownership.

Unanimity in case of complete markets

When markets are complete grad $F(\bar{y})$ belongs to the convex cone generated by the vectors $\text{grad}_r\, v^i(\bar{p}, \bar{r}^i)$ for $i \in \mathbb{I}_0$ at an equilibrium if and only if $\lambda = (1, \beta(1), \ldots, \beta(T))$; the uniquely determined discount factors. Using Lemma 3.6.C we get the following corollary showing that the partial preordering of production plans, and corresponding spot net revenue vectors, can be extended to a complete preordering when all owners agree on the evaluation of net income. In particular this will be the case if asset markets are complete.

Corollary 3.6.E *Under the same assumptions as in* Theorem 3.6.D. *If* $\dim \langle W \rangle = T$, *so that markets are complete, then the convex cone generated by the vectors* $\text{grad}_r\, v^i(\bar{p}, \bar{r}^i)$, $i \in \mathbb{I}_0$, *is the ray generated by the uniquely determined state prices* $(1, \beta(1), \ldots, \beta(T))$. *If* $\bar{y} \in Y$ *then there is no production plan preferred to* $\bar{y}$ *if and only if*

$$\sum_{t\in\mathbb{T}} \beta(t)\bar{p}(t)\bar{y}(t) \geq \sum_{t\in\mathbb{T}} \beta(t)\bar{p}(t)y(t) \quad \textit{for } y \in Y$$

3.7 Money and the Existence of a Monetary Equilibrium

An economy with money

Let us enlarge the set of goods with one good, good $L+1$, which can be stored. This commodity has no intrinsic value to the consumers; it does not enter their utility functions. In an economy where there are no other means of transferring purchasing power but, possibly, to store good $L+1$, it appears that good $L+1$ might still be valuable to the consumers. We will refer to good $L+1$ as money.

We assume that each consumer has some initial endowment of money. It would be convenient to assign the price 1 to money but this presumes that money will, in equilibrium, have a positive price. Hence it is preferable to allow money to have any (nonnegative) price, $q(t)$, at date t, for $t \in \mathbb{T}$.

Let $\mathcal{E} = (C^i, u^i, e^i)_{i\in\mathbb{I}}$ be an *economy with money*, that is, an economy where each consumer besides his initial endowment of commodities, also has an initial endowment of money, $\hat{m}^i \geq 0$, for $i \in \mathbb{I}$ at date 0 and where $\Sigma_{i\in I}\hat{m}^i = M > 0$. Consumer i's Problem is

$$\text{Max}_{c^i}\, u^i(c^i(0), c^i(1), ..., c^i(T)) \quad \text{s. to } c^i \in C^i$$
$$\begin{cases} p(0)c^i(0) + q(0)m^i(0) \leq p(0)e^i(0) + q(0)\hat{m}^i \\ p(t)c^i(t) + q(t)m^i(t) \leq p(t)e^i(t) + q(t)m^i(t-1) \quad \text{for } t \in \mathbb{T}_1 \end{cases}$$
3.7.A

At date t, $0 < t \leq T$ the consumer enters with his money holdings from date $t-1$ and has the amount $p(t)e^i(t) + q(t)m^i(t-1)$ to spend in the spot-market. The consumer can use this amount to buy goods or to acquire money to be carried over to date $t+1$.

(Non)existence of a monetary equilibrium

The definition of a monetary equilibrium is now straightforward.

Definition 3.7.A *A* **monetary equilibrium** *is a tuple* $((\bar{c}^i, \bar{m}^i)_{i\in\mathbb{I}}, p, q)$ *such that:*

(a) *for* $i \in I$, $(\bar{c}^i, \bar{m}^i)$ *is a solution to Consumer* i*'s Problem* 3.7.A *given the prices* $p \in (\mathbb{R}^L_{++})^{T+1}$ *and* $q \in \mathbb{R}^{T+1}_+$

(b) *for* $t \in \mathbb{T}_1$, $\Sigma_{i\in\mathbb{I}}\bar{m}^i(t) = \Sigma_{i\in\mathbb{I}}\bar{m}^i(t-1) = M$

(c) *for* $t \in \mathbb{T}$, $\Sigma_{i\in\mathbb{I}}\bar{c}^i(t) = \Sigma_{i\in\mathbb{I}}e^i(t)$

Condition (**b**) implies that the demand for money is equal to M, at each date.

Proposition 3.7.B *Let* $\mathcal{E}$ *be an economy with money and let* $((\bar{c}^i, \bar{m}^i)_{i\in\mathbb{I}}, p, q)$ *be a monetary equilibrium for* $\mathcal{E}$. *Then* $q(t) = 0$, *for* $t \in \mathbb{T}$.

Proof: The market balance condition for money at date T, implies that for some consumer $i \in \mathbb{I}$, say consumer 1, $\bar{m}^i(T) > 0$. Since $(\bar{c}^1, \bar{m}^1)$ solves Consumer 1's Problem 3.7.A, this can be the case only if $q(T) = 0$.

We can now apply the same reasoning to date $T-1$. Again for some consumer $i \in \mathbb{I}$, say consumer 1, $\bar{m}^i(T-1) > 0$. Since $(\bar{c}^1, \bar{m}^1)$ solves Consumer 1's Problem 3.7.A, this can be the case only if $q(T-1) = 0$.

Applying the same reasoning to dates $T-2$, $T-3, \ldots, 1, 0$ we get

$$q(t) = 0 \quad \text{for } t \in \mathbb{T}$$

□

Proposition 3.7.B shows that it is not possible to extend the Arrow-Debreu model to include (fiat) money, with a positive price.

There is an alternative way of stating the result. Let us refer to a monetary equilibrium where the price of money is 1, at each date, as a *normalized* monetary equilibrium. The conclusion is, in this case, that there does not exist a normalized monetary equilibrium.

What distinguishes money, as defined here, from a nominal asset, that can be traded at every date, but pays dividends only at the last date? A nominal asset is available in 0 net supply. Hence any claim to T-crowns is matched by a corresponding obligation to deliver T-crowns, and it follows that the net deliveries of T-crowns sum to 0 in equilibrium. With money, some agents are needed, who at the last date are willing to hold money, and accordingly refrain from consumption.

Temporary equilibrium and overlapping generations economies

There are, at least, two ways to introduce money in an economy. On the one hand, one could give up the requirement that markets for commodities and money should balance at *every* date. Allowing for, perhaps incorrect, expectations about future prices at date 0, equality of supply and demand is required *only* at date 0; the current date. This is the approach taken in *temporary equilibrium models*. In these models a last date of the economy is avoided by postulating an unknown future. Under reasonable assumptions there are consumers who want to carry purchasing power forward. This gives a demand for money at the current date.

On the other hand, we could retain the assumption of perfect foresight and market balance at each date and relax the assumption of a finite time horizon. Instead an infinity of dates is postulated. Again there may, at any date, be consumers in the economy who want to carry purchasing power forward. Such consumers will demand money and so again money might get a positive value at an equilibrium. This is the approach taken in the study of *overlapping generations economies*.

Summary

In this chapter we have introduced assets as a means for the agents to transfer income between dates. We showed that the asset structure generated a subspace of income transfers and that with real assets this subspace would in general depend on the spot prices of goods.

With complete asset markets each net income vector at future dates was attainable and with incomplete asset markets the subspace of attainable net income vectors was seen to be a proper subspace and some net income vector at future dates was not attainable independent of the agents ability to pay at date 0.

If there is to be no arbitrage then assets have to be priced in accordance with a set of discount factors, which were then seen to be the date 0 price of spot income for the different future dates. We also showed that with complete asset markets the vector of discount factor was uniquely determined if there was to be no arbitrage.

A Radner equilibrium was seen to be essentially an equilibrium relative to a subspace of income transfers, where the subspace was the linear subspace spanned by the assets (at the equilibrium prices). At a Radner equilibrium the portfolios necessary to obtain the desired net income vector were specified.

We showed that for an economy with production and a given production plan the action of the producer in the asset market was irrelevant and there was no need for the consumers to trade in the stock if the producer's spot net revenue vector belonged to the subspace generated by the other assets. We also showed that with incomplete asset markets the evaluation of production plans, given spot prices for the goods, by different owners would in general be different.

Money was shown to differ from assets because money acquired at some date gave no guarantee of increased purchasing power at some future date.

Exercises

Exercise 3.A Consider an economy with three dates and one nominal asset. Let

$$V = \begin{bmatrix} 1 \\ 1 \end{bmatrix}$$

be the value dividend matrix.

(a) Let

$$\begin{cases} \beta = (1, \beta(1), \beta(2)) = (1, \frac{4}{5}, \frac{1}{5}) \\ \beta' = (1, \beta'(1), \beta'(2)) = (1, \frac{1}{5}, \frac{4}{5}) \end{cases}$$

Does β define a date 0 price q such that (V, q) is arbitrage free? Does β' define a price q such that (V, q) is arbitrage free?

(b) At what prices date 0 prices, q, is (V, q) arbitrage free?

(c) What will determine the asset price, q, in a Radner equilibrium?

Exercise 3.B Consider an economy with three dates and one nominal asset. Let

$$V = \begin{bmatrix} 1 \\ -1 \end{bmatrix}$$

be the value dividend matrix.

(a) At what a date 0 prices q is (V, q) arbitrage free?

(b) Let q be a price such that (V, q) is arbitrage free. Find all the discount factors, $\beta = (1, \beta(1), \beta(2))$, which induce the price q for the assets.

(c) What will determine the date 0 asset price q in a Radner equilibrium? Under what circumstances would the asset price be positive in equilibrium?

Exercise 3.C Consider an economy with T dates and an asset structure with $J = T + 1$ assets and assume that assets $1, \ldots, T$ are linearly independent. Let the date 0 prices for assets $1, \ldots, T$ be determined by discount factors $\beta = (1, \beta(1), \ldots, \beta(T))$.

(a) Consider asset $T + 1$. Show that there is a portfolio, θ^{T+1}, of the first T assets giving the same future dividend vector as asset $T+1$, that is, a replicating portfolio.

(b) Calculate the date 0 cost of the replicating portfolio from (a).

(c) Show that the date 0 price calculated in (b) for asset $T + 1$ is the same as the date 0 price induced by the discount factors

(d) Is the asset structure (V, q) arbitrage free with the date 0 price q equal to the cost of asset $T + 1$ as calculated in (b)?

(e) Assume that trade in asset $T+1$ is forbidden. How does this change the subspace of net income vectors?

Exercise 3.D Consider an economy with two goods, three dates and two real assets, so that $L = 2$, $T = 2$ and $J = 2$, and with dividend matrix V given by

	Asset	
Date	1	2
1	$\begin{bmatrix} 0 \\ 0 \end{bmatrix}$	$\begin{bmatrix} 1 \\ 0 \end{bmatrix}$
2	$\begin{bmatrix} 1 \\ -1 \end{bmatrix}$	$\begin{bmatrix} 0 \\ 0 \end{bmatrix}$

(a) For which future commodity prices does the value dividend matrix have rank 2?

(b) For which future commodity prices does the value dividend matrix have rank 1?

(c) Let $\bar{p} \in \mathbb{R}^4_{++}$ be future commodity prices such that the value dividend matrix has rank 2. Show that there is an open set $U_{\bar{p}}$ with $\bar{p} \in U_{\bar{p}}$ such that the value dividend matrix has rank 2 for $p \in U_{\bar{p}}$.

(d) Let $\langle W(p) \rangle$ be the extended subspace of income transfers. Find a basis for the orthogonal subspace, $\langle W(p) \rangle^{\perp}$, which varies differentiably with $p \in U_{\bar{p}}$ if $U_{\bar{p}}$ is small enough so that $\operatorname{rank} V(p) = 2$ for $p \in U_{\bar{p}}$.

Exercise 3.E Consider an economy with three dates and 2 nominal assets with dividend matrix

	Asset	
Date	1	2
1	1	2
2	2	1

(a) Assume that short-selling is not allowed. Illustrate in a diagram the future net income vectors that can be achieved by buying non-negative quantities of the assets.

(b) Show that the set of net income vectors from (a) is a convex cone, K.

(c) Let (q_1, q_2) be the date 0 prices of the assets. Assume that the Consumer Net Income Vector Problem

$$\text{Max}_{(r,\theta)} \; v(p, r) \quad \text{s. to } r = W\theta \text{ and } \theta \in \mathbb{R}^2_+$$

has a solution $(\bar{r}, \bar{\theta})$ with $\bar{r}$ belonging to the interior of K. Show that the gradient of v evaluated at $(p, \bar{r})$ defines discount factors so that

$$\begin{aligned} D_{r(0)}v(p,\bar{r})q_1 &= D_{r(1)}v(p,\bar{r}) \cdot 1 + D_{r(2)}v(p,\bar{r}) \cdot 2 \\ D_{r(0)}v(p,\bar{r})q_2 &= D_{r(1)}v(p,\bar{r}) \cdot 2 + D_{r(2)}v(p,\bar{r}) \cdot 1 \end{aligned}$$

Exercise 3.F Let $\mathcal{E} = (C, u, e)$ be an economy with $T = 2$, $L = 2$ and a single consumer satisfying the Maintained Assumptions.

(a) Let $\mathbb{M} = \{0\}$. Find necessary and sufficient conditions for $p = (p(0), p(1), p(2)) \in \mathbb{R}^6_{++}$ to be a spot-market equilibrium price system (for a spot-market equilibrium relative to $\mathbb{M}$). Denote the set of such spot-market equilibrium price system by P.

(b) Let $p \in P$ and let $\operatorname{grad} v(p, 0, 0, 0)$ be the gradient of the indirect utility function. Characterize all the linear subspaces $\mathbb{M}$ such that (e, p) is a spot-market equilibrium relative to $\mathbb{M}$.

(c) Extend the economy with a real asset structure with dividend matrix A. Let (e, p) and $(e, \tilde{p})$ be two spot-market equilibria with $p \neq \tilde{p}$. Show that (e, p) and $(e, \tilde{p})$ can be extended to Radner equilibria in a unique way and that the date 0 asset prices are the same for both extensions. (*Hint:* Use the result from Exercise2.D)

(d) Assume that the economy is extended instead with a nominal dividend matrix V. Are the conclusions from (d) still true?

(e) "For economies with a single consumer the distinction between complete and incomplete markets is not important". Use your conclusions from (c) and (d) to comment on this.

Exercise 3.G Consider an economy with $T = 2$, where there is single asset, so that $J = 1$. The asset is a bond paying nothing at date 1 and one 2-crown at date 2. However, assume for the moment that the bond can be traded also at date 1. Show that the bond induces new assets, 2 and 3 so that the dividend matrix is

	Asset		
Date	1	2	3
1	0	$-q(1)$	$q(1)$
2	1	1	0

where $q(1)$ is the date 1 price of the bond in 1-crowns.

(a) Is the asset market complete if the bond can be traded only at date 0?

(b) Assume that, rather than trading the bond at date 1, consumers can at date 0 trade the induced assets and that a Radner equilibrium has resulted in in date 0 prices for the assets q_1, q_2 and q_3. Can the original asset be replicated by the induced assets? Is the asset market complete if (i) $q(1) > 0$ (ii) $q(1) = 0$

(c) Assume that in the Radner equilibrium from (b) $q(1) > 0$. Let $\beta = (1, \beta(1), \beta(2))$ be discount factors in the equilibrium. Find the date 0 prices of the assets 1, 2 and 3.

(d) Assume that the date 0 price of asset 3, that is, q_3 is positive. To interpret the equilibrium as an equilibrium with retrading of the bond the price q_2 ought to be 0. Show that there is a Radner equilibrium differing from the given one only in portfolio choices, $q(1), q_2$ and q_3 where $q_2 = 0$.

Exercise 3.H Consider an economy with one good, $L = 1$, three dates, $T = 2$, two assets, $J = 2$, two consumers, $\mathbb{I} = \{a, b\}$ and a single producer. The production set is

$$Y = \{y \in \mathbb{R}^3 \mid F(y) = y(1) + y(2) + y(0),\ y(0) \leq 0 \text{ and } y(1), y(2) \geq 0\}$$

which does not satisfy the assumption (Y1). The assets and initial endowments are

	Assets		Init. end.	
Date	1	2	e^a	e^b
0			2	2
1	1	0	2	0
2	0	1	0	2

The consumers have utility functions

$$u^a(c) = \frac{1}{3}\ln c(0) + \frac{1}{3}\ln c(1) + \frac{1}{3}\ln c(2)$$

$$u^b(c) = \frac{1}{3}\ln c(0) + \frac{1}{3}\ln c(1) + \frac{1}{3}\ln c(2)$$

and own initial shares of the producer $\alpha^a = 1/3$ and $\alpha^b = 2/3$.

(a) Disregard the assets and initial shares. Find a Walras equilibrium for the economy.

(b) Interpret the forward prices from (a) as spot-prices and find the net income vectors in the equilibrium, $\bar{r}^{\alpha}$ and $\bar{r}^{b}$, taking into account the initial ownership of the producer. What are the discount factors in the equilibrium?

(c) Find a Radner equilibrium with a given production plan $y = (-4/3, 2/3, 2/3)$ where there is no trade in the stock so that $\bar{\gamma}^a = \bar{\gamma}^b = 0$ and where the producer chooses the portfolio $\bar{\theta}^0 = 0$. Find portfolios $\bar{\theta}^a$ and $\bar{\theta}^b$, needed to obtain the net income vectors from (b). What is the future spot net revenue vector in the equilibrium and what is the equilibrium stock price?

(d) Modify the Radner equilibrium with a given production plan $y = (-4/3, 2/3, 2/3)$, from (c), so that $\bar{\gamma}^a = -1/3$ and $\bar{\gamma}^b = 1/3$ and let the producer choose the portfolio $\bar{\theta}^0 = (-2/3, -2/3)$. How should the consumers adjust their portfolios to obtain the income vectors from (b)?

Exercise 3.I Consider an economy as in Exercise 3.H but without the two assets and where the consumers can not trade in the stock.

(a) Find a Radner equilibrium for the economy given the production plan $y = (-4/3, 2/3, 2/3)$ and equilibrium prices $\bar{p} = (1, 1, 1)$. Find the indirect utility functions, v^a and v^b, for the consumers.

(b) Calculate the gradient of F at the given production plan and the gradient of the consumers indirect utility functions at the equilibrium consumptions. Let $\bar{r}^a$ and $\bar{r}^b$ be the net income vectors in the equilibrium. Does $\operatorname{grad} F$ belong to the convex cone induced by $\operatorname{grad}_r v^a(\bar{p}, \bar{r}^a)$ and $\operatorname{grad}_r v^b(\bar{p}, \bar{r}^b)$? Does there exist a production plan preferred by both consumers?

(c) Assume that the initial endowments were instead $e^a = (2, 2, 0)$ and $e^b = (2, 2, 0)$. Does $\operatorname{grad} F$ belong to the convex cone induced by $\operatorname{grad}_r v^a(\bar{p}, \bar{r}^a)$ and $\operatorname{grad}_r v^b(\bar{p}, \bar{r}^b)$. Does there exist a production plan preferred to $y = (-4/3, 2/3, 2/3)$ by both consumers?

(d) With the endowments as in (c) which is the best production plan for consumer a? For consumer b?

Chapter 4

ECONOMIES WITH UNCERTAINTY

Introduction

As was pointed out in Chapter 3, the reasoning, results and definitions, regarding economies over time, *carry over* to the case of economies with uncertainty.

In Chapter 3 we studied study economies over time but with no uncertainty. With *uncertainty* we are forced to consider at least two dates. At the first date, the consumers are uncertain about some event that will obtain at the second date. Still agents must make plans at the first date regarding consumption and/or production at the second date.

Uncertainty is introduced in the form of an (elementary) event, or *state of nature (state of the world)*, chosen by *Nature*. At the first date the agents do not know which event Nature has chosen. Hence the plans of the agents and the contracts for deliveries between the agents or to and from the market must be formulated as *contingency* plans to the extent that they involve deliveries at the second date. Once it is revealed, at the second date, which state Nature has chosen the uncertainty is resolved and some of the deliveries are carried out.

A simple example of such a contingent delivery is a fire insurance for a house. At the first date the insurance is paid independent of which unknown event is chosen by Nature at date 1; "fire" or "no fire". A delivery from the insurance company to the insured in the form of money to enable the insured to rebuild the house takes place only if Nature chooses the event "fire". The basic idea in the extension of the Arrow-Debreu model to encompass uncertainty is to allow, for each of the goods, contracts for delivery contingent on each of the events that may be chosen by Nature. The feature that the delivery of goods is contingent on the state of nature is often, somewhat inappropriately, referred to by saying that the economy has "contingent commodities". Such commodities are discussed in Section 4.2 along with the definition of complete contingency markets and Walras equilibria. A discussion of Radner equilibria for the general model of time and uncertainty is given in Section 4.9.

In an economy with contingent commodities the preferences of the consumers will depict not only their taste concerning the goods but also their *beliefs* regarding the event chosen by Nature. We consider, in Section 4.3, the assumptions on preferences; monotonicity and convexity, with this extended interpretation in mind.

The latter part of the chapter, Section 4.5 to Section 4.9 concerns the generalization of the Arrow-Debreu model to many dates and uncertainty. This is included for the sake of completeness and although it makes it possible to discuss assets traded at many dates; frequent trading, the later chapters do not rely on the exposition here. The reader should find the definitions and interpretations straightforward after having dealt in detail with economies over time and the two date-uncertainty model.

The first task is to find a way to describe the varying information between agents and/or over time. This is done in Section 4.5 where we introduce event trees. An event tree is a convenient way to describe the evolution of information over time when the information is common among all the agents. With less than full information consumption plans and production plans have to satisfy certain informational restrictions. This is the topic of Section 4.6. When the time varying but common information of the agents is described by an event tree these informational restrictions are taken care of automatically. If information differs between agents the agents may be able to distinguish, at some date, different elementary events. In this case there is an alternative description of the informational restrictions which is convenient. This is the subject of the latter part of Section 4.6.

Finally, Sections 4.7 to 4.9 show how we can generalize the earlier definitions and results to economies over several dates under uncertainty.

4.1 Different Kinds of Uncertainty

A first task is to delimit the type of uncertainty we want to study.

We may distinguish uncertainty according to the possibilities of the agents to influence the outcome. Consider, for example,

- *weather conditions*: these seem (with the present state of technology) to be outside the influence of the agents.
- *the price of a stock*: if there are many agents then the influence of any single agent on the price of the stock is negligible but, of course, the agents' actions collectively determine the price.
- *theft of a bicycle*: here the owner can influence the probability of the event "theft" by installing different kinds of locks on the bicycle.

The type of uncertainty we want to consider is uncertainty which can not be influenced by the agents separately. To a single agent the uncertainty of the price of a stock price may be beyond the influence of the agent and our analysis will apply. This kind of uncertainty is often insurable. At the other extreme end is uncertainty like theft of a bicycle which is more difficult to insure against. In this case the agent's actions to a large extent determine whether the event occurs or not. Such a situation is characterized by "moral hazard". If a bicycle is heavily insured the insured has an incentive to neglect to take actions which prevents the bike from being stolen.

4.2 Contracts for Contingent Delivery

In Chapter 2 we made a distinction between a good; characterized by its physical properties and a commodity; a certain good delivered at a specific date. By assuming a single location for delivery we were able to ignore the place of delivery. We will continue to do so here as well.

Contingent commodities

In the beginning of the fifties, see Arrow [1953, 1964] and Debreu [1953, 1959, 1983], the concept of commodity was extended to *contingent commodities*. This made it possible to extend the Arrow-Debreu economy to take uncertainty into account.

Although the idea of a contingent commodity may seem far-fetched for someone who is not familiar with the intricacies of economics the idea behind it is familiar enough. Consider a consumer who buys insurance for her house at date 0. Assume that the insurance contract specifies that the consumer shall receive a copy of the original house in case the house burns down (and nothing otherwise).

Table 4.2.A: *Fire insurance*

Date	Event	
0	{0	Insurance premium
1	Fire	Copy of original house
	No fire	Nothing

At date 0 the consumer pays some price for the insurance. This date 0 cost of the insurance is not recovered whatever event Nature chooses. Having bought the insurance the consumer owns a "risky prospect" which is described by the part of Table 4.2.A referring to date 1.

Planned and realized deliveries

The consumer receives a copy of the original house only if the house burns down so the delivery of the house is *contingent* on the state chosen by Nature. At date 0 there are *planned deliveries* to the consumer; nothing if the house does not burn down and a copy of the original house if the house burns down. The *realized deliveries* to the consumer at date 1 depend on which event Nature has chosen and will be either nothing, if the house does not burn down, or a copy of the original house, if the house burns down.

Two Dates and Uncertainty

Let us now generalize the idea from the insurance example and apply it to each possible state and each good. Let

$$\mathbb{S} = \{0, 1, 2, ..., S\}$$

be the set of the elementary event at date 0 and the possible states at date 1; $1, 2, ..., S$, and let

$$\mathbb{S}_1 = \{1, 2, ..., S\}$$

be the states of Nature at date 1. Nature chooses one, and only one, state in $\mathbb{S}_1$ and at date 0, the time of planning Nature's choice is not known to the agents. At date 1 Nature's choice is revealed to all the agents. We thus have common and complete information at date 1.

The consumer problem

The consumers' initial endowment is contingent on the state of Nature as shown in Table 4.2.B but the consumer knows at date 0 for every state, which goods bundle she will receive.

Table 4.2.B: *Endowment, consumption and prices*

Date	State	Init. Endow.	Cons.	Prices
0	0	$e(0)$	$c(0)$	$P(0)$
	1	$e(1)$	$c(1)$	$P(1)$
1	2	$e(2)$	$c(2)$	$P(2)$
	$\vdots$	$\vdots$	$\vdots$	$\vdots$
	S	$e(S)$	$c(S)$	$P(S)$

In the simplest interpretation of the model there are prices, in 0-crowns to be paid at date 0 for each (contingent) commodity, that can be delivered at date 1. Cf. the interpretation over time in Section 2.2. At date 0 the consumers can buy goods for consumption at that date at the prices $P(0)$ but the consumer may also at date 0 make plans for her consumption, at date 1. A consumption plan will usually depend on the choice made by Nature which is revealed only at date 1.

The consumer pays $P_l(s)$ 0-crowns at date 0 to have delivered 1 unit of good l if state s occurs. The delivery of 1 unit of good l is realized only if state s occurs. Good l and the event s define a contingent commodity which can conveniently be thought of as "good $l-$if state s". This interpretation of the model is referred to as "complete contingency markets" at date 0. The Consumer Problem in this setting is

$$\begin{aligned} \text{Max}_c \quad & u(c(0), c(1), ..., c(S)) \quad \text{s. to } c \in C \text{ and} \\ & P(0)c(0) \ + \ P(1)c(1) \ + \ ... \ + \ P(S)c(S) \ \leq \ w \qquad \mathbf{4.2.A} \\ & \text{with } w \ = \ P(0)e(0) \ + \ P(1)e(1) \ + \ ... \ + \ P(S)e(S) \end{aligned}$$

Formally this is the same as the Consumer Problem over time. We have only substituted the time-state index, s, for the time index t. In the interpretation the consumer sells the initial endowment at date 0 and obtains the wealth w which is then used to buy a consumption (plan).

With the *consumption plan* $(c(0), c(1), ..., c(S))$ the consumer will realize the consumption $(c(0), c(s)) \in (\mathbb{R}^L_{++})^2$, if Nature chooses $s \in \mathbb{S}_1$. The *realized deliveries to* (positive) and *from* (negative) the consumer are $c(0) - e(0)$ at date 0 and $c(s) - e(s)$ at date 1.

The price for sure delivery

A consumer who wants the sure delivery of 1 unit of some good $l \in \mathbb{L}$ can secure this delivery by buying 1 unit of each of the commodities: "good l if state 1", "good l if state 2",..., "good l if state s". Thus good l gives rise to S contingent commodities, "good l if state s", for $s \in \mathbb{S}$. A *futures contract* for the delivery of one unit of good l or a *sure delivery* of good l is obtained by buying 1 unit of each of the contingent commodities.

Let $\hat{P}_l$ be the price for sure delivery at date 1 of one unit of good l. $\hat{P}_l$ is the sum of the prices of "good l- if state s" where the summation is over $s \in \mathbb{S}_1$. Thus

$$\hat{P}_l \ = \ \textstyle\sum_{s \in \mathbb{S}_1} P_l(s) \quad \text{for } l \in \mathbb{L}$$

Walras equilibrium

Let us now consider an economy $\mathcal{E} = (C^i, u^i, e^i)_{i \in \mathbb{I}}$. A *Walras equilibrium* for $\mathcal{E}$ is an allocation and a price system, $((\bar{c}^i)_{i \in \mathbb{I}}, P)$ where, for $i \in \mathbb{I}$, $\bar{c}^i$ solves the Consumer Problem 4.2.A. Since $(\bar{c}^i)_{i \in \mathbb{I}}$ is an allocation we have market balance which means

$$\begin{cases} \sum_{i \in \mathbb{I}} \bar{c}^i(0) &= \sum_{i \in \mathbb{I}} e^i(0) \\ \sum_{i \in \mathbb{I}} \bar{c}^i(1) &= \sum_{i \in \mathbb{I}} e^i(1) \\ & \vdots \\ \sum_{i \in \mathbb{I}} \bar{c}^i(S) &= \sum_{i \in \mathbb{I}} e^i(S) \end{cases}$$

These conditions ensure that markets balance at date 0 but the market balance conditions also ensure that the date 0 trade in the contingent commodities is such that no matter which state Nature chooses at date 1 the realized deliveries, $\Sigma_{i \in \mathbb{I}}(\bar{c}^i(s) - e^i(s))$ sum to 0. Hence the consumers' plans are compatible for each choice by Nature.

4.3 Preferences and Beliefs

In this section we consider some of the assumptions regarding the consumers' preferences and their interpretation, in case of uncertainty. The preferences of the consumer depict, on the one hand, her taste regarding the different goods and, on the other hand, an evaluation of the likelihood of the different states at date 1. We will limit ourselves to the case of two goods and three events at date 1, so that $L = 2$ and $S = 3$.

Table 4.3.A: *Monotonicity of preferences*

		Consumption	
Date	State	$\bar{c}$	$\hat{c}$
0	0	$(2,3)$	$(2,3)$
1	1	$(2,2)$	$(2,2)$
	2	$(2,2)$	$(2,2)$
	3	$(2,2)$	$(3,3)$

Monotonicity of preferences

Let us consider a consumer who compares the consumptions $\bar{c}, \hat{c} \in \mathbb{R}^8$, as in Table 4.3.A.

If the utility function is strongly monotone then $u(\hat{c}) > u(\bar{c})$. This is reasonable unless the consumer regards state 3 not to be possible and hence believes that Nature will not choose this state at date 0.

Convexity and consumption variability across states

Consider a consumer who compares the consumptions $\bar{c}, \hat{c}$ and $\tilde{c} \in \mathbb{R}^8$, given in Table 4.3.B.

Table 4.3.B: *Preferences and consumption variability across states*

		Consumption		
Date	State	$\bar{c}$	$\hat{c}$	$\tilde{c}$
0	0	$(2,3)$	$(2,3)$	$(2,3)$
1	1	$(2,2)$	$(2,2)$	$(2,2)$
	2	$(2,2)$	$(3,1)$	$(1,3)$
	3	$(2,2)$	$(1,3)$	$(3,1)$

The consumptions $\hat{c}$ and $\tilde{c}$ differ only for the states 2 and 3. Both $\hat{c}$ and $\tilde{c}$ are, speaking vaguely, more spread out than $\bar{c}$ which involves a sure consumption at date 1. If the consumer's utility function satisfies strong convexity then

$$u(\bar{c}) = u(\frac{1}{2}\hat{c} + \frac{1}{2}\tilde{c}) > \min(u(\hat{c}), u(\tilde{c}))$$

This indicates that strict convexity of preferences (corresponding to strict *quasi*-concavity for the utility function) implies an *aversion* to variability of the consumption across states at date 1. If $u(\hat{c}) = u(\tilde{c})$ then $\bar{c}$ is preferred to both $\hat{c}$ and $\tilde{c}$.

Preferences and probability judgements

To see how the consumer's preferences entails *beliefs* about the state of Nature turn to Table 4.3.C.

There the consumption $\hat{c}$ involves large deliveries of goods to the consumer, if Nature chooses state 2, and the consumption $\bar{c}$ involves the same large deliveries to the consumer, if Nature chooses state 3. The consumptions are otherwise equal. If the consumer prefers $\hat{c}$ to $\bar{c}$ it is tempting to conclude that this is because the consumer regards state 2 as more likely than state 3. This conclusion is warranted unless there is an interplay between the valuation of the goods bundle received and the state. For example, let the goods be two kinds of nutrients, let state 2 be "warm winter" and state 3 be "cold winter".

Table 4.3.C: *Preferences and probablility judgements*

		Consumption	
Date	State	$\bar{c}$	$\hat{c}$
0	0	$(2,3)$	$(2,3)$
1	1	$(2,2)$	$(2,2)$
	2	$(2,2)$	$(4,5)$
	3	$(4,5)$	$(2,2)$

Then the consumer may regard state 2, "warm winter" as more likely than state 3, "cold winter". Still the consumer may prefer the consumption $\bar{c}$ to $\hat{c}$ since her need for nutrition is larger when there is a cold winter.

Separation of taste and beliefs

For many problems it is useful to have separation of beliefs from tastes. Assume that a consumer has a utility function of the form

$$u(c(0), c(1), c(2), c(3)) =$$
$$= \pi(1)\tilde{u}(c(0), c(1)) + \pi(2)\tilde{u}(c(0), c(2)) + \pi(3)\tilde{u}(c(0), c(3)) \qquad \textbf{4.3.A}$$

with $\pi(1), \pi(2), \pi(3) \geq 0$ and $\pi(1) + \pi(2) + \pi(3) = 1$ and where $\tilde{u} : (\mathbb{R}^L_{++})^2 \longrightarrow \mathbb{R}$ is a function satisfying the Maintained Assumptions. It is close at hand to interpret $\pi(1), \pi(2)$ and $\pi(3)$ as (subjective) probabilities. There is a *separation* of the *beliefs* (probability judgements) and taste or the evaluation of the *consequences*; in this case the commodity bundle actually received. When the consumer has a utility function of the form given in 4.3.A then $\tilde{u}$ is the utility function over alternatives with sure delivery at date 1 and a date 0 consumption which is independent of Nature's choice. Consider the consumptions in Table 4.3.D assuming that $c' = \bar{c}(1) = \bar{c}(2) = \bar{c}(3)$, $c'' = \hat{c}(1) = \hat{c}(2) = \hat{c}(3)$ and $\bar{c}(0) = \hat{c}(0) = c_0$.

Table 4.3.D: *Preferences for sure deliveries*

		Consumption					Consumption	
Date	State	$\bar{c}$	$\hat{c}$		Date	State	$\bar{c}$	$\hat{c}$
0	{ 0	$\bar{c}(0)$	$\hat{c}(0)$	or	0	{ 0	$\bar{c}(0)$	$\hat{c}(0)$
	1	$\bar{c}(1)$	$\hat{c}(1)$			1	c'	c''
1	{ 2	$\bar{c}(2)$	$\hat{c}(2)$		1	{ 2	c'	c''
	3	$\bar{c}(3)$	$\hat{c}(3)$			3	c'	c''

A simple calculation, using 4.3.A yields

$$u(c) \;=\; u(c(0), \bar{c}(1), \bar{c}(2), \bar{c}(3)) \;>\; u(c(0), \hat{c}(1), \hat{c}(2), \hat{c}(3)) \;=\; u(\hat{c})$$

if and only if

$$\tilde{u}(c_0, c') \;>\; \tilde{u}(c_0, c'')$$

Let us now turn to the consumptions in Table 4.3.E, which differ only for the states 1 and 2.
Indeed we have, if $\tilde{u}(c_0, c') - \tilde{u}(c_0, c'') > 0$,

$$u(\bar{c}) = u(c_0, c', c'', c(3)) \;>\; u(c_0, c'', c', c(3)) \;=\; u(\hat{c})$$

if and only if

$$\pi(1)\tilde{u}(c_0, c') \,+\, \pi(2)\tilde{u}(c_0, c'') \;>\; \pi(1)\tilde{u}(c_0, c'') \,+\, \pi(2)\tilde{u}(c_0, c')$$

if and only if

$$[\pi(1) - \pi(2)]\,[\tilde{u}(c_0, c') - \tilde{u}(c_0, c'')] \;>\; 0$$

if and only if

$$\pi(1) - \pi(2) \;>\; 0$$

which shows that the interpretation of $\pi(s), s = 1, 2, 3,$ as (subjective) probabilities makes sense. If $\tilde{u}(c_0, c') > \tilde{u}(c_0, c'')$ then the consumer ought to prefer $\bar{c}$ to $\hat{c}$ if and only if $\pi(1) > \pi(2)$.

Table 4.3.E: *Implied (subjective) probabilities*

		Consumption	
Date	State	$\bar{c}$	$\hat{c}$
0	0	$c(0)$	$c(0)$
1	1	c'	c''
	2	c''	c'
	3	$c(3)$	$c(3)$

The indirect utility function

With uncertainty the *indirect utility function* describes the consumer's preferences over *contingent spot income*, for given spot prices. On the one hand, the consumer evaluates date 0 income and date 1 income. On the other hand, her preferences shows how she is willing to substitute spot income in different states of Nature at date 1. The indirect utility function also depicts the consumer´s taste as well as her probability judgements, in much the same way as the consumer's (direct) utility function.

4.4 Analogy to Economies over Time: Terminology in Case of Uncertainty

The purpose of this section is to reinterpret some of the results, from Chapter 2 and 3 in an economy with uncertainty and two dates.

We have already seen in Section 4.2 that the interpretation of the Arrow-Debreu economy over time where all the exchanges for current and forward deliveries take place at date 0 carries over to the case of uncertainty and two dates.

From Chapter 2 we know that there is another interpretation where consumers trade in future spot income at date 0. The spot income deliveries are carried out at the future dates. Then the consumers engage in exchange of goods on the spot markets. In Chapter 3, Section 3.1 we saw that this interpretation was similar to the interpretation of a Radner equilibrium for an economy with Arrow-Debreu assets. The key feature was that the date 0 prices of the Arrow-Debreu assets equalled the date 0 prices of spot income. We take the Radner equilibrium as a the starting point for a second interpretation of the Arrow-Debreu model with uncertainty and two dates.

Arrow-Debreu assets and state prices

At date 0 the consumers trade in the Arrow-Debreu assets and in the goods, to be delivered at date 0. The s'th Arrow-Debreu asset gives the holder 1 s–crown, if state s occurs and nothing otherwise. The, date 0 price of 1 unit of income to be received contingent on the occurrence of state s is, $\beta(s)$, for $s \in \mathbb{S}_1$, where $\beta(s)$ is the date 0 price of the s'th Arrow-Debreu asset. For the case of uncertainty the prices, $\beta(s), s \in \mathbb{S}_1$, are referred to as *state prices* of (spot) income. The notation is given in Table 4.4.A.

Table 4.4.A: *The Arrow-Debreu assets*

		Asset					
Date	State	1	2	$\cdots$	s	$\cdots$	S
0	0	$-\beta(1)$	$-\beta(2)$	$\cdots$	$-\beta(s)$	$\cdots$	$-\beta(S)$
1	1	1	0		0		0
	2	0	1		0		0
	$\vdots$	$\vdots$	$\vdots$	$\cdots$	$\vdots$	$\cdots$	$\vdots$
	s	0	0		1		0
	$\vdots$	$\vdots$	$\vdots$		$\vdots$		$\vdots$
	S	0	0		0		1

A consumer enters date 1 with a portfolio of the assets and having realized her date 0 consumption. Her portfolio gives her a date 1 net income vector of contingent income.

By assumption, the consumer has correct expectations about the goods prices that will prevail on the date 1 spot-market if Nature chooses the state $s \in \mathbb{S}_1$. Once s is chosen each of the assets, except asset s, is of no value. The net income that the consumer can spend in the spot-market is the number of units of the s'th asset that she holds and the gross income she can spend is the sum of the spot value of her initial endowment in the realized state and the value of her net income. The fundamental difference in the interpretation of the model with merely time and with time *and* uncertainty is that with time *all* planned deliveries are carried out whereas with uncertainty only those contingent on the realized state are carried out. Note, however, that the market balance conditions for the case of uncertainty ensure that we will get market balance no matter what state Nature chooses.

Time consistency

In the interpretation with spot-markets there arises a problem of *time consistency* of the consumer's choice both for the case of time and for the case of uncertainty but the problem is more pronounced in case of uncertainty. It arises since a consumer takes decisions at two different points in time. At date 0; the date of planning, the consumer plans in accordance with the utility function

$$u(c) = u(c(0), c(1), c(2), ..., c(S))$$

Let $\bar{c} = (\bar{c}(0), \bar{c}(1), \bar{c}(2), ..., \bar{c}(S))$ be a solution to the Consumer Problem given the correctly anticipated spot-market prices. At date 0 the amount(s) $\bar{c}(0)$ is consumed. At date 1the state of Nature is revealed, say $\hat{s}$, and the consumer may trade at the correctly foreseen spot-prices $p(\hat{s})$. The trade in assets or spot-income at date 0 gives the consumer a net income $r(\hat{s})$ and her budget restriction in the spot-market at date 1 is

$$p(\hat{s})(c(\hat{s}) - e(\hat{s})) \leq r(\hat{s}) \quad \text{or} \quad p(\hat{s})c(\hat{s}) \leq p(\hat{s})e(\hat{s}) + r(\hat{s})$$

At date 0 the consumer planned to choose the consumption $\bar{c}(\hat{s})$ if state $\hat{s}$ occurred. Will this plan be consistent with her choice in the spot-market at date 1? This depends on the utility function used over consumption in state $\hat{s}$. Assume that the consumer uses the utility function $\hat{u} : \mathbb{R}^L \longrightarrow \mathbb{R}$ given by

$$\hat{u}(c(1)) = u(\bar{c}(0), \bar{c}(1), \bar{c}(2), ..., c(\hat{s}), ..., \bar{c}(S))$$

that is, the date 1 utility function for the value of the realized consumption at date 0 and the *planned* values of consumption in the states $s \in \mathbb{S}_1$, for $s \neq \hat{s}$. Then it is easy to see that $\bar{c}(\hat{s})$ is a solution to, the consumers spot market problem

$$\begin{aligned} \text{Max}_{c(\hat{s})} \quad & u(\bar{c}(0), \bar{c}(1), \bar{c}(2), ..., c(\hat{s}), ..., \bar{c}(S)) \quad \text{s. to } c(\hat{s}) \in \mathbb{R}^L_{++} \text{and} \\ & p(\hat{s})(c(\hat{s}) - e(\hat{s})) \leq r(\hat{s}) \end{aligned}$$

Thus her choice in the spot market at date 1 will agree with her planned choice from date 0.

Time consistency and separable utility

Note that, if the utility function is separable over time *and* states, so that

$$u(c) = u(c(0), c(1), c(2), ..., c(S)) = \tilde{u}_0(c(0)) + \textstyle\sum_{s \in \mathbb{S}_1} \tilde{u}_s(c(s))$$

for some functions $\tilde{u}_s : \mathbb{R}^L \longrightarrow \mathbb{R}, s \in \mathbb{S}$, then the utility function to be applied, once is realized, is simply $u_{\hat{s}}(\cdot)$ which is independent of the consumption at date 0 and the planned consumptions in state $s \neq \hat{s}$.

State prices and redundant assets

Although the interpretation of the Arrow-Debreu model over time and under uncertainty are analogous there are some results which make sense only over time or only under uncertainty. The example of spanning through the frequent trading of a single asset from Section 3.4 does not make sense for the case of two dates and uncertainty. (As will be seen it does make sense once we introduce more dates along with uncertainty.) An option does not make sense in an economy over time where there is no uncertainty but as the following example shows our earlier results carry over to the pricing of an option when there are complete markets.

Example 4.4.A Pricing of an option with complete markets

Assume that there is given a stock and a bond, with values as in Table 4.4.B, and a call option on the stock available at date 0 with strike price 45. The option gives the holder the right, but not the obligation, to buy 1 unit of the stock at date 1 for the price 45. What should be the price of the option, q_3, at date 0?

Table 4.4.B: *The option, stock and bond*

		Asset		
Date	State	Stock	Bond	Option
Date 0	0	-50	-18	$-q_3$
Date 1	1	40	20	0
	2	60	20	15

The option is not exercised in state 1 since its strike price is 45. The value of the option in state 2 is the difference between the strike price and the value of the stock, that is, $60 - 45 = 15$.
Since the date 1 dividend vectors of the stock and the bond are linearly independent the rank of the dividend matrix, excluding the option, is 2 and we have complete markets. To check whether the dividend matrix and the date 0 prices exclude arbitrage we solve

$$\begin{aligned} -50 + 40\beta(1) + 60\beta(2) &= 0 \\ -18 + 20\beta(1) + 20\beta(2) &= 0 \end{aligned}$$

This has the unique solution $(\beta(1), \beta(2)) = (2/10,\ 7/10)$. The existence of positive state-prices implies that there are no arbitrage possibilities.

Using the state prices to price the option

To avoid arbitrage the date 0 price of the option should be chosen so that the value of its extended dividend vector is 0 when evaluated with the state prices. Hence q_3 should satisfy

$$-q_3 + 0 \cdot \beta(1) + 15 \cdot \beta(2) = 0$$

which gives $q_3 = 15(7/10) = 21/2$.

A replicating portfolio

Another way to arrive at the same conclusion is to find a portfolio, $\theta = (\theta_1, \theta_2)$, of the bond and the stock replicating the date 1 dividend vector of the option. This leads to the following equation

$$\begin{bmatrix} 40 \\ 60 \end{bmatrix} \theta_1 + \begin{bmatrix} 20 \\ 20 \end{bmatrix} \theta_2 = \begin{bmatrix} 0 \\ 15 \end{bmatrix}$$

The solution is $(\theta_1, \theta_2) = (3/4, \ -3/2)$. The date 0 price of the option should equal the date 0 cost of this portfolio. Thus

$$q_3 = q_1\theta_1 + q_2\theta_2 = \frac{3}{4}50 - \frac{3}{2}18 = \frac{21}{2}$$

Note that the bond gives the same number of $s-$crowns, independently of whether state 1 or state 2 is realized. The sum $\beta(1) + \beta(2) = 9/10$ can be interpreted as the time discount factor. It is the date 0 cost of the portfolio $(\theta_1, \theta_2) = (0, \ 1/20)$ with the *sure* date 1 dividend vector giving 1 s-crown in state s, for $s = 1, 2$.

4.5 Increasing Information over Time

In this section we introduce partitions and filtrations in order to describe differences in information which might arise, for example, over time. We restrict the discussion to finite sets although many of the concepts make good sense also for sets which are not finite.

Partitions and Information

Definition 4.5.A *Let $\mathbb{S}$ be a finite set and let $\mathcal{P} = \{\sigma_1, \sigma_2, ..., \sigma_K\}$ be a family of subsets of $\mathbb{S}$, so that $\sigma_k \subset \mathbb{S}$, for $k \in \mathbb{K} = \{1, 2, ..., K\}$. $\mathcal{P}$ is a* **partition** *of $\mathbb{S}$ if:*

(i) $\sigma_k \neq \emptyset$ *for* $k \in \mathbb{K}$

(ii) $\sigma_k \cap \sigma_{k'} = \emptyset$ *for* $k \neq k'$ *and* $k, k' \in \mathbb{K}$

(iii) $\cup_{k \in \mathbb{K}} \sigma_k = \mathbb{S}$.

Assume that Nature chooses $\bar{s} \in \mathbb{S}$. If an agents information is given by a partition $\mathcal{P}$, then there is precisely one of the subsets, say $\bar{\sigma}$, in $\mathcal{P}$ that owns $\bar{s}$. The agent is unable to say which element of $\bar{\sigma}$ has been chosen, but knows that it belongs to $\bar{\sigma}$.

The situation where the agent has *full* or *complete information* is the case where the partition is the set of all singletons, that is, $\mathcal{P} = \{\{1\}, \{2\}, ..., \{S\}\}$. The other extreme case is when the agent is *totally uniformed* which corresponds to the partition $\mathcal{P} = \{\mathbb{S}\}$.

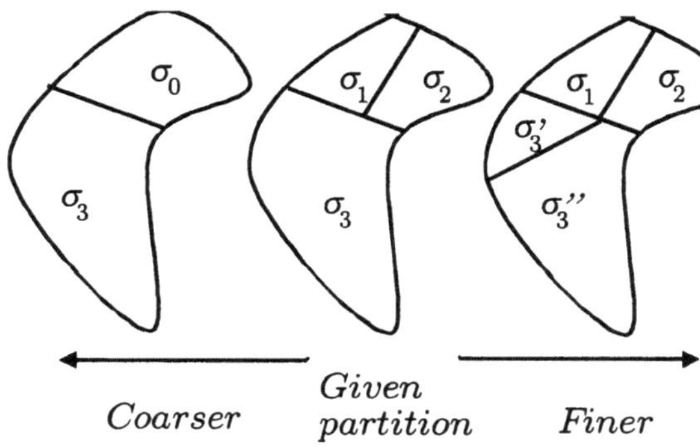

Figure 4.5.A: *By partioning sets from a given partition we get a finer partition and taking union of sets of the original partition we get a coarser partition*

Finer and coarser partitions

Let $\mathbb{S}$ be a finite set and let

$$\begin{aligned} \mathcal{P} &= \{\sigma_1, \sigma_2, ..., \sigma_K\} \\ \mathcal{P}' &= \{\sigma'_1, \sigma'_2, ..., \sigma'_{K'}\} \end{aligned}$$

be two partitions of $\mathbb{S}$. The partition $\mathcal{P}'$ is *at least as fine as* $\mathcal{P}$ if $\mathcal{P} = \mathcal{P}'$ or $\mathcal{P}'$ is derived from $\mathcal{P}$ by partitioning one or more subsets of $\mathcal{P}$. The partition $\mathcal{P}'$ is *at least as coarse as* $\mathcal{P}$ if $\mathcal{P}$ is *at least as fine as* $\mathcal{P}'$. If $\mathcal{P}'$ is at least as coarse as $\mathcal{P}$ then $\mathcal{P} = \mathcal{P}'$ or $\mathcal{P}'$ is derived from $\mathcal{P}$ by taking the union(s) of some of the subsets of $\mathcal{P}$. Cf. Figure 4.5.A. Finer partitions correspond to more information as will be evident below.

Event Trees

With two dates or more time enters in an essential way into the decision problems of consumers and producers. Information will usually increase with time. An *event tree* is a convenient way to describe the increasing information over time. The following example conveys the idea.

The states of the world; elementary events at date 3

Consider an economy extending over the four dates 0, 1, 2 and 3. An agent does not know at date 0 about the weather at the future dates. Assume that at each future date the weather can be *su*nny or *ra*iny. The future "weather history" is then given by a triple like (su, ra, su), indicating *su*nny, at date 1, *ra*iny at date 2, and *su*nny at date 3. A "state of the world" is a "weather history". Hence

$$\mathbb{S} = \left\{ \begin{array}{l} (su, su, su), (su, su, ra), (su, ra, su), (su, ra, ra), \\ \qquad (ra, su, su), (ra, su, ra), (ra, ra, su), (ra, ra, ra) \end{array} \right\}$$

Of course the complete "weather history" will not be known until date 3. The one-point subsets of $\mathbb{S}$, the states of the world, are the *elementary events at date* 3. These one-point subsets are often identified with the elements of $\mathbb{S}$. (We will always assume full information, at the final date.) In Figure 4.5.B we have shown the event tree and the corresponding partitions. Figure 4.5.C is the same event tree with the partition suppressed.

Intuition suggests that "rain at date 2" should be an event. How should this event be expressed using elements from $\mathbb{S}$? Those "weather histories" where there is rain at date 2, belong to the set

$$\{(su, ra, ra), (su, ra, su), (ra, ra, su), (ra, ra, ra), \}$$

and the event "rain, at date 2" corresponds to Nature choosing an element in this set. Note that at date 2 the agent will not know whether Nature has chosen, say, (su, ra, ra) or (su, ra, su), but the agent will still be able to infer at date 2 whether "rain at date 2" has occurred or not.

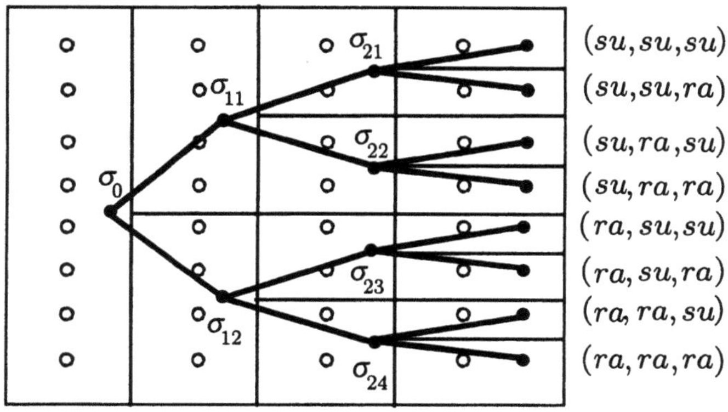

Figure 4.5.B: *The sequence of successively finer partitions forms a filtration, which is described by an event tree. Each date t node of the tree corresponds to a set in the partition for date t*

Elementary events at date 0, 1 and 2

At date 0 the agent does not know anything about the state chosen by Nature. Hence

$$\mathcal{P}_0 = \{\sigma_0\} = \{\mathbb{S}\}$$

is the partition describing the agents information at date 0 and $\sigma_0 = \mathbb{S}$ is the only *elementary event at date* 0. This event corresponds to the *node* (or vertex) at date 0 of the *event tree*, cf. Figure 4.5.C. At date 1 the agent is able to infer whether "rain, at date 1" or "sun, at date 1" has occurred. The partition describing the information at date 1 is given by a partition $\mathcal{P}_1$ which is finer than $\mathcal{P}_0$,

$$\begin{aligned}
\mathcal{P}_1 &= \{\sigma_{11}, \sigma_{12}\} \text{ where} \\
&\sigma_{11} = \{(su, su, su), (su, su, ra), (su, ra, su), (su, ra, ra)\} \\
&\sigma_{12} = \{(ra, su, su), (ra, su, ra), (ra, ra, su), (ra, ra, ra), \}
\end{aligned}$$

where σ_{11} and σ_{12} are the *elementary events at date* 1. These two elementary events at date 1 correspond to "sun at date 1" and "rain at date 1".

At date 2 the agent knows the weather history up to that date and the corresponding elementary events at date 2 are the elements of the partition $\mathcal{P}_2$ which is finer than $\mathcal{P}_1$,

$$\begin{aligned}\mathcal{P}_2 &= \{\sigma_{21}, \sigma_{22}, \sigma_{23}, \sigma_{24}\} \quad \text{where} \\ \sigma_{21} &= \{(su, su, su), (su, su, ra)\} \\ \sigma_{22} &= \{(su, ra, su), (su, ra, ra)\} \\ \sigma_{23} &= \{(ra, su, su), (ra, su, ra)\} \\ \sigma_{24} &= \{(ra, ra, su), (ra, ra, ra)\}\end{aligned}$$

Hence $\sigma_{21}, \sigma_{22}, \sigma_{23}$ and σ_{24} are the *elementary events at date* 2 corresponding to the events that the "weather history" for date 1 and 2 has been $(su, su), (su, ra), (ra, su)$ or (ra, ra).

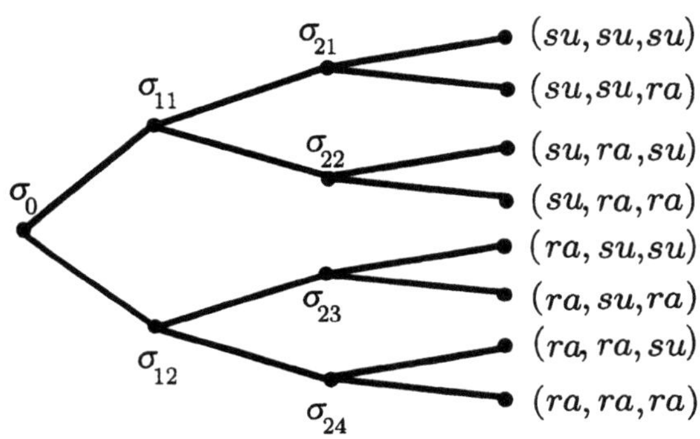

Figure 4.5.C: *To each elementary event at the final date corresponds a realized path in the event tree*

An example of an event, which is not an elementary event at date 2 is the event $\sigma_{21} \cup \sigma_{22}$ corresponding to the event "sun at date 1 and sun or rain at date 2". (This event, however, was an elementary event, at date 1.)

Finally, at date 3 the agent knows the whole "weather history", that is, which element of $\mathbb{S}$ Nature has chosen.

Filtrations, nodes and arcs

Note that in the sequence of partitions $\mathcal{P}_0$, $\mathcal{P}_1$, $\mathcal{P}_2$ and $\mathcal{P}_3$, each partition (except $\mathcal{P}_0$) is finer than the preceding one which reflects the increasing information over time. The increasing fineness of the partitions means that the agents do no forget. For example, at date 2 the agents remember whether it was sunny or rainy at date 1. A sequence of partitions is a *filtration* if each partition apart from the initial partition is at least as fine as the preceding one.

Any filtration of a finite set can be illustrated in an event tree. Each node of the event tree corresponds to a subset of one of the partitions. The *arcs*, leading forward from a given node connects the node with other nodes; the *immediate successors*, which correspond to subsets in the partition of the given node. Some of the nodes do not have any successors. These are the *terminal nodes.* Each node, except the *initial node*, also has a unique *predecessor.*

Example 4.5.A Event tree with two dates and uncertainty

The model from Section 4.2 gives rise to a particularly simple event tree. At date 0, there is no information and the corresponding partition is $\mathcal{P}_0 = \{\mathbb{S}_1\}$. At date 1, there is complete information and $\mathcal{P}_1$ is the partition of $\{\mathbb{S}_1\}$ into singletons; the finest partition of $\{\mathbb{S}_1\}$. We used $\mathbb{S} = \{0, 1, ..., S\}$ to denote the nodes of the event tree.

□

Example 4.5.B Event tree for an economy over time

The economy considered in Chapter 2 where there was no uncertainty but several dates can be described by a trivial event tree where there is a single state of Nature and a single node at each date and the partition is at each date the trivial one.

Event Tree; a General Definition*

An event tree is an instance of an oriented graph. We begin by giving an abstract definition.

Definition 4.5.B *An* **event tree** *is a pair,* $(\mathbb{P}, A)$ *where*

(i) $\mathbb{P}$ *is a non-empty finite set*

(ii) A *is a set of pairs of points of* $\mathbb{P}$; *that is,* $A \subset \mathbb{P} \times \mathbb{P}$

(iii) *there is a unique* $\sigma_0 \in \mathbb{P}$ *such that* $(\sigma', \sigma_0) \notin A$, *for* $\sigma' \in \mathbb{P}$

(iv) *if* $\sigma' \neq \sigma_0$ *then there exists a unique* $\sigma \in \Sigma$ *such that* $(\sigma, \sigma') \in A$

(v) *there is no cycle in* A : *if* $(\sigma_1, \sigma_2), (\sigma_2, \sigma_3), ..., (\sigma_k, \sigma_{k+1}) \in A$ *then* $\sigma_1 \neq \sigma_{k+1}$

The set $\mathbb{P}$ is the set of *nodes* in the tree and A the set of *arcs*. If $(\sigma, \sigma') \in A$ then σ is *the* (immediate) *predecessor* of σ' and σ' is *a* (immediate) *successor* of σ. Condition **(iii)** asserts that there is a unique *initial node*; a node having no predecessor while **(iv)** asserts that every other node has a predecessor. Condition **(v)** asserts that one can not by following a sequence of arcs return to the starting point. Note that **(v)** implies: if $(\sigma, \sigma') \in A$ then $(\sigma', \sigma) \notin A$. A node σ such that $(\sigma, \sigma') \notin A$ for each $\sigma' \in \mathbb{P}$ is a *terminal node*.

Event tree induced by a filtration

Let us now see how a filtration of an arbitrary, but finite, set induces an event tree. Let $\mathcal{P}_0, \mathcal{P}_1, \ldots, \mathcal{P}_T$ be a sequence of partitions of $\mathbb{S}$, where, for $t \in \mathbb{T}\backslash\{T\}$, the partition $\mathcal{P}_{t+1}$ is finer than $\mathcal{P}_t$, the partition at date 0 is $\mathcal{P}_0 = \{\mathbb{S}\}$ and $\mathcal{P}_T$ is the finest partition of $\mathbb{S}$.

We construct an event tree by letting the nodes be the elements in

$$\mathbb{P} = \cup_{t \in \mathbb{T}} \mathcal{P}(t)$$

and the arcs

$$\{(\sigma, \sigma') \in \mathbb{P} \times \mathbb{P} \mid \sigma' \subset \sigma \text{ and for some } t \in \mathbb{T}, \sigma \in \mathcal{P}_t \text{ and } \sigma' \in \mathcal{P}_{t+1}\}$$

The terminal nodes are the elements of $\mathcal{P}_T$ corresponding to the elements of $\mathbb{S}$. The initial node is $\mathbb{S}$.

Once the state of nature is known a particular terminal node is singled out. From the terminal node there is a unique *path* back to the initial node. The nodes on this path define for every date a unique elementary event at that date. This sequence of nodes or elementary events defines the *realized path, realization* or *trajectory*, determined by Nature's choice of the terminal node of the path.

4.6 Consumption Plans and Information

In this section we study the consumer problem in an economy with uncertainty and an arbitrary, but finite, number of dates and states. There are essentially two ways to define the commodity space.

First Description

The first one is the simplest and is sufficient as long as the information of the consumers is the same. With this approach the informational restrictions, which are implied by the varying information over time, are automatically taken care of.

The commodity space

In this case the commodity space is the linear vector space of functions defined on the nodes of an event tree.

Recall that $\mathbb{R}^n$ can be viewed as a set of functions. A vector in $\mathbb{R}^n$ is then a function

$$\{1, 2, ..., n\} \longrightarrow \mathbb{R}$$

With the standard notation, the n-tuple $(x_1, x_2, ..., x_n)$ gives the values of the function for $1, 2, ..., n$. As usual vectors can be added, multiplied by scalars etc. If x and y are functions then the functions $x + y$ and λx, where λ is a scalar, are defined by

$$\begin{aligned} (x+y)(j) &= x(j)+y(j) \quad \text{for } j \in \{1, 2, ..., n\} \\ (\lambda x)(j) &= \lambda x(j) \quad \text{for } j \in \{1, 2, ..., n\} \end{aligned} \qquad \textbf{4.6.A}$$

The crucial point is that the domain of the functions is unimportant. Addition and multiplication with a scalar takes place in the range of the functions by operations on the *values* of the function(s).

Clearly then we can substitute for the "index" set, $\{1, 2, ..., n\}$, any (non-empty) set and the functions defined on this set will be a linear vector space with addition and multiplication by a scalar defined through the obvious generalization of 4.6.A.

Let an event tree be defined by the partitions $\mathcal{P}_0, \mathcal{P}_1, ..., \mathcal{P}_T$. The nodes of the tree is the set $\mathbb{P} = \cup_{t \in \mathbb{T}} \mathcal{P}(t)$ and the set of future nodes is denoted $\mathbb{P}_1$. The set of functions

$$X = \{x \mid x : \mathbb{P} \longrightarrow \mathbb{R}^L\}$$

is a linear vector space. We take this set as the commodity space of an economy. To get a vector in the commodity space we attach to each node of the event tree a vector of $\mathbb{R}^L$. Vectors (functions) in X are added, by adding the corresponding vectors of $\mathbb{R}^L$ node by node and multiplication by a real number amounts to multiplying the vectors of $\mathbb{R}^L$ at each of the nodes.

Consumption, endowment and net trade

By our Maintained Assumptions the consumption set is the set of vectors in the commodity space with each component positive. Hence a consumption plan belonging to the consumption set is a function

$$c : \mathbb{P} \longrightarrow \mathbb{R}^L_{++}$$

For most purposes, it is convenient to think of the vectors as functions defined on the event tree but it is also possible to get a description resembling the one in Section 4.2 by enumerating the nodes, in a suitable way. See Table 4.6.A.

From Table 4.6.A it is clear that the dimension of the commodity space is the number of nodes, $\#\mathbb{P}$, multiplied by the number of goods, that is, $L \times \#\mathbb{P}$.

A consumer's endowment is a vector in the consumption set

$$e : \mathbb{P} \longrightarrow \mathbb{R}^L_{++}$$

and given a consumption the corresponding net trade is the vector

$$c - e : \mathbb{P} \longrightarrow \mathbb{R}^L$$

The net trade at σ, $c(\sigma) - e(\sigma)$, shows the deliveries *to* (positive) or *from* (negative) the consumer of the L goods, contingent on the event σ being on the realized path.

Table 4.6.A: *Consumption and endowment; first description*

Date	Event	Cons.	Endowm.
0	σ_0	$c(\sigma_0)$	$e(\sigma_0)$
1	σ_{11}	$c(\sigma_{11})$	$e(\sigma_{11})$
	$\vdots$	$\vdots$	$\vdots$
	$\sigma_{1K(1)}$	$c(\sigma_{1K(1)})$	$e(\sigma_{1K(1)})$
	$\vdots$	$\vdots$	$\vdots$
T	σ_{T1}	$c(\sigma_{T1})$	$e(\sigma_{T1})$
	$\vdots$	$\vdots$	$\vdots$
	$\sigma_{TK(T)}$	$c(\sigma_{TK(T)})$	$e(\sigma_{TK(T)})$

The consumer's utility function is defined on C. As in Section 4.2, the consumer's preferences depicts her taste regarding the goods, time preference as well as her beliefs about the likelihood of different states of Nature.

Realized consumption and net trade

Let a consumption plan, c, be given and let Nature choose a state, $\bar{s}$. The chosen state (terminal node) determines the realized path and the nodes (elementary events) on the realized path. This, in turn, determines the *realized consumption* and the *realized net trade*. Thus the realized consumption, given $\bar{s}$, is a function of time only and the same is true for the realized net trade.

Second Description*

The second approach is more explicit about the informational restrictions, pertaining to initial endowment, feasible consumptions and feasible production plans. The second approach can easily allow for asymmetric information between agents over time.

It is implicit in our first description of the consumer, that a consumption must satisfy certain informational restrictions.

Consider the event tree in Figure 4.5.C. At date 0, the consumer is not allowed to *plan* to consume different amount of, say, good 1, at date 2 contingent on whether Nature has chosen (su, su, ra) or (su, su, su). This is a reasonable restriction since it is not possible at date 2 to ascertain which of the two states (su, su, ra) or (su, su, su) that has been chosen by Nature. Technically speaking, the singleton sets are not (elementary) events, at date 2. Thus contracts contingent on these (set of) states of the world are not allowed.

The commodity space; second description

The informational restrictions, which were automatically taken care of in our first description, are made explicit if we proceed as follows. Take as the commodity space the set of functions

$$\hat{X} = \{\hat{x} \mid \hat{x} : \mathbb{S} \times \mathbb{T} \longrightarrow \mathbb{R}^L\}$$

$\hat{X}$ is a vector space of dimension $\#(\mathbb{S} \times \mathbb{T}) \times L$. The set $\mathbb{S} \times \mathbb{T}$ can be visualized as in Figure 4.5.B where each of the unfilled small circles corresponds to an element of $\mathbb{S} \times \mathbb{T}$.

Informational restrictions

A function $\hat{x} \in \hat{X}$ gives rise to the functions

$$\hat{x}(\cdot, t) : \mathbb{S} \longrightarrow \mathbb{R}^L \quad \text{for } t \in \mathbb{T}$$

The set of functions satisfying the informational requirements consists of the functions such that, for $t \in \mathbb{T}$, the function $\hat{x}(\cdot, t)$ is *measurable with respect to the partition* $\mathcal{P}(t)$. By definition this is the case if and only if $\hat{x}(\cdot, t)$ is constant on each date t node, for $t \in \mathbb{T}$, so that $s, s' \in \sigma$ implies $\hat{x}(s, t) = \hat{x}(s', t)$ for σ a t-node.

Relation between the two descriptions

Thus, if $c : \mathbb{P} \longrightarrow \mathbb{R}^L$ is a consumption, using the first description above, and $\hat{c} : \mathbb{S} \times \mathbb{T} \longrightarrow \mathbb{R}^L$ is a consumption, using the present description then c and $\hat{c}$ describe the "same" consumption if, for $t \in \mathbb{T}$,

$$\hat{c}(s,t) = c(\sigma_{tk}) \quad \text{for } s \in \sigma_{tk} \text{ and } \sigma_{tk} \in \mathcal{P}_t$$

It is then seen that it is superfluous to give *all* the values of the function $\hat{c}(\cdot,:)$, since the measurability requirement implies that $\hat{c}(\cdot,:)$ is constant on each node. This was what made possible the first description above. With the present description the initial endowment and the net trade will also be functions defined on $\mathbb{S} \times \mathbb{T}$. The measurability requirement for the initial endowment is natural; if the initial endowment was not a measurable function then it would provide more information than the corresponding partition for some date.

Table 4.6.B: *Relating the two ways to describe the commodity space*

	First Description	Second Description
Consumption	$c(\sigma_{tk})$	$\hat{c}(s,t)$ for $s \in \sigma_{tk}$
Endowment	$e(\sigma_{tk})$	$\hat{e}(s,t)$ for $s \in \sigma_{tk}$
Net trade	$c(\sigma_{tk}) - e(\sigma_{tk})$	$\hat{c}(s,t) - \hat{e}(s,t)$ for $s \in \sigma_{tk}$
Prices	$P(\sigma_{tk})$	$\sum_{s\in\sigma_{tk}} \hat{P}(s,t)$
Net Expenditure	$P(\sigma_{tk})\,[c(\sigma_{tk}) - e(\sigma_{tk})]$	$\sum_{s\in\sigma_{tk}} \hat{P}(s,t)\,[\hat{c}(s,t) - \hat{e}(s,t)]$

We relate the two ways of describing the consumer in Table 4.6.B. (Disregard, for the moment, the last two lines.)

4.7 Walras Equilibrium

In this section we discuss shortly, for the sake of completeness, the interpretation of the general model with time and uncertainty for the case of complete contingent markets. We do this for the two different ways of describing the commodity space. As will be seen the reinterpretation consists essentially of substituting the date-event index for the time index and the reader will by now be able to fill in missing details.

Equilibrium with Complete Contingent Markets

Date 0 prices

Let us start by using the first description from above of the consumer.

Assume that at date 0 each commodity can be bought and sold for future delivery at date t contingent on each (elementary) event at date t. The price, in σ_0-crowns of good l to be delivered at date t contingent on the elementary event $\sigma \in \mathcal{P}_t$, is $P_l(\sigma)$. The price for the sure delivery at date t of good l is the sum, $\Sigma_{\sigma \in \mathcal{P}_t} P_l(\sigma)$. Thus a date 0 price system is a function

$$P : \mathbb{P} \longrightarrow \mathbb{R}^L_{++}$$

The consumer problem

The consumer's wealth is the value of her initial endowment. The consumer uses this wealth to buy commodities, at date 0. Apart from notation, the Consumer Problem is the same as in Chapter 2.

$$\text{Max}_c \quad u\left[(c(\sigma))_{\sigma \in \mathbb{P}}\right] \quad \text{s. to } c \in C \text{ and}$$

$$\textstyle\sum_{\sigma \in \mathbb{P}} P(\sigma) c(\sigma) \ \leq \ w$$

where $w = \sum_{\sigma \in \mathbb{P}} P(\sigma) e(\sigma)$. The budget restriction can be written as a restriction on the consumer's net trade.

$$\textstyle\sum_{\sigma \in \mathbb{P}} P(\sigma) \left[c(\sigma) - e(\sigma)\right] \ \leq \ 0$$

which has the interpretation that the date 0 value of the planned deliveries *to* the consumer and *from* the consumer should be nonpositive.

Walras equilibrium

Let us now consider an economy with I consumers described by their consumption sets, initial endowments and utility functions. A *Walras equilibrium* is an allocation and a price system $\left[(\bar{c}^i)_{i \in \mathbb{I}}, (P(\sigma))_{\sigma \in \mathbb{P}}\right]$, where, for $i \in \mathbb{I}$, the consumption $\bar{c}^i = (\bar{c}^i(\sigma))_{\sigma \in \mathbf{P}}$ is a solution to the Consumer Problem.

Since $((\bar{c}^i(\sigma))_{\sigma \in \mathbf{P}})_{i \in \mathbb{I}}$ is an allocation, the consumptions are individually feasible and give market balance, that is,

$$\sum_{i \in \mathbb{I}} c^i(\sigma) = \sum_{i \in \mathbb{I}} e^i(\sigma) \quad \text{for every } \sigma \in \mathbb{P}$$

The choice by Nature determines a path which is revealed gradually over time. The measurability restrictions, which are automatically satisfied with the first description of the consumer, ensure that it is possible to determine at each date which deliveries are to be carried out at that date. The *realized consumptions* are determined by the state of the world defining the realized path.

Using the second description of the commodity space*

If we use the second description then the commodity space will be different. The price system is in this case a function

$$\hat{P} : \mathbb{S} \times \mathbb{T} \longrightarrow \mathbb{R}^L_{++}$$

Consider a consumer and a net trade for the consumer corresponding to a node $\sigma_{tk} \in \mathcal{P}_t$. The value of the net trade, at this node, is

$$\sum_{s \in \sigma_{tk}} \hat{P}(s,t)\,[\hat{c}(s,t) - \hat{e}(s,t)]$$

The measurability restrictions imply that $[\hat{c}(s,t) - \hat{e}(s,t)]$ is a constant vector in $\mathbb{R}^L$ as s varies over σ_{tk}. Hence the value of the net trade can be written

$$\left[\sum_{s \in \sigma_{tk}} \hat{P}(s,t)\right] [\hat{c}(s,t) - \hat{e}(s,t)]$$

The conditions

$$\hat{c}(s,t) = c(\sigma_{tk}) \quad \text{for } t \in \mathbb{T},\ s \in \sigma_{tk} \text{ and } \sigma_{tk} \in \mathcal{P}_t$$

$$\sum_{s \in \sigma_{tk}} \hat{P}(s,t) = P(\sigma_{tk}) \quad \text{for } t \in \mathbb{T} \text{ and } \sigma_{tk} \in \mathcal{P}_t$$

show how prices and the net trade, using the first description, are related to prices and the net trade, using the second description. In particular the values of $\hat{P}$ for $s \in \sigma_{tk}$, are of no importance as long as they sum to $P(\sigma_{tk})$.

We have indicated in Table 4.6.B, how prices and net expenditures are related for the two descriptions of the consumer.

4.8 Assets, State Prices and Arbitrage

The purpose of this section is to see how the concepts from economies over time carry over to economies with several dates and uncertainty. In particular how frequent trading of a few assets may increase the dimension of the subspace of income transfers.

Nominal assets, dividend and extended dividend vectors

Almost all the reasoning from Chapter 3 carries over to the present situation. A *nominal asset*, v, is defined by its dividends over the nodes apart from the initial node. The dividend vector is a function

$$v : \mathbb{P}_1 \longrightarrow \mathbb{R}$$

and the value, $v(\sigma) \in \mathbb{R}$, gives the number of $\sigma-$crowns to be delivered *to* or *from* the holder of one unit of the asset if the event $\sigma \in \mathbb{P}_1$ occurs. Extending the dividend vector with the date 0 price of the asset in 0-crowns we get the extended dividend vector, which is thus a function

$$w : \mathbb{P} \longrightarrow \mathbb{R}$$

By enumerating the nodes in an orderly way we can illustrate the (extended) dividend vector in a matrix as before. See Table 4.8.A. The assets corresponding to the Arrow-Debreu assets in the economy over time are the $\#\mathbb{P}_1$ assets where the asset defined by $\sigma \in \mathbb{P}_1$ pays one $\sigma-$crowns if σ occurs and nothing otherwise.

Real assets

A *real asset* is defined by its future real dividend vector, a. Thus, its dividends are given by a function

$$a : \mathbb{P}_1 \longrightarrow \mathbb{R}^L$$

where $a(\sigma) \in \mathbb{R}^L$ gives the deliveries of goods to or from the consumer if the event σ occurs. Given the spot prices $(p(\sigma))_{\sigma \in \mathbb{P}_1}$, the *value dividend vector* is generated by taking for each node $\sigma \in \mathbb{P}_1$ the scalar product of $p(\sigma)$ and $a(\sigma)$. Thus, $p(\sigma)a(\sigma)$ is the amount of $\sigma-$crowns which can be claimed (or which has to be paid) by an owner of one unit of the asset if σ occurs.

Table 4.8.A: *State prices and dividend vector*

Date	State	State Prices	Asset j
0	σ_0	1	$-q_j$
1	σ_{11}	$\beta(\sigma_{11})$	$v^j(\sigma_{11})$
	$\vdots$	$\vdots$	$\vdots$
	$\sigma_{1K(1)}$	$\beta(\sigma_{1K(1)})$	$v^j(\sigma_{1K(1)})$
	$\vdots$	$\vdots$	$\vdots$
T	σ_{T1}	$\beta(\sigma_{T1})$	$v^j(\sigma_{T1})$
	$\vdots$	$\vdots$	$\vdots$
	$\sigma_{TK(T)}$	$\beta(\sigma_{TK(T)})$	$v^j(\sigma_{TK(T)})$

From Table 4.8.A it should be clear that the former definitions of *complete* and *incomplete asset markets* apply to the present situation.

State prices and time discount factors

For the case of several dates and uncertainty the notion of a state price is somewhat inappropriate since the prices referred to are the date 0 prices for delivery of 1 unit of income in *event* σ. Although "event prices" would be a better terminology we will stick to "state prices" since this appears to be the accepted terminology. Thus a *state price vector* is a function

$$\beta : \mathbb{P} \longrightarrow \mathbb{R}_{++} \quad \text{with } \beta(\sigma_0) \; = \; 1$$

Theorem 3.2.B, from Chapter 3, still applies. Hence if the assets are priced so that there is no arbitrage then there exists a state price vector such that the prices of the assets are determined as the discounted values, using the state prices of the (future) dividends. Furthermore, asset markets are complete if and only if the state price vector is uniquely determined. A state price vector, β, gives rise to *time discount factors.* The price, in $0-$crowns of a *sure* delivery of a $t-$crown, $t \in \mathbb{T}$, is the sum of the prices of σ−crowns for $\sigma \in \mathcal{P}_t$. Hence the (time) discount factor, from t to 0, is $\sum_{\sigma \in \mathcal{P}_t} \beta(\sigma)$.

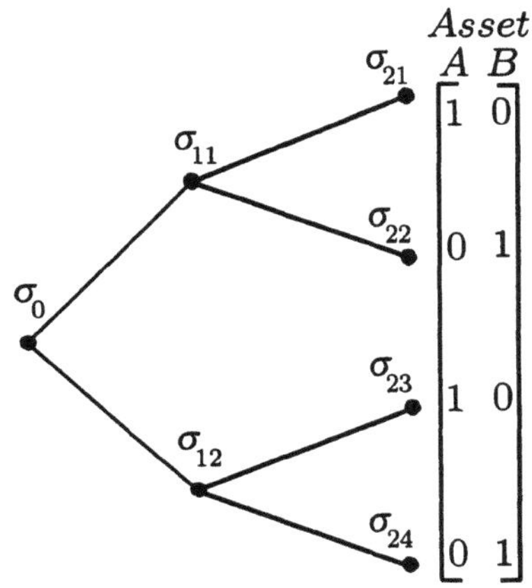

Figure 4.8.A: *Dividends of the given assets in the example of frequent trading*

The remarks above indicate that all of the analysis in Chapter 3 applies to the present situation.

Frequently traded assets

With several dates and uncertainty the possibility of assets that are traded at several dates reappears. (Cf. Chapter 3, Section 3.4). Below we will give an example of such a *frequently traded* asset. If the subspace of income transfers is generated by assets traded not only at the initial node, its dimension may vary also in the case where the assets are nominal.

Spanning through Frequent Trading

In this section we will give an example of how with frequent trading a few assets may suffice to give complete markets.

The initial assets

We consider a simple example of an economy with uncertainty extending over dates 0, 1 and 2 with two initial assets. The dividend vectors of the initial assets are given in Table 4.8.B. There it is seen that the initial assets pay dividends only at date 2. Figure 4.8.A shows the event tree for the economy. The maximal number of successors to any node in the tree is 2. This is the *spanning number* of the tree. It is equal to the number of initial assets.

The dividends may also be given in an extended dividend matrix as in Table 4.8.B

Table 4.8.B: *The original assets*

Date	State	Asset A	Asset B	State Price
0	σ_0	$q_A(\sigma_0)$	$q_B(\sigma_0)$	1
1	σ_{11}	0	0	$\beta(\sigma_{11})$
	σ_{12}	0	0	$\beta(\sigma_{12})$
2	σ_{21}	1	0	$\beta(\sigma_{21})$
	σ_{22}	0	1	$\beta(\sigma_{22})$
	σ_{23}	1	0	$\beta(\sigma_{23})$
	σ_{24}	0	1	$\beta(\sigma_{24})$

Induced assets

Assume that the initial assets can be traded, at both date 0 and date 1. The initial assets generate new assets, which will be referred to as induced assets. These induced assets are given in Table 4.8.C. (For the moment disregard the frame.) Columns are indexes as in Table 4.8.C. Asset 1 and 2 can be used to transfer income from date 0 to date 1 and the other assets to transfer income from date 1 to date 2.

Assume that these 6 induced assets can all be traded at date 0. For example, asset 3 costs 0 σ_0-crowns, but the buyer is obliged to deliver $q_A(\sigma_{11})$ σ_{11}-crowns if node σ_{11} is reached and receives 1 σ_{21}-crown, if the node σ_{21} is reached.

Table 4.8.C: *The induced assets*

1	2	3	4	5	6
$-q_A(\sigma_0)$	$-q_B(\sigma_0)$	0	0	0	0
$q_A(\sigma_{11})$	$q_B(\sigma_{11})$	$-q_A(\sigma_{11})$	$-q_B(\sigma_{11})$	0	0
$q_A(\sigma_{12})$	$q_B(\sigma_{12})$	0	0	$-q_A(\sigma_{12})$	$-q_B(\sigma_{12})$
0	0	1	0	0	0
0	0	0	1	0	0
0	0	0	0	1	0
0	0	0	0	0	1

Whether the actual trade takes place at date 0, or at date 1 (at node σ_{11}) does not matter as long as price expectations are correct. In view of this the 6×6 submatrix formed by the last 6 rows and the columns in Table 4.8.C is taken to be the date 0 value dividend matrix, which will be denoted V. The extended value dividend matrix, W, is the full 7×6 matrix in Table 4.8.C.

Replicating the initial assets

The initial assets, A and B, can be replicated by the induced assets. Let $\theta = (\theta_1, \theta_2, ..., \theta_6)$ denote a portfolio of the 6 induced assets. It is then easy to see that asset A is replicated by the portfolio $(1, 0, 1, 0, 1, 0)$ and asset B is replicated by the portfolio $(0, 1, 0, 1, 0, 1)$. Hence if the assets 1,2,...,6 are available then the assets A and B are redundant.

Let $\theta = (\theta_1, \theta_2, ..., \theta_6)$ be a portfolio of the induced assets. θ determines a portfolio in the original assets at the different nodes and reveals how the consumer adjusts her portfolio of the original assets over time.

Table 4.8.D shows the portfolio in the original assets, which the consumer holds "leaving" the node. (The portfolio as the consumer "enters", say date 1, is (θ_1, θ_2). Thus the consumer arrives att node σ_{11} with the portfolio (θ_1, θ_2) and leaves with (θ_3, θ_4) of the assets A and B implying buying or selling $(\theta_3, \theta_4) - (\theta_1, \theta_2)$ at the node.

Table 4.8.D: *Portfolio in the original assets*

Date	State	Asset A	Asset B
0	σ_0	θ_1	θ_2
1	σ_{11}	θ_3	θ_4
	σ_{12}	θ_5	θ_6

No arbitrage and state prices

Assume in the remaining part of the example that we are considering asset prices of both the original and the induced assets from an equilibrium so that there is no arbitrage. Then there exists a positive state price vector as given in Table 4.8.B such that the value of each of the assets is 0. For example, consider asset 3. We have

$$\beta(\sigma_{11}) \cdot -q_A(\sigma_{11}) + \beta(\sigma_{21}) \cdot 1 \;=\; 0 \quad \text{or} \quad q_A(\sigma_{11}) \;=\; \frac{\beta(\sigma_{21})}{\beta(\sigma_{11})}$$

Note that $\beta(\sigma_{21})$ is the price of σ_{21}-crowns, in σ_0-crowns and $1/\beta(\sigma_{11})$ is the price of σ_0-crowns in σ_{11}-crowns. Thus $q_A(\sigma_{11})$ is the price of σ_{21}−crowns, in σ_{11}−crowns. In particular it follows that the asset prices, at each date, are positive.

Income transfers between date 1 and 2

Let us first check which income transfers are possible between date 1 and 2, if the event σ_{12} occurs. The (extended) dividend matrix from this node is the matrix we get from W by deleting all columns except column 5 and 6 together with the rows corresponding to $\sigma_0, \sigma_{11}, \sigma_{21}$ and σ_{22}. This matrix is given in Table 4.8.E.

The lower 2×2 submatrix of this matrix, will be denoted $V_{\sigma_{12}}$. Clearly this matrix has full rank and allows any income vector, to nodes σ_{23} and σ_{24}.
A similar reasoning applies to node σ_{11}. Hence there will always be "complete asset markets from date 1" and onwards.

Table 4.8.E: *Income transfers, date 1 and 2*

Date	State	5	6
1	σ_{12}	$-q_A(\sigma_{12})$	$-q_B(\sigma_{12})$
2	σ_{23}	1	0
	σ_{24}	0	1

Income transfers between date 0 and date 1

We can now concentrate on the possibilities of transferring income between date 0 and date 1. If V has full rank then, in particular, a consumer may achieve any income vector $(r(\sigma_{11}), r(\sigma_{12}))$ (disregarding the σ_0 cost) by buying a suitable portfolio $\theta = (\theta_1, \theta_2, 0, ..., 0)$ at date 0 and selling the same portfolio at date 1.

V has full rank if and only if $V\theta = 0$ has the unique solution $\theta = 0$. Let V_1 denote the submatrix of V, which is framed in Table 4.8.C. It is easy to see that

$$V\theta = 0 \quad \text{if and only if} \quad \begin{cases} V_1 \begin{bmatrix} \theta_1 \\ \theta_2 \end{bmatrix} = 0 \quad \text{and} \\ \theta_3 = \theta_4 = \theta_5 = \theta_6 = 0 \end{cases}$$

Clearly $V\theta = 0$ has the unique solution $\theta = 0$ if and only if rank $V_1 = 2$. Hence, in an equilibrium, the possibilities to transfer income from date 0 to the date 1 nodes determine whether asset markets are complete or not.

The matrix V_1 is from the no-arbitrage condition with any state price vector β

$$\begin{bmatrix} q_A(\sigma_{11}) & q_B(\sigma_{11}) \\ q_A(\sigma_{12}) & q_B(\sigma_{12}) \end{bmatrix} = \begin{bmatrix} \dfrac{\beta(\sigma_{21})}{\beta(\sigma_{11})} & \dfrac{\beta(\sigma_{22})}{\beta(\sigma_{11})} \\ \dfrac{\beta(\sigma_{23})}{\beta(\sigma_{12})} & \dfrac{\beta(\sigma_{24})}{\beta(\sigma_{12})} \end{bmatrix}$$

Thus the rank is 2 if and only if the state prices, at the nodes following σ_{11}, are not proportional to the state prices at the nodes following σ_{12}. If asset markets are incomplete then they will be proportional for any state price vector, consistent with no arbitrage.

Complete markets and the spanning number

We have shown that, except for a small set of asset prices, the 2 initial assets are enough to span a subspace of income transfers of full dimension. In this case we get complete asset markets from the assumption that the assets can be traded at each date and in each event. In general, it turns out that one needs at least as many assets as the spanning number of the tree to obtain a subspace of income transfers of full dimension.

The rank of the dividend matrix

Consider an economy with the initial assets, as above, and allow for retrading of the assets. The "drop in rank" problem now arises even with nominal assets. The economy can have a spot market equilibrium relative to $\mathbb{M}$, where $\mathbb{M}$ is a homogenous hyperplane with normal given by some state prices $(\beta(\sigma))_{\sigma\in\mathbb{P}}$. If the state prices generates a matrix, V_1, as above, with rank 1 then the subspace spanned by the assets will be a proper linear subspace of $\mathbb{M}$.

In the example we considered nominal assets. Kreps [1982] has studied the case of an economy with, initially, at least as many real assets as the spanning number of the event tree. For almost all such asset structures the economy has a Radner equilibrium with a subspace of income transfers of full dimension in case the initial assets can be traded at each node.

4.9 Radner Equilibrium

We will now show how the definition of a Radner equilibrium should be modified in order to apply to an economy with uncertainty and several dates.

Economy with an asset structure

The definition of an economy with an asset structure is the same as before but we have to change the description of the asset structure somewhat.

Definition 4.9.A *An* **economy with an asset structure** *is an economy and a dividend matrix. It will be denoted* $\mathcal{E} = \{(C^i, u^i, e^i)_{i\in\mathbb{I}}, (v^j)_{j\in\mathbb{J}}\}$.

The only novelty to be allowed for is the asset structure. This is given by the J assets. Initially the economy may have fewer than J assets but J is the number of assets which are available, taking into account that some or all of the original assets assets may be traded at several nodes. It will be clear from the context if nominal or real assets are considered. What matters to the consumers is the linear subspace of income transfers spanned by the assets. As before let

$$w^j : \mathbb{P} \longrightarrow \mathbb{R}$$

denote the extended dividend vector of asset $j \in \mathbb{J}$. A portfolio $\theta \in \mathbb{R}^J$ gives a consumer the net income vector

$$r : \mathbb{P} \longrightarrow \mathbb{R} \text{ where } r = \sum_{j\in\mathbb{J}} w^j \theta_j$$

and the delivery of σ-crowns, to or from the consumer, at node σ is $r(\sigma) = \sum_{j\in\mathbb{J}} w^j(\sigma)\theta_j$.

Definition of a Radner equilibrium

Proposition 3.5.B, in Chapter 3, shows that if the set $\mathbb{M}$, in the definition of a spot market equilibrium relative to a subspace, is generated by an asset structure, then we can define portfolios and get a corresponding Radner equilibrium.

As in Chapter 3 a Radner equilibrium is an equilibrium relative to a subspace supplemented by a specification of the consumers' portfolios. We only have to substitute the date-event index for the time index. The relation between the asset structure and the subspace of income transfers may be less obvious than in Chapter 3. Some care must be taken in the interpretation of the date 0 prices of assets given by the vector $q \in \mathbb{R}^J$

since q gives the date 0 prices also for the induced assets. For an asset, that is bought at date 2 and sold at date 3, the date 0 price should be 0 in order for the interpretation to make sense. Cf. Exercise 3.G.

Thus a Radner equilibrium is a tuple of consumptions, portfolios, spot prices for goods and asset prices such that $((\bar{c}^i)_{i\in\mathbb{I}}, p)$ is an equilibrium relative to $\mathbb{M} = \langle W \rangle$ and the portfolios finance the consumptions plans, that is, $(\bar{r}^i(\sigma))_{\sigma\in\mathbb{P}} = [p(\sigma)(\bar{c}^i(\sigma) - e^i(\sigma))]_{\sigma\in\mathbb{P}} = W\bar{\theta}^i$ for some portfolio $\bar{\theta}^i$.

A detailed definition of a Radner equilibrium*

For the sake of completeness we spell out the conditions in the definition of a Radner equilibrium in detail below.

Definition 4.9.B *A* **Radner equilibrium** *for an economy with an asset structure,* $\mathcal{E} = \{(C^i, u^i, e^i)_{i\in\mathbb{I}}, (v^j)_{j\in\mathbb{J}}\}$, *is a tuple*

$$((\bar{c}^i, \bar{\theta}^i)_{i\in\mathbb{I}}, p, q)$$

where

- $p \in (\mathbb{R}^L_{++})^{\#\mathbb{P}}$ *are the commodity prices,*
- $q \in \mathbb{R}^J$ *are the asset prices and,*
- *for* $i \in I$, $\bar{c}^i \in C^i$ *is a consumption plan and* $\bar{\theta}^i \in \mathbb{R}^J$ *a portfolio,*

and where $((\bar{c}^i, \bar{\theta}^i)_{i\in\mathbb{I}}, p, q)$ *satisfies:*

(a) *for* $i \in I$, $\bar{\theta}^i$ *solves*

$$\text{Max}_{\theta^i}\; v^i(p, (r^i(\sigma))_{\sigma\in\mathbb{P}}) \quad s.\ to\ r^i(\sigma) = \sum_{j\in\mathbb{J}} w^j(\sigma)\bar{\theta}^i_j \quad for\ \sigma \in \mathbb{P}$$

(b) *for* $i \in I$, $\bar{c}^i$ *solves, with* $\bar{r}^i(\sigma) = \sum_{j\in\mathbb{J}} w^j(\sigma)\bar{\theta}^i_j$ *for* $\sigma \in \mathbb{P}$,

$$\text{Max}_{c^i}\quad u^i((c^i(\sigma))_{\sigma\in\mathbb{P}}) \quad s.\ to\ c^i \in C^i\ and$$
$$p(\sigma)(c^i(\sigma) - e^i(\sigma)) \leq \bar{r}^i(\sigma) \quad for\ \sigma \in \mathbb{P}$$

(c)

$$\begin{cases} \sum_{i\in\mathbb{I}} \bar{c}^i(\sigma) = \sum_{i\in\mathbb{I}} e^i(\sigma) \text{ for } \sigma \in \mathbb{P} & \textit{(Market Balance for Goods at Each Date, in Each Event)} \\ \sum_{i\in\mathbb{I}} \bar{\theta}^i = 0 & \textit{(Market Balance for Assets)} \end{cases}$$

The **rank** *of a Radner equilibrium is the dimension of the subspace spanned by the functions* $(w^j)_{j\in\mathbb{J}}$.

Summary

In this chapter we introduced uncertainty, which was taken to be beyond the influence of the consumers.

The simple model with two dates, 0 and 1, where uncertainty was resolved at date 1 was seen to be formally equivalent to an economy over time. The novel feature with uncertainty was the clear distinction between the *plan* of an agent, concerning commodities or spot income, and the *realized* deliveries.

The interpretation of commodities as contingent commodities led to a reconsideration of the assumptions on preferences, which were seen to depict taste and time preferences as well as probability judgements concerning the events.

In the latter part of the chapter we considered the extension of the model to more than two dates. The increasing information over time was described by an event tree, which in turn reflected the feature that the partitions giving the elementary events at each date became finer over time.

The general model using an event tree to represent uncertainty was seen to encompass the models with time, studied in Chapter 2, and the model with two dates and uncertainty studied in the first part of this chapter.

The introduction of uncertainty along with time gave rise to some new phenomena. The time consistency problem, which was present already in the economy involving only time, became more pronounced with uncertainty and many dates.

With uncertainty new types of assets, like options, were of interest. The fact that asset markets could be active at many future dates made it possible to trade assets which paid dividends only at the last date, not only at date 0, but at all dates but the last. The frequent trading of a few assets was seen to bring about a large subspace of income transfers. With frequent trading there was a possibility of a "drop in rank" of the value dividend matrix also with nominal assets.

Exercises

Exercise 4.A Consider a consumer and the following consumptions:

		Consumption				
Date	State	A	B	C	D	E
0	0	$(1,1)$	$(1,1)$	$(1,1)$	$(1,1)$	$(1,1)$
1	1	$(1,1)$	$(3,3)$	$(2,2)$	$(\frac{3}{2},\frac{3}{2})$	$(2,1)$
	2	$(2,2)$	$(2,2)$	$(1,1)$	$(\frac{3}{2},\frac{3}{2})$	$(2,1)$

(a) What is the commodity space?

(b) What can be said about the preferences regarding the consumptions A,...,E using only that the utility function is increasing in each argument. Assume that the consumer regards state 2 as impossible. What are the consequences for the preferences? Will the utility function satisfy the Maintained assumptions in this case?

(c) What can be said about the preferences concerning the consumptions A,...,E using only that the utility function is increasing in each argument and strictly quasi-concave?

(d) Assume that the consumer prefers consumption A to consumption C. What would you infer about the consumer's probability judgements? If state 2 entails the consumer being "sick" and good 1 is "penicillin" would you still consider that inference valid?

Exercise 4.B Consider a consumer satisfying the Maintained Assumptions acting in an economy with two dates, $T = 1$, two states at date 1, $S = 2$, and 1 good, $L = 1$. Let $e = (e(0), e(1), e(2))$ be the consumers initial endowment and assume that the consumer has the utility function $u : \mathbb{R}^3_{++} \longrightarrow \mathbb{R}$ where

$$u(c) = \tilde{u}(c(0)) + \pi(1)\tilde{u}(c(1)) + \pi(2)\tilde{u}(c(2))$$

where $\tilde{u} : \mathbb{R}_{++} \longrightarrow \mathbb{R}$, $\pi(1) + \pi(2) = 1$ and $\pi(1), \pi(2) > 0$.

(a) Show that the indirect utility function is

$$\begin{aligned} v(p,r) &= \tilde{u}\left(e(0) + \frac{r(0)}{p(0)}\right) \\ &+\pi(1)\tilde{u}\left(e(1) + \frac{r(1)}{p(1)}\right) + \pi(2)\tilde{u}\left(e(2) + \frac{r(2)}{p(2)}\right) \end{aligned}$$

(b) Consider the Consumer Net Income Vector Problem with complete markets and state prices $\beta = (1, \beta(1), \beta(2))$

$$\begin{aligned} \text{Max}_r \quad v(p,r) \quad \text{s. to } r \in E(p) \quad \text{and} \\ r(0) + \beta(1)r(1) + \beta(2)r(2) = 0 \end{aligned}$$

and derive the first order conditions satisfied by a solution, $\bar{r}$.

(c) Let us say that the (implicit) future goods prices, induced by the date 1 spot-prices and the state prices, are actuarially fair if they are proportional to $(\pi(1), \pi(2))$. Examine the condition from (b) to find a condition on $(r(1), r(2))$ implying that future goods prices are actuarially fair.

Exercise 4.C Consider an economy satisfying the Maintained Assumptions with two consumer $\mathbb{I} = \{a, b\}$ with two dates, $T = 1$, two states at date 1, $S = 2$, 1 good, $L = 1$ and two assets, $J = 2$. Let $e^i = (e^i(0), e^i(1), e^i(2))$, for $i = a, b$, be the consumers' initial endowment and assume that the consumers have utility functions $u^i : \mathbb{R}^3_{++} \longrightarrow \mathbb{R}$, for $i = a, b$, where

$$u^i(c) = \tilde{u}(c(0)) + \pi(1)\tilde{u}(c(1)) + \pi(2)\tilde{u}(c(2)) \text{ for } i = a, b$$

and $\pi(s) = 1/2$ for $s = 1, 2$. The initial endowments and assets are given by

		Initial end.		Assets	
Date	State	e^a	e^b	1	2
0	$\{$ 0	1	1	$-\beta(1)$	$-\beta(2)$
1	$\{$ 1	1	0	1	0
	2	0	1	0	1

Thus the economy has *individual risk* but no aggregate risk since $e^a(1) + e^b(1) = e^a(2) + e^b(2)$.

(a) Show that the economy has a Radner equilibrium with $p(1) = p(2) = 1$, $c^a = (1, 1/2, 1/2)$ and $e^b = (1, 1/2, 1/2)$. Find the state prices $(1, \beta(1), \beta(2))$ and the date 0 spot price for the good.

(b) Show that $\bar{r}^a = (0, -1/2, 1/2)$ and $r^b = (0, 1/2, -1/2)$ are solutions to the Consumer Net Income Vector Problem for consumer a and b.

(c) Show that the discounted equilibrium prices (the implicit date 0 prices for forward delivery) are actuarially fair. (Cf. Exercise 4.B)

Exercise 4.D Consider a Radner equilibrium in an economy with $T = 1, L = 1, S = 2$ and complete markets. Assume that the economy has a risk-neutral consumer with

$$u(c) = c(0) + \pi(1)c(1) + \pi(2)c(2)$$

for $c \in \mathbb{R}^3_{++}$ whose equilibrium consumption belongs to $\mathbb{R}^3_{++}$. Show that the implicit future goods prices are actuarially fair. (Cf. Exercise 4.B)

Exercise 4.E Let an economy have only the two assets as given in Figure 4.8.A in Section 4.8 and thus disregard the possibility of frequent trading. In a Radner equilibrium there is a unique discount factor between date 0 and date 2 but not between date 0 and date 1. Explain!

Exercise 4.F Consider an economy with an asset structure where the consumers $\mathbb{I} = \{a, b\}$ satisfy the Maintained Assumptions, $T = 1, L = 1, S$ and J arbitrary. Assume that there is a Radner equilibrium for the economy with net income vectors $\bar{r}^a$ and $\bar{r}^b$. Let V be the dividend matrix and

$$W = \begin{bmatrix} -q \\ V \end{bmatrix}$$

the extended dividend matrix in the Radner equilibrium.

(a) W is arbitrage free. Why?

(b) Assume that there are state prices $(1, \tilde{\beta})$ and $(1, \hat{\beta})$. Show that the projections of $\tilde{\beta}$ and $\hat{\beta}$ on $\langle V \rangle$ are equal. Denote this common vector in $\langle V \rangle$ by β_V.

(c) Let θ be a portfolio. Show that $q^\theta = \beta_V \cdot v^\theta$ where v^θ is the dividend vector and q^θ the date 0 price for the portfolio θ.

(d) Consider two consumers, a and b, with net income vectors $\bar{r}^a$ and $\bar{r}^b$. Let $Dv^i(p, \bar{r}^i) = (D_0 v^i(p, \bar{r}^i), \text{grad}_1 v^i(p, \bar{r}^i))$ for $i = a, b$ so that $grad_1 v^i(p, \bar{r}^i)$ is the vector of date 1 marginal utilities of income. Show that there positive numbers α^a, α^b and vectors $\beta_\perp^a$, $\beta_\perp^b$such that

$$\begin{aligned} \text{grad}_1 v^a(p, \bar{r}^a) &= \alpha^a \beta_V + \beta_\perp^a \quad \text{with } \beta_\perp^a \in \langle V \rangle^\perp \\ \text{grad}_1 v^b(p, \bar{r}^b) &= \alpha^b \beta_V + \beta_\perp^b \quad \text{with } \beta_\perp^b \in \langle V \rangle^\perp \end{aligned}$$

Hence, for $r' \in \langle V \rangle$,

$$\text{grad}_1 v^a(p, \bar{r}^a) \cdot r' = \text{grad}_1 v^b(p, \bar{r}^b) \cdot r'$$

so that consumer a's and b's evaluation of net income vectors in $\langle V \rangle$ agree. (*Hint:* Use that $Dv^i(p, \bar{r}^i)$ is proportional to some vector $(1, \beta^i) \in \langle W \rangle^\perp$ for $i = a, b$.)

(e) If $\beta_\perp^a$ is not proportional to $\beta_\perp^b$, as given in (d), then there will be some (future) net income vector in $\mathbb{R}^S$ which is evaluated differently by the consumers. True or false?

(f) Use the result from (d) to show that the consumers' evaluation of portfolios is the same, that is, for $\theta \in \mathbb{R}^J$

$$\text{grad}_r v^a(p, \bar{r}^a) \cdot V\theta \geq 0 \text{ if and only if } \text{grad}_r v^b(p, \bar{r}^b) \cdot V\theta \geq 0$$

(g) Choose arbitrary positive probabilities for the states, $\pi(s)$ for $s \in \mathbb{S}_1$, and consider the vector with components $\beta_V(s)/\pi(s) = \gamma(s)$. Use the result from (c) to show that for any portfolio θ

$$q_\theta = E_\pi(\gamma)E_\pi(v^\theta) + cov_\pi(\gamma, v^\theta)$$

where $E_\pi(\cdot)$ denotes the expectation with respect to π and $cov_\pi(\cdot, :)$ denotes the covariance with respect to π.

Exercise 4.G Consider an economy with an asset structure. In the text we have implicitly assumed that each consumer $i \in \mathbb{I}$ has an initial portfolio $\hat{\theta}^i = 0$. Assume that we allow for $\hat{\theta}^i \neq 0$, for $i \in \mathbb{I}$.

(a) Does the Consumer Problem in such an economy make sense?

(b) Assume that $\sum_{i \in \mathbb{I}} \hat{\theta}^i = 0$. Could the definition of a Radner equilibrium be modified to allow for this? What should be the relation of the net income vectors and the portfolios in equilibrium?

(c) Assume that $\sum_{i \in \mathbb{I}} \hat{\theta}^i \neq 0$. For example, assume that $\hat{\theta}^i_j = 0$ for $i \in \mathbb{I}$, $j \in \mathbb{J}$ except that $\hat{\theta}^1_1 > 0$. Can there exist a Radner equilibrium (reasonably defined) for the economy?

Exercise 4.H Consider an economy with an asset structure with two dates, $T = 1$, and S states with where the date 0 price of asset $j \in \mathbb{J}$ in a Radner equilibrium is q_j. Let $(1, \beta(1), \ldots, \beta(S)) \in \mathbb{R}^{S+1}_{++}$ be state prices.

(a) Assume that $(1, 1, \ldots, 1) \in \langle V \rangle$. What is the date 0 price of 1 unit of sure income at date 1? Why is it reasonable to refer to that price as the time discount factor?

(b) Use the vector β to define define numbers $\pi(s) = \beta(s)/\sum_{s \in \mathbb{S}_1} \beta(s)$ which are positive and sum to 1. What is the interpretation of $\sum_{s \in \mathbb{S}_1} \pi(s)v^j(s)$ for $j \in \mathbb{J}$?

(c) Show that $q_j = \left[\sum_{s \in \mathbf{S}_1} \beta(s)\right] E_\pi(v^j)$ where E_π denotes the expectation with respect to π. Interpret!

The probabilities given by π, as defined in (b), are often referred to as *risk-neutral probabilities.*

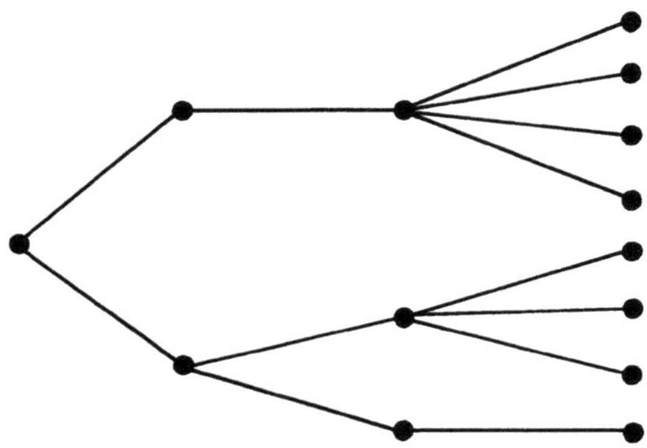

Figure A: *An event tree over 4 dates*

Exercise 4.I Consider the event tree in Figure A.

(a) Find a set of states of the world and a filtration such that the given event tree is compatible with the filtration.

(b) For each of the elementary events at date 3 find the realized path. Explain why for any event tree the set of terminal nodes can be chosen as states of the world.

(c) Assume that there are two nominal assets traded at every node which is not terminal and which pay dividends only at some terminal nodes. The numbers $q(\sigma)$, $q'(\sigma)$ denote prices if the node is not a terminal node and the dividend at σ in case σ is a terminal node. Choose a suitable enumeration of the nodes and find the extended dividend matrix.

(d) What is the spanning number of the tree?

(e) Is the asset market complete? Describe the possible incompleteness of the asset market.

(f) Assume that the elementary events at date 3 are assigned the same probabilities. Find the probabilities for the elementary events at other dates.

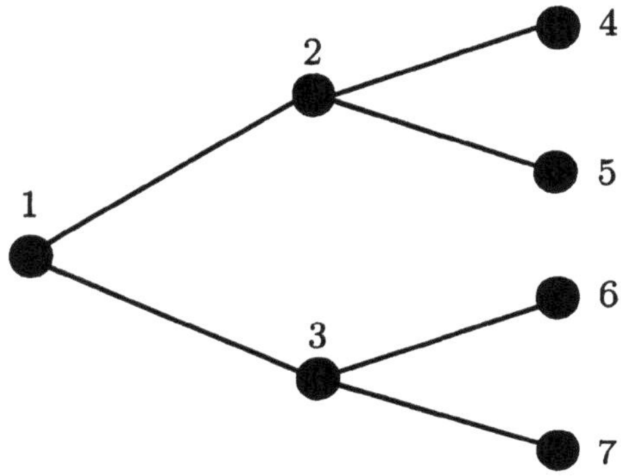

Figure B: *For* Exercise 4.J

Exercise 4.J (Exercise 4.H continued) Consider an economy with three dates, $T = 2$, and 4 elementary events at date 2 as illustrated in Figure B. Assume that there are unique state prices $\beta(\sigma)$, for $\sigma \in \mathbb{P}_1$ where $\mathbb{P}_1$ is the set of future nodes. Let j be an asset traded at date 0 and date 1 paying dividends at every future node

(a) What is the time discount factor between date 0 and date 1?

(b) What are the time discount factors between date 1 and date 2 corresponding to nodes 2 and 3? What is the time discount factor between date 0 and date 2?

(c) Show, as in Exercise 4.H that the price, q_j, of asset j is

$$q_j = \left[\sum_{\sigma \in \mathbb{P}_1} \beta(\sigma)\right] \sum_{\sigma \in \mathbb{P}_1} \pi(\sigma) v^j(\sigma)$$

where $\pi(\sigma)$ are suitably defined probabilities. What is the interpretation?

(d) Let $\mathcal{P}_1$ be the date 1 nodes and $\mathcal{P}_2$ the date 2 nodes. Show that

$$q_j = \left[\sum_{\sigma \in \mathcal{P}_1} \beta(\sigma)\right] \sum_{\sigma \in \mathcal{P}_1} \pi_1(\sigma) v^j(\sigma) + \left[\sum_{\sigma \in \mathcal{P}_2} \beta(\sigma)\right] \sum_{\sigma \in \mathcal{P}_2} \pi_2(\sigma) v^j(\sigma)$$

where $\pi_1(\sigma)$ and $\pi_2(\sigma)$ are suitably defined probability measures. What is the interpretation?

(e) Would the formulas in (c) and (d) hold in case the state prices were not unique?

Exercise 4.K Assume that we give up the assumption of a finite time horizon in the construction of an event tree. For example, tossing a (possibly unfair) coin gives rise to a construct like the one indicated in Figure C. What should be taken as states of the world? How should the definition of an event tree be modified in order to allow for such a construct? Is it still true that each state of the world can be identified with a unique realized path? What should be the filtration associated with the tree?

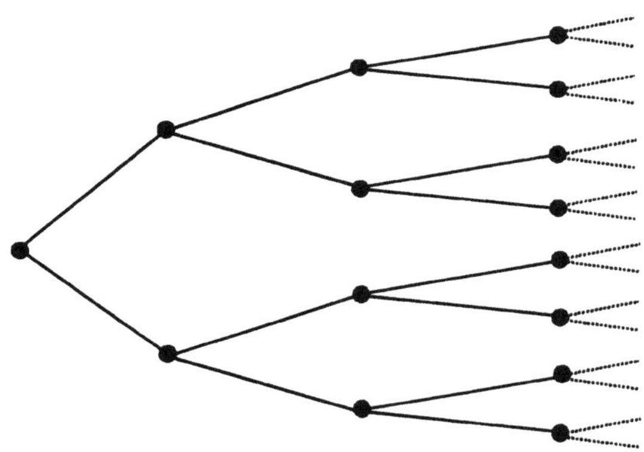

Figure C: *An infinite "event tree" for the Black-Scholes and the Lucas model*

In the exercise below we consider an "event tree" extending into the indefinite future as described in Exercise 4.K.

Exercise 4.L (Black and Scholes [1973]) In an economy a (Radner) equilibrium has resulted in prices of a stock equal to $(q^S(\sigma))_{\sigma\in\mathbb{P}}$ and a constant bond price equal to q^B for $\sigma \in \mathbb{P}$. Given the elementary event (node) at the current date there are only two successors at the next date and at every node prices at the successors to this node are given by the table below.

Date	Sucessor	Asset 1	2	State price
current		q^B	$q^S(\sigma)$	1
next	*up*	1	$q^S(\sigma)U$	β_u
	down	1	$q^S(\sigma)D$	β_d

Assume that $0 < q^B D < 1 < q^B U$

(a) Find the state prices between the current date and the next date as functions of q^B and $q^S(\sigma)$. Note that the state prices, relevant for a node and its immediate successors are constant over time.

(b) Use the current-next date prices to find the date 0 prices for income at date 2. (*Hint:* Discount in two steps.)

(c) Normalize the state prices β_u, β_d between date current date and the next date by defining

$$\pi = \frac{\beta_u}{\beta_u + \beta_d}$$
$$1 - \pi = \frac{\beta_d}{\beta_u + \beta_d}$$

and interpret π as the probability of an increase in the price of stock.

(d) Find the probabilities for the elementary events at date 2.

(e) Consider the stock price at date 2. Find its distribution with respect to the probability measure given by π.

(f) Consider the state price at date 2. Find its distribution with respect to the probability measure given by π.

(g) Consider a European call option (an option that can be exercised only at the date of expiration) expiring at date 2 with exercise price $q_0^S UD$ where q_0^S is the date 0 price of the stock. Use the state prices to find the date 0 price of the option.

Exercise 4.M (Continuation of Exercise 4.L)

(a) Consider the "down" node at date 1. Find a portfolio, say θ_d, replicating the date 2 dividends of the option, given in part (g) of Exercise 4.L, from this node.

(b) Consider the "up" node at date 1. Find a portfolio, say θ_u replicating the date 2 dividends of the option from this node.

(c) From (a) and (b) you have the portfolios θ_u and θ_d. Find a date 0 portfolio, say θ_0, whose dividends at date 1 equals the cost of θ_u, if Nature chooses "up", and θ_d if Nature chooses "down".

(d) Show that the date 0 price of the portfolio θ_0 equals the price of the option found in (g) of Exercise 4.L.

Exercise 4.N (Lucas [1978] model of asset pricing). Consider an economy with infinitely many future dates and with uncertainty described by the "event tree" in Figure C with the probability of an "up"-step being p at each node, with $0 < p < 1$. The economy has a single good, $L = 1$, one consumer and a single producer. The producer has chosen a production plan $(\bar{y}(\sigma))_{\sigma \in \mathbb{P}}$ where $\bar{y}(\sigma) > 0$ for $\sigma \in \mathbb{P}$ (inputs were used before the current date). The consumer owns all of the stock which is taken to be one unit but has no initial endowment otherwise. At each node the real dividends $\bar{y}(\sigma)$ accrue to the consumer provided she holds on to the stock. There is, for $\sigma \in \mathbb{P}$, a bond yielding 1 unit of sure income at the next date available in 0 supply. Let $q^B(\sigma)$ be the price of the bond and $q^S(\sigma)$ the postdividend price of the stock for $\sigma \in \mathbb{P}$. Assume that the spot-price for the good is 1 at each node. The consumer has a utility function u which is separable over states and over time and which is given by a time and state invariant function $\tilde{u}$. Thus

$$u(c(\sigma)_{\sigma \in \mathbb{P}}) = \sum_{t=0}^{\infty} \lambda^t \sum_{\sigma \in \mathcal{P}_t} \tilde{u}(c(\sigma) P_\sigma$$

where $\tilde{u} : \mathbb{R}_{++} \longrightarrow \mathbb{R}$ is a differentiable, strictly concave function with $D\tilde{u}$ everywhere positive, $0 < \lambda < 1$ is a subjective discount factor and P_σ is the probability of the elementary event given by σ.

(a) Assume that there are date 0 state prices $\beta(\sigma)$ for $\sigma \in \mathbb{P}$. How are these prices related to the spot-prices for the good Determine the consumer's date 0 wealth if she can sell the date 0 and the future dividends at date 0.

(b) Formulate the budget restriction in the Consumer Problem given the wealth from (a) and the date 0 prices $\beta(\sigma)$ for consumption at node σ. Show that the solution to the Consumer Problem is unique.

(c) Assume that $\bar{c} = (\bar{c}(\sigma)_{\sigma \in \mathbb{P}})$ is a solution to the Consumer Problem from (b). Show that there is $\alpha \in \mathbb{R}$ such that

$$\lambda^t D\tilde{u}(\bar{c}(\sigma))P_\sigma - \alpha\beta(\sigma) \;=\; 0 \quad \text{for } t \in T \text{ and } \sigma \in \mathcal{P}_t$$

(*Hint:* Consider the consumer Problem with consumption fixed at $(\bar{c}(\sigma)_{\sigma \in \mathbb{P}})$ except for two nodes $\hat{\sigma}$ and σ'. Argue that $\bar{c}(\hat{\sigma})$ and $\bar{c}(\sigma')$ is a solution to the induced Consumer Problem in two variables.)

(d) Consider the problem of finding $(\beta(\sigma)_{\sigma \in \mathbb{P}})$, with $\beta(\sigma_0) = 1$ at the initial node, such that the solution to the Consumer Problem from (b), $\bar{c} = (\bar{c}(\sigma)_{\sigma \in \mathbb{P}})$, equals $(\bar{y}(\sigma))_{\sigma \in \mathbb{P}}$. Use the marginal conditions from (c) to show that

$$\bar{\beta}(\sigma) \;=\; \frac{1}{\alpha}\lambda^t D\tilde{u}(\bar{y}(\sigma))P_\sigma \quad \text{for } t \in T \text{ and } \sigma \in \mathcal{P}_t$$

with $\alpha = D\tilde{u}(\bar{y}(\sigma_0))$ achieves this.

(e) Assume that the (postdividend) price of the stock at node σ is the value of the future dividends. Check that the date 0 portfolio $\theta = (1, 0)$, that is, one unit of the stock and 0 units of the bond allows the consumer to consume $\bar{y}(\sigma_0)$ at date 0 and the amount $\bar{y}(\sigma)$ at both date 1 nodes and also to acquire the portfolio $\theta_\sigma = (1, 0)$ at both date 1 nodes.

Chapter 5

EXISTENCE AND DETERMINACY

Introduction

General equilibrium theory aims at providing an explanation of the relative values of commodities. In the theory of incomplete asset markets this explanation is extended by the explicit study of income transfers, between dates or states, to the pricing of assets.

The basic assumption underlying the theory is that using a reasonable equilibrium concept we will observe prices and choices which have at least some similarity with what the theory predicts.

If there is to be such an agreement between theory and reality, then it is a necessary condition that the model is consistent. Thus one has to prove that the model, under a set of reasonable assumptions, possesses at least one equilibrium. Existence of equilibrium does not imply that the model is satisfactory, but non-existence of equilibrium definitely disqualifies the model from any serious consideration as an explanation to observed phenomena.

A second theme concerns the determinacy of equilibrium. In general equilibrium theory it is shown that it is to much to hope for a unique equilibrium. Still a model which has a very large set of equilibria is not satisfactory as an explanation of relative values. For economies over time or under uncertainty a multitude of equilibria also makes the assumption of perfect foresight particularly dubious.

For the general equilibrium model with incomplete markets, Radner [1972] demonstrated the existence of a Radner equilibrium for the case where the agents are restricted to choose portfolios in some compact, convex set. Often such a "compactification" is innocuous, since it can in the end be removed while the actions of the agents stay bounded. An explicit example by Hart [1975] shows that for the case of incomplete markets and real assets the bounds on the asset trade can not be removed in equilibrium, without affecting the equilibrium allocation. Hart's example inspired much research into the existence problem and forced the introduction of new tools to study the existence problem, cf. Duffie and Shafer [1985], but his example turned out to in the end to be atypical. A further example by Polemarchakis and Ku [1990] indicated that the existence problem might be more serious when the economy also has derivative assets like options. In Section 5.1 and at the end of 5.2 we give simplified versions of Hart's example and the example by Polemarchakis and Ku.

For economies with nominal assets existence of a Radner equilibrium is less problematic and can be proved using a fixed point, or fixed point like, theorem. Cf. Werner [1985] and Werner [1987].

Debreu [1970] showed that typically an exchange economy will have a finite and odd number of equilibria. His contribution is noteworthy for the introduction of methods from differential topology into economics. Extending these methods to economies with incomplete markets it is has been shown that an economy with real assets, typically, has a (non-zero) finite number of Radner equilibria. Cf. Duffie and Shafer [1985, 1986] and Hirsch et al. [1990]

For economies with nominal assets and incomplete asset markets the situation is less satisfactory. Here the economy will typically have continuum of equilibria. This was first noticed by Cass [1985] and his result was extended by Geanakoplos and Mas-Colell [1989] as well as by Balasko and Cass [1989]. In the first part of Section 5.4 we give an example of indeterminacy and the last part of Section 5.4 contains a statement of a general theorem on indeterminacy with nominal assets.

5.1 Real Assets and Existence of a Radner Equilibrium

In this section we study a simplified version of a famous counterexample by Hart [1975], which we will somewhat inappropriately refer to as "Hart's example". Hart's original example was for an economy with two dates and uncertainty as described in Exercise 5.D.

An Example of Non-Existence

Hart showed that the restrictions on asset trade, as introduced by Radner, were not innocuous. He gave an explicit example of a class of economies with real assets, whose members did not have a Radner equilibrium. His example also showed that asset trade in an equilibrium has to be larger the closer the economy is to an economy which fails to have a Radner equilibrium.

Hart's example was the starting point for a number of contributions to the existence of equilibrium with incomplete asset markets. It turned out that to prove existence of equilibrium new tools had to be used. There is at present no proof based merely on a fixed point, or a fixed point like, theorem.

Subsequent results, which we will discuss in Section 5.2, showed that the economies in Hart's example were the exception rather than the rule and that "most" economies, satisfying standard assumptions, do have a Radner equilibrium.

The economy

We study an economy over time where $\mathbb{T} = \{0, 1\}$, $\mathbb{L} = \{1, 2\}$ and $\mathbb{I} = \{a, b\}$. The consumers have time separable utility functions

Date	Utility a	Utility b
0	$\alpha^a \ln c_1^a(0) + \beta^a \ln c_2^a(0)$	$\alpha^b \ln c_1^b(0) + \beta^b \ln c_2^b(0)$
1	$+\ \alpha^a \ln c_1^a(1) + \beta^a \ln c_2^a(1)$	$+\ \alpha^b \ln c_1^b(1) + \beta^b \ln c_2^b(1)$

and endowments

Date	e^a	e^b
0	$\begin{bmatrix} (1-\varepsilon)h \\ (1-\varepsilon)h \end{bmatrix}$	$\begin{bmatrix} \varepsilon \\ \varepsilon \end{bmatrix}$
1	$\begin{bmatrix} \varepsilon h \\ \varepsilon h \end{bmatrix}$	$\begin{bmatrix} 1-\varepsilon \\ 1-\varepsilon \end{bmatrix}$

Thus for $h = 1$ the total endowment is the same at both dates and equal to 1 for both goods.

Asset, spot prices and discount factor

The economy has a single asset which can be traded at the current date. Its price is q and the owner of one unit of the asset receives 1 unit of the first good and is obliged to deliver 1 unit of the second good, at date 1. This is illustrated below along with the notation for spot prices and the discount factor.

Date	Asset 1	Spot Prices	Discount factor
0	$-q$	$\begin{bmatrix} p_1(0) \\ p_2(0) \end{bmatrix}$	$\beta(0) = 1$
1	$\begin{bmatrix} 1 \\ -1 \end{bmatrix}$	$\begin{bmatrix} p_1(1) \\ p_2(1) \end{bmatrix}$	$\beta(1)$

We have a complete asset market if and only if $p_1(1) \neq p_2(1)$. We can now state

Theorem 5.1.A *Let $\mathcal{E}$ be the economy with an asset structure as defined above. Assume*

(i) $\alpha^a + \alpha^b = 1, \alpha^a > \alpha^b, \alpha^a + \beta^a = 1$ *and* $\alpha^b + \beta^b = 1$

(ii) $0 < \varepsilon < 1/2$

If $h = 1$ then $\mathcal{E}$ does not have a Radner equilibrium.

As will be seen a Radner equilibrium fails to exist since a Radner equilibrium has either rank 1 or 0. If the rank is 1 the only candidates for equilibrium prices at date 1 are equal prices for both goods but at equal prices there is a "drop in rank" of the value dividend matrix so that its rank is 0. On the other hand if the rank is assumed to be 0 then date 1 goods prices are not equal which implies that the rank is 1.

The corollary below shows that for $h \neq 1$ but close to 1, the economy has a Radner equilibrium of rank 1, but consumers have to engage in very large trades in the asset, in order to achieve the desired income transfers between date 0 and date 1. In the limit the desired income transfers can not be achieved through the asset trade.

Corollary 5.1.B *Under the assumptions of* Theorem 5.1.A *the economy has a unique Radner equilibrium for $h > 1$. Consumer a's demand for the asset, in the equilibrium, tends to infinity as h tends to* 1.

Wealth and income transfers

It is easy to see that if there is a Radner equilibrium then there is one where spot prices are normalized at each date to sum to one, that is, $p_1(0) + p_2(0) = 1$ and $p_1(1) + p_2(1) = 1$. We restrict attention to such equilibria. We can calculate the value of the components of the initial endowment which is given below along with the notation for the net income vectors.

Date	Value init. end. a	Value init. end. b	r^a	r^b
0	$\begin{bmatrix} p_1(0)(1-\varepsilon)h) \\ p_2(0)(1-\varepsilon)h \end{bmatrix}$	$\begin{bmatrix} p_1(0)\varepsilon \\ p_2(0)\varepsilon \end{bmatrix}$	$r^a(0)$	$r^b(0)$
1	$\begin{bmatrix} p_1(1)\varepsilon h \\ p_2(1)\varepsilon h \end{bmatrix}$	$\begin{bmatrix} p_1(1)(1-\varepsilon) \\ p_2(1)(1-\varepsilon) \end{bmatrix}$	$r^a(1)$	$r^b(1)$

The normalization of prices makes it easy to calculate the spot income accruing from the initial endowment

Date	Spot value init. end. a	Spot value init. end. b
0	$w^a(0) = (1-\varepsilon)h$	$w^b(0) = \varepsilon$
1	$w^a(1) = \varepsilon h$	$w^b(1) = (1-\varepsilon)$

Hence the (discounted) wealth is $\hat{w}^a = (1-\varepsilon)h + \beta(1)\varepsilon h$ for consumer a and $\hat{w}^b = \varepsilon + \beta(1)(1-\varepsilon)$ for consumer b.

Example 2.6.A in Chapter 2 shows that if, say, consumer a is only restricted in his choice of the net income vector, r^a, by the restriction $r^a(0) + \beta(1)r^a(1) = 0$ then consumer a will choose r^a so that the discounted gross spot income, $\beta(t)\left[w^a(t) + r^a(t)\right]$, $t = 1,2$ is the same for both dates. (This depends on the fact that the sum of the coefficients, for each date, in the utility function are equal.) Hence we get the following demand for net income vectors in case the asset markets are complete.

$$\begin{bmatrix} r^a(0) \\ \\ r^a(1) \end{bmatrix} = \begin{bmatrix} \frac{1}{2}\frac{\hat{w}^a}{1} & - w^a(0) \\ \\ \frac{1}{2}\frac{\hat{w}^a}{\beta(1)} & - w^a(1) \end{bmatrix} = \begin{bmatrix} -\frac{1}{2}(1-\varepsilon)h + \frac{1}{2}\beta(1)\varepsilon h \\ \\ \frac{1}{2}\frac{(1-\varepsilon)h}{\beta(1)} - \frac{1}{2}\varepsilon h \end{bmatrix}$$

$$\begin{bmatrix} r^b(0) \\ \\ r^b(1) \end{bmatrix} = \begin{bmatrix} \frac{1}{2}\frac{\hat{w}^b}{1} & - w^b(0) \\ \\ \frac{1}{2}\frac{\hat{w}^b}{\beta(1)} & - w^b(1) \end{bmatrix} = \begin{bmatrix} \frac{1}{2}\frac{\beta(1)(1-\varepsilon)}{1} - \frac{1}{2}\varepsilon \\ \\ \frac{1}{2}\frac{\varepsilon}{\beta(1)} - \frac{1}{2}(1-\varepsilon) \end{bmatrix} \qquad \mathbf{5.1.A}$$

We can now prove Theorem 5.1.A.

Proof of Theorem 5.1.A:

Step 1: *There is no Radner equilibrium with* $\dim\langle V\rangle = 1$ *when* $h = 1$. Assume, in order to arrive at a contradiction, that there is a Radner equilibrium of rank 1. If $\dim\langle V\rangle = 1$ then we have complete markets and there is a unique date 0 price for date 1 income, $\beta(1)$. In the equilibrium $\beta(1)$ is determined by the balance condition for net income at date 1, the condition $r^a(1) + r^b(1) = 0$ or

$$\frac{1}{2}\left[\frac{(1-\varepsilon)h}{\beta(1)} - \varepsilon h\right] + \frac{1}{2}\left[\frac{\varepsilon}{\beta(1)} - (1-\varepsilon)\right] = 0$$

which has the unique solution

$$\beta(1) = \frac{\varepsilon + (1-\varepsilon)h}{\varepsilon h + (1-\varepsilon)}$$

Recall that $0 < \varepsilon < 1/2$. Hence $h > 1$ implies $\beta(1) > 1$. Furthermore, the normalization of the spot prices results in $\beta(1)$ being independent of the spot prices.

Using the Walrasian demand the market balance conditions at date 1 are

$$\begin{aligned} \alpha^a \frac{1}{2} \frac{\hat{w}^a}{\beta(1) p_1(1)} + \alpha^b \frac{1}{2} \frac{\hat{w}^b}{\beta(1) p_1(1)} &= \varepsilon h + (1 - \varepsilon) \\ \beta^a \frac{1}{2} \frac{\hat{w}^a}{\beta(1) p_2(1)} + \beta^b \frac{1}{2} \frac{\hat{w}^b}{\beta(1) p_2(1)} &= \varepsilon h + (1 - \varepsilon) \end{aligned}$$

which gives

$$\begin{aligned} p_1(1) &= \frac{1}{2\beta(1)} \frac{\alpha^a \hat{w}^a + \alpha^b \hat{w}^b}{\varepsilon h + (1 - \varepsilon)} \\ p_2(1) &= \frac{1}{2\beta(1)} \frac{\beta^a \hat{w}^a + \beta^b \hat{w}^b}{\varepsilon h + (1 - \varepsilon)} \end{aligned}$$

or, substituting the values for $\hat{w}^a$ and $\hat{w}^b$,

$$\begin{aligned} p_1(1) &= \frac{1}{2} \frac{\alpha^a \left[(1 - \varepsilon) + \beta(1)\varepsilon\right] h + \alpha^b \left[\varepsilon + \beta(1)(1 - \varepsilon)\right]}{\varepsilon + (1 - \varepsilon) h} \\ p_2(1) &= \frac{1}{2} \frac{\beta^a \left[(1 - \varepsilon) + \beta(1)\varepsilon\right] h + \beta^b \left[\varepsilon + \beta(1)(1 - \varepsilon)\right]}{\varepsilon + (1 - \varepsilon) h} \end{aligned} \qquad \textbf{5.1.B}$$

It is now easy to see that $h = 1$ implies $\beta(1) = 1$ and $p_1(1) = p_2(1) = 1/2$. Hence $\dim \langle V \rangle = 0$ which contradicts our assumption $\dim \langle V \rangle = 1$.

Step 2: *There is no Radner equilibrium with* $\dim \langle V \rangle = 0$ *when* $h = 1$. Assume, in order to arrive at a contradiction, that there is a Radner equilibrium of rank 0. Note that $\dim \langle V \rangle = 0$ if and only if the normalized spot prices at date 1 are both equal to 1/2. On the other hand, the equilibrium prices at date 1 give market balance at date 1. Using the pure spot market demand we have the market balance conditions

$$\begin{aligned} \alpha^a \frac{\varepsilon + r^a(1)}{p_1(1)} + \alpha^b \frac{(1 - \varepsilon) + r^b(1)}{p_1(1)} &= 1 \\ \beta^a \frac{\varepsilon + r^a(1)}{p_2(1)} + \beta^b \frac{(1 - \varepsilon) + r^b(1)}{p_2(1)} &= 1 \end{aligned}$$

Since $r^a(0) = r^b(0) = r^a(1) = r^b(1) = 0$ it follows that

$$p_1(1) = \alpha^a\varepsilon + \alpha^b(1-\varepsilon)$$
$$p_2(1) = \beta^a\varepsilon + \beta^b(1-\varepsilon)$$

Since $\alpha^a > \alpha^b$, $\beta^a < \beta^b$ and $\varepsilon \in]0, 1/2[$ we have $p_1(1) \neq p_2(1)$ which implies that the subspace of income transfers $\langle V \rangle$ has dimension 1, in contradiction to our assumption.

Step 3: *There is no Radner equilibrium when* $h = 1$.
The Radner equilibrium $\langle V \rangle$ has rank 1 or 0 and since we have shown that there is no Radner equilibrium where the of ranks 0 and no Radner equilibrium of rank 1 it follows that there does not exist a Radner equilibrium for the economy.

□

Using the result from relation 5.1.B it is now easy to prove the corollary.

Proof of Corollary 5.1.B: First consider a fixed $h > 1$. The economy has a Walras equilibrium and the Walras equilibrium prices are given by 5.1.B. Hence

$$p_1(1) - p_2(1) = \frac{1}{2}\frac{1}{\varepsilon + (1-\varepsilon)h}\left[(\alpha^a - \beta^a)\left[(1-\varepsilon) + \beta(1)\varepsilon\right]h + (\alpha^b - \beta^b)\left[\varepsilon + \beta(1)(1-\varepsilon)\right]\right]$$

which implies that $p_1(1) - p_2(1) > 0$. However, this implies that the Walras equilibrium allocation can be realized as Radner equilibrium allocation with the given asset structure. The date 0 price of the asset, q, is determined by $\beta(1), p_1(1), p_2(1)$ so that $q = \beta(1)\left[p_1(1) - p_2(1)\right]$. From 5.1.A we get the demand for net income at date 1 for consumer a

$$r^a(1) = \frac{1}{2}\frac{(1-\varepsilon)h}{\beta(1)} - \frac{1}{2}\varepsilon h$$

which implies that consumer a's asset demand is

$$\theta^a = \frac{r^a(1)}{p_1(1) - p_2(1)}$$

Now let h tend to 1. Then $\beta(1)$ tends to 1, $p_1(1) - p_2(1)$ tends to 0, $r^a(1)$ tends to the positive number $(1/2) - \varepsilon$ and thus θ^a tends to infinity.

5.2 Existence of a Pseudo Equilibrium

Hart's example shows that one can not hope for a general existence theorem for Radner equilibrium when the economy has real assets.

On the other hand, the situation encountered in Hart's example might be rare so that there would "almost always" exist a Radner equilibrium. To make precise the "almost always" one needs to introduce measure theoretic considerations but, as will be seen, a knowledge of the notion of a set of measure 0 is sufficient. We will state the general result for an economy with two dates, S states of the world, L goods and J assets

Pseudo equilibrium

Rather than trying to prove an "almost always" existence theorem, the strategy has been to weaken the equilibrium concept and to prove the existence of a *pseudo equilibrium*. A pseudo equilibrium is a particular instance of a spot-market equilibrium.

The idea behind the pseudo equilibrium concept is simple. The difficulty in Hart's example stems from the, possible, drop in rank of the dividend matrix. Instead of demanding that the consumers should choose their net income vectors in the subspace spanned by the dividend matrix, they are allowed to choose in a related subspace. Let us momentarily assume that the *real* $SL \times J$ dividend matrix has rank $J < S$. Then the value dividend matrix will have rank at most J and for most prices the rank will be equal to J. A pseudo equilibrium is a spot-market equilibrium relative to a subset $\mathbb{M}$, where $\mathbb{M}$ is a subspace of dimension J. The novelty is that the subspace $\mathbb{M}$ is required to contain, possibly as a proper subspace, the linear subspace spanned by the column vectors of the value dividend matrix, at the equilibrium prices. Thus, if the dividend matrix has rank J at a pseudo equilibrium, then the linear subspace spanned by the column vectors coincides with $\mathbb{M}$.

It has been shown, Duffie and Shafer [1985,1986] that a pseudo equilibrium exists for a large class of economies. For the proof Duffie and Shafer use results from differential topology, which were first introduced into economics by Debreu [1970] and further applied by Balasko [1988],

Dierker [1974] and Mas-Colell[1985]. There is, at present, no proof using a fixed point theorem or a similar result. Another approach to the existence problem is taken by Hirsh et al. [1990].

The Coincidence of Pseudo Equilibria and Radner Equilibria*

The introduction of pseudo equilibria makes it possible to prove existence of a Radner equilibrium for a large class of economies. If the rank of the value dividend matrix is maximal, J, at a pseudo equilibrium then the pseudo equilibrium is also a Radner equilibrium. How should we make precise that this occurs "almost always"?

Exceptional and generic sets

To this end, choose some utility functions, $u^1, ..., u^I$. Consider the class of economies with an asset structure which arise as endowments $(e^i)_{i \in \mathbb{I}}$ are varied along with the real dividend matrix, A. We get a mapping

$$(e^1, ..., e^I, A) \longrightarrow \mathcal{E}(e, A)$$

taking the initial endowments and the real dividend matrix to an economy, with these initial endowments and this real dividend matrix.

Now we may, for each (e, A), consider the pseudo equilibria of $\mathcal{E}(e, A)$ and the Radner equilibria of the same economy. Every Radner equilibrium is also a pseudo equilibrium, so the issue to be resolved is the following. Is each pseudo equilibrium of $\mathcal{E}(e, A)$, also a Radner equilibrium, or equivalently: is the value dividend matrix of rank J at each pseudo equilibrium of $\mathcal{E}(e, A)$?

Since e is a vector in $\mathbb{R}_{++}^{L(S+1)I}$ and the matrix A is given by a point in $\mathbb{R}^{LSJ}$, the pair (e, A) belongs to $\mathbb{R}_{++}^{L(S+1)I} \times \mathbb{R}^{L(S+1)I+LSJ}$. Let us consider our problem as (e, A) belongs to a cube $Q \subset \mathbb{R}_{++}^{L(S+1)I} \times \mathbb{R}^{L(S+1)I+LSJ}$ with side 1. The *exceptional set*, restricted to Q, is the set

$$E = \left\{ (e, A) \in Q \;\middle|\; \begin{array}{c} \mathcal{E}(e, A) \text{ has a pseudo equilibrium with rank} \\ \text{of the value dividend matrix less than } J \end{array} \right\}$$

It is easy to calculate the volume, or (Lesbegue) measure, of a cube. One simply multiplies the lengths of its sides. Since each side of Q has length 1, the measure of Q is 1.

Given any $\varepsilon > 0$, there exists cubes $(Q_n)_{n\in\mathbb{N}}$, contained in Q, such that the $\sum_{n\in\mathbb{N}} vol(Q_n) < \varepsilon$, where $vol(Q_n)$ is the volume of Q_n, for $n \in \mathbb{N}$, and such that $E \subset \cup_{n\in\mathbb{N}} Q_n$. This implies that the exceptional set, E, is contained in a set of measure smaller than ε and thus E itself has measure smaller than ε. Since this is true for each $\varepsilon > 0$, the set E has, by definition, measure 0.

One may think of the measure on Q as a probability measure. The probability that a pair (e, A), chosen at random in Q, also belongs to E is 0.

Note that E will contain two types of points. If (e, A) is such that $\operatorname{rank} A < J$ then the rank of the value dividend matrix is trivially less than J. The more interesting points (e, A) in E are those where $\operatorname{rank} A = J$ but where, for some pseudo equilibrium for $\mathcal{E}(e, A)$, the rank of the value dividend matrix is less than J.

The set $Q \backslash E$ is the set of economies, where each pseudo equilibrium is also a Radner equilibrium. $Q \backslash E$ has measure 1 and is a *generic set.*

One can show that E is closed set. Hence its complement $Q \backslash E$ is an open set. Thus given an economy in $Q \backslash E$, sufficiently small variations of the endowments and real dividend matrix, will leave us with an economy in $Q \backslash E$.

We have given the discussion for the cube Q. The reasoning extends to the whole space of economies $\mathbb{R}_{++}^{L(S+1)I} \times \mathbb{R}^{L(S+1)I+LSJ}$ since this space can be covered by countably many cubes. The exceptional set for the whole space is contained in the union of the exceptional sets for these cubes. Since the measure of a union of a countable collection of sets of measure 0 is again 0 we get the result that the measure of the exceptional set is 0. The probability interpretation is however lost.

A consequence of the results above is that Hart's example of an economy, which does not have a Radner equilibrium, makes use of an economy which is a atypical. Typically, or generically, an economy with real assets does have a Radner equilibrium. This was also suggested by the variation of the parameter h, in Hart's example. Only for the value $h = 1$ did the economy fail to have a Radner equilibrium.

Generic set; a formal definition*

It is convenient to have a formal definition of a generic set in $\mathbb{R}^n$. The concept of a subset of measure 0 makes good sense in $\mathbb{R}^n$ but the probability interpretation, suggested above, fails for general subsets of $\mathbb{R}^n$ since such a set might not have finite measure.

Definition 5.2.A *Let $Z \subset \mathbb{R}^n$ be an non-empty open set. A subset $G \subset Z$ is a* **generic set** *in Z if*

(a) *G is open*

(b) *the complement of G in Z, the set $Z \backslash G$, has measure* 0

If G is a generic set in Z then we will refer to the complement of G in Z, that is, the set $Z \backslash G$, as an **exceptional set**.

The Determinacy of Pseudo Equilibria*

General equilibrium theory aims at explaining the resource allocation and prices in an economy, as the outcome of utility maximizing individuals together with the condition of compatible actions, that is, market balance.

It would be very satisfying if, for a large class of economies, equilibrium did exist and was unique. This has turned out to be too much to hope for but for Walras equilibrium; corresponding to complete asset markets, it has been shown that the number of equilibria, generically, is finite and odd. Cf. Debreu [1970], Dierker [1974] or Balasko[1988]. Consideration of examples in an Edgeworth's box suggest the existence of economies with a continuum of Walras equilibria. However careful consideration of these examples also suggests that typically there is a finite number of equilibria.

The situation with incomplete asset markets and real assets, is similar to the one for Walras equilibria. One can show that for a generic set in endowments and the real dividend matrix, there is a finite set of equilibria. This is in sharp contrast with the results for economies with nominal assets and incomplete asset markets, to be studied in Section 5.3 and 5.4.

A General Existence Theorem. Real Assets*

We can now state a general theorem. Existing proofs make use of mathematical results which go beyond the exposition here. A proof can be found in Duffie and Shafer [1985]

Theorem 5.2.B *Let* $\mathbb{E} = \{(C^i, u^i, e^i)_{i \in \mathbb{I}}, A\}$ *be a family of economies with real assets parametrized by the initial endowments and the real asset structure* $(e, A) \in \mathbb{R}^{L(S+1)I}_{++} \times \mathbb{R}^{LSJ}$ *where* $0 \leq J < S$. *Let* $\mathcal{E}(e, A)$ *be the economy induced by* (e, A). *Then*

(a) $\mathcal{E}(e, A) \in \mathbb{E}$ *has a spot-market equilibrium relative to a subspace* $\mathbb{M}$ *of dimension* J *such that* $\langle V(p) \rangle \subset \mathbb{M}$ *where* $V(p)$ *is the value dividend matrix at the equilibrium prices*

(b) *there is a generic set of endowments and real dividend matrices* $\mathbb{D} \subset \mathbb{R}^{L(S+1)I}_{++} \times \mathbb{R}^{LSJ}$ *such that* $(e, A) \in \mathbb{D}$ *implies* $\langle V(p) \rangle = \mathbb{M}$ *for each equilibrium of* $\mathcal{E}(e, A)$.

Part (a) asserts that the economy $\mathcal{E}(e, A)$ has a pseudo equilibrium. Part (b) that generically each of the pseudo equilibria is a Radner equilibrium.

A Robust Counterexample to Existence. Real Assets

In this section we will see that if we allow for "assets" whose dividends depends on the spot-prices, but not necessarily as linear functions of the spot-prices, then the problem of non-existence may be aggravated.

The "drop in rank problem" in Hart's example

Hart's example in Section 5.1 shows that there are economies which do not have a Radner equilibrium. Consider the value dividend matrix of the economy studied in the example. The owner of one unit of the asset receives the amount $p_1(1) - p_2(1)$ of 1-crowns at date 1. Hence the asset can be used to transfer purchasing power in equilibrium unless the spot prices happen to be equal at date 1. If we take $\mathbb{R}^2_{++}$ as the ambient set of prices, then the set of prices for which the "drop in rank" problem occurs is a ray emanating from the origin. Hence its volume (area) is 0 so that generically, in the date 1 prices, the value dividend matrix has rank 1. This indicates that the problem of non-existence could be more serious if the "drop in rank" problem occurred for a larger subset of prices.

With real assets defined as in Chapter 3 the value dividend matrix is generated by spot prices and real dividend vectors and under weak conditions on the real dividend matrix, the resulting value dividend matrix will have full rank for a generic set of prices.

An example by Polemarchakis and Ku [1990] shows that if we allow for a more general asset structure the "drop in rank" problem may be robust so that non-existence can not be remedied by varying the economy slightly. We now give an example similar to the one in Polemarchakis and Ku to convey the idea.

An example in the vein of Polemarchakis and Ku.

We study an economy with an asset structure with no uncertainty where $\mathbb{T} = \{0,1\}$, $\mathbb{L} = \{1,2\}$ and $\mathbb{I} = \{a,b\}$. The consumers have time separable utility functions

Date	Utility a	Utility b
0	$\alpha \ln c_1^a(0) + \alpha \ln c_2^a(0)$	$\beta \ln c_1^b(0) + \beta \ln c_2^b(0)$
1	$+ 4 \ln c_1^a(1) + \ln c_2^a(1)$	$+ \ln c_1^b(1) + 2 \ln c_2^b(1)$

and their endowments are $e^a = e^b = (1,1,1,1)$. The economy has a single "optionlike" asset with date 0 price q and dividend $\max(p_1(1) - ap_2(1), 0)$ at date 1 where $a > 0$. We then have the following proposition.

Proposition 5.2.C *If* $a = 1$, $\beta = 1$ *and* $\alpha \geq 5$ *then the economy does not have a Radner equilibrium*

Proof: Assume that $((\bar{c}^a, \bar{\theta}^a), (\bar{c}^b, \bar{\theta}^b), \bar{p}, \bar{q})$ is a Radner equilibrium. Then either $\bar{p}_1(1) - \bar{p}_2(1) \leq 0$ and the rank of the dividend matrix is 0 or

$\bar{p}_1(1) - \bar{p}_2(1) > 0$ in which case the rank is 1 and the asset market is complete.

Case $\bar{p}_1(1) - \bar{p}_2(1) \leq 0$. In this case we have a pure spot market equilibrium. Without loss of generality we can normalize spot-prices so that $\bar{p}_1(1) + \bar{p}_2(1) = 1$. The date 1 incomes are then $w^a(1) = w^b(1) = 1$ and the market balance condition at date 1 is

$$\frac{4}{5}\frac{w^a(1)}{\bar{p}_1(1)} + \frac{1}{3}\frac{w^b(1)}{\bar{p}_1(1)} = 2$$
$$\frac{1}{5}\frac{w^a(1)}{\bar{p}_2(1)} + \frac{2}{3}\frac{w^b(1)}{\bar{p}_2(1)} = 2$$

which gives $\bar{p}_1(1) = 17/30$ and $\bar{p}_2(1) = 13/30$ so that $\bar{p}_1(1) - \bar{p}_2(1) > 0$, contradicting $\bar{p}_1(1) - \bar{p}_2(1) \leq 0$.

Case $\bar{p}_1(1) - \bar{p}_2(1) > 0$. In this case the asset market is complete so there is a $\beta(1)$ such that

$$(\bar{p}_1(0), \bar{p}_2(0), \beta(1)\bar{p}_1(1), \beta(1)\bar{p}_2(1))$$

are Walras equilibrium prices. The wealths are

$$\hat{w}^a = \hat{w}^b = \bar{p}_1(0) + \bar{p}_2(0) + \beta(1)\bar{p}_1(1) + \beta(1)\bar{p}_2(1)$$

which, noting that the sum of the coefficients in the utility functions are $5 + 2\alpha$ and $3 + 2\beta$, respectively, implies the following market balance conditions for date 1

$$\frac{4}{5+2\alpha}\frac{\hat{w}^a}{\beta(1)\bar{p}_1(1)} + \frac{1}{3+2\beta}\frac{\hat{w}^b}{\beta(1)\bar{p}_1(1)} = 2$$
$$\frac{1}{5+2\alpha}\frac{\hat{w}^a}{\beta(1)\bar{p}_2(1)} + \frac{2}{3+2\beta}\frac{\hat{w}^b}{\beta(1)\bar{p}_2(1)} = 2$$

Then, with $\hat{w} = \hat{w}^a = \hat{w}^b$,

$$[\bar{p}_1(1) - \bar{p}_2(1)] = \frac{1}{2\beta(1)}\hat{w}\left(\frac{3}{5+2\alpha} - \frac{1}{3+2\beta}\right)$$

Hence if $\beta = 1$ and

$$\frac{3}{5+2\alpha} - \frac{1}{5} \leq 0 \quad \text{or} \quad \alpha \geq 5$$

then $\bar{p}_1(1) - \bar{p}_2(1) \leq 0$ which contradicts our assumption that $\bar{p}_1(1) - \bar{p}_2(1) > 0$. Hence the economy can not have a Radner equilibrium where $\bar{p}_1(1) - \bar{p}_2(1) > 0$.

In an equilibrium $\bar{p}_1(1) - \bar{p}_2(1)$ is positive or nonpositive and each case leads to a contradiction. Hence there does not exist a Radner equilibrium for the economy.

□

Clearly, if $\beta = 1$ and $\alpha > 5$ then slight variations in endowments or the parameter a will not restore equilibrium since Walrasian demand, pure spot market demand and equilibrium spot-prices are continuous in the endowments.

Krasa [1989] pointed out that the original example of Polemarchakis and Ku [1990], formulated for an economy with two dates and uncertainty, relied on the feature that total initial endowment does not vary enough over the states of the world. He showed, for given utility functions and options as assets, that the fraction of economies having at least one Radner equilibrium tends to 1 as the variation in the total initial endowment increases. Since the example above is for an economy over time, there is indeed very little variation in total endowments at date 1. If the economy extended over several dates and the single asset was to pay $\max(p_1(t) - ap_2(t), 0)$ at each future date then, by an argument similar to Krasa's, enough variation in the total endowment over dates would result in price variations large enough to prevent the dividends from the asset to be the 0 vector and a pseudo equilibrium would necessarily be a Radner equilibrium of rank 1.

5.3 Nominal Assets and Existence of a Radner Equilibrium

In Section 5.1 we saw that an economy with real assets may fail to have a Radner equilibrium. The reason for this was brought out by Hart's example, which showed that there might be a drop in rank of the value dividend matrix, at the potential equilibrium prices. The drop in rank corresponds to an abrupt change in the possibilities of transferring income between states or dates. Since the elements in the value dividend matrix were linear functions of the spot-prices the drop in rank occurred for very few prices. The example of Polemarchakis and Ku shows that the *linear* dependence is essential.

For an economy with two dates and nominal assets the value dividend matrix is constant and the only source of variation in the possibilities of income transfers is from the variation of asset prices. Hence, under the Maintained Assumptions, it is to be expected that a Radner equilibrium does exist and furthermore that existence can be proved by means of a fixed point, or fixed point like, theorem.

We state the following theorem without proof.

Theorem 5.3.A *Let* $\mathcal{E} = ((C^i, u^i, e^i)_{i\in\mathbb{I}}, V)$ *be an economy with a nominal asset structure. There exists a Radner equilibrium,* $((\bar{c}^i, \bar{\theta}^i)_{i\in\mathbb{I}}, \bar{p}, \bar{q})$, *for* $\mathcal{E}$.

Werner [1985] has proved a theorem like Theorem 5.3.A, using the excess demand correspondence. Werner [1987] also proved a more general result, allowing for incomplete preferences, using methods originating in Mas-Colell [1974] and Mas-Colell and Gale [1975].

5.4 Indeterminacy of Radner Equilibrium with Nominal Assets

We have argued that existence of equilibrium is a necessary condition for a model to be a candidate as an explanation of prices and the allocation of resources, in an economy. On the other hand, it is desirable that the model should not have "too many" equilibria, since then prices and equilibrium actions are seriously indeterminate. This, in turn, implies that the model can only predict that the outcome belongs to a fairly large set. As pointed out earlier, it is too much to hope for uniqueness of equilibrium, and a finite number of equilibria is considered satisfactory.

With nominal assets, the existence of an equilibrium is ensured under weak assumptions on the economy but typically an economy with nominal assets will have a continuum of equilibria, so that there is indeterminacy. The indeterminacy concerns both prices and equilibrium actions. We begin by giving an example and go on to discuss the implications for resource allocation, in an economy with nominal assets.

An Example of Indeterminacy with Nominal Assets

Consider an economy $\mathcal{E}$ under uncertainty with an asset structure where $\mathbb{S} = \{1, 2\}, L = 1, \mathbb{I} = \{a, b, \}$ and 1 nominal asset, as described in Table 5.4.A.

Table 5.4.A: *Notation for the example with nominal assets*

Date	State	End. a	End. b	Cons.	Asset
0	0	4	4	$c^i(0)$	$-q$
1	1	6	3	$c^i(1)$	1
	2	3	6	$c^i(2)$	1

For $i \in \{a, b\}$, the consumption set is $C^i = \mathbb{R}^3_{++}$ and the utility function $u^i : C^i \longrightarrow \mathbb{R}$, is the same for both consumers,

$$u^i(c(0), c(1), c(2)) = \ln c(0) + \frac{1}{2} \ln c(1) + \frac{1}{2} \ln c(2)$$

The Consumer Problem is (dropping the index for the consumer)

$$\text{Max}_{(c,\theta)} \quad \ln c(0) + \frac{1}{2} \ln c(1) + \frac{1}{2} \ln c(2) \quad \text{s. to } c \in C$$

$$\left\{ \begin{array}{rcl} p(0)(c(0) - e(0) & \leq & r(0) \\ p(1)(c(1) - e(1)) & \leq & r(1) \\ p(2)(c(2) - e(2)) & \leq & r(2) \end{array} \right. \quad \text{and} \quad r = \begin{bmatrix} -q \\ 1 \\ 1 \end{bmatrix} \theta \qquad \textbf{5.4.A}$$

It is easy to see that (c, θ) satisfies the restrictions in 5.4.A at prices $(p(0), p(1), p(2), q)$ if and only if $(c, (q/p(0))\theta)$ satisfies the restrictions in 5.4.A at prices $(1, (q/p(0))p(1), (q/p(0))p(2), 1)$ and it follows that $\left[(\bar{c}^i, \bar{\theta}^i)_{i \in \mathbb{I}}, (p(0), p(1), p(2), q)\right]$ is a Radner equilibrium if and only if

$$(\bar{c}^i, \frac{q}{p(0)}\bar{\theta}^i)_{i\in\mathbb{I}}, (1, \frac{q}{p(0)}p(1), \frac{q}{p(0)}p(2), 1)$$

is a Radner equilibrium. Hence we may normalize prices so that $p(0) = q = 1$ and make a corresponding variation of the equilibrium portfolios but retain the same equilibrium consumptions. Since our interest is in Radner equilibria, where the equilibrium *allocations* differ we will from here on restrict attention to price systems $(p(0), p(1), p(2), q)$, where $p(0) = q = 1$.

Equilibrium portfolio choices

The Consumer Problem reduces to a choice of a portfolio by substitution in 5.4.A

$$\text{Max}_\theta \ \ln(e(0) - \theta) + \frac{1}{2}\ln\left(e(1) + \frac{\theta}{p(1)}\right) + \frac{1}{2}\ln\left(e(2) + \frac{\theta}{p(2)}\right) \quad \textbf{5.4.B}$$

$\bar{\theta}$ is a solution to the portfolio problem 5.4.B if and only if $\bar{\theta}$ satisfies the first-order condition for a maximum

$$\frac{1}{e(0)-\theta} - \frac{1}{2}\frac{1}{p(1)e(1)+\theta} - \frac{1}{2}\frac{1}{p(2)e(2)+\theta} = 0$$

Let $\bar{\theta}^a$ and $\bar{\theta}^b$ be the portfolios chosen in a Radner equilibrium. Then $\bar{\theta}^a + \bar{\theta}^b = 0$ by the market balance condition for the asset market and the following conditions are satisfied, with $\theta = \bar{\theta}^a = -\bar{\theta}^b$,

$$\begin{aligned} F_1(p(1), p(2), \theta) &= \frac{1}{4-\theta} - \frac{1}{2}\frac{1}{6p(1)+\theta} - \frac{1}{2}\frac{1}{3p(2)+\theta} = 0 \\ F_2(p(1), p(2), \theta) &= \frac{1}{4+\theta} - \frac{1}{2}\frac{1}{3p(1)-\theta} - \frac{1}{2}\frac{1}{6p(2)-\theta} = 0 \end{aligned} \quad \textbf{5.4.C}$$

Conversely, if θ satisfies 5.4.C then $-\theta$ satisfies 5.4.C and θ, $-\theta$ solve the consumers' portfolio problem.

If the asset market is balanced then, using the budget restrictions, it is seen that the each of the commodity markets also balances. It follows that if $(p(1), p(2), \bar{\theta})$ is a solution to 5.4.C, then $(\bar{c}^i, \bar{\theta}^i)_{i\in\mathbb{I}}, (1, p(1), p(2), 1)$, where $\bar{\theta}^a = \bar{\theta}$, $\bar{\theta}^b = -\bar{\theta}$ and

$$\bar{c}^a = \begin{bmatrix} e^a(0) - \bar{\theta}^a \\ e^a(1) + \dfrac{\bar{\theta}^a}{p(1)} \\ e^a(2) + \dfrac{\bar{\theta}^a}{p(2)} \end{bmatrix} \quad \text{and} \quad \bar{c}^b = \begin{bmatrix} e^b(0) - \bar{\theta}^b \\ e^b(1) + \dfrac{\bar{\theta}^b}{p(1)} \\ e^b(2) + \dfrac{\bar{\theta}^b}{p(2)} \end{bmatrix} \qquad \textbf{5.4.D}$$

is a Radner equilibrium. (Hence $\theta \in \,]-4, +4[$, since otherwise one of the equilibrium consumptions would fail to be positive in at least one coordinate.)

A continuum of equilibria

Substituting $p(1) = p(2) = 1$ and $\theta = 0$ into 5.4.C it is found that these values satisfy 5.4.C. We will show that for each θ close to 0 there is a unique solution $(p(1), p(2))$, close to $(1, 1)$. By the Implicit Function Theorem, Theorem C in Appendix A we can solve (locally) for $(p(1), p(2))$ as functions of θ if the Jacobian with respect to $p(1)$ and $p(2)$, evaluated at $(p(1), p(2), \theta) = (1, 1, 0)$, has full rank. This matrix is

$$\begin{bmatrix} \dfrac{\partial F_1(1,1,0)}{\partial p(1)} & \dfrac{\partial F_1(1,1,0)}{\partial p(2)} \\ \dfrac{\partial F_2(1,1,0)}{\partial p(1)} & \dfrac{\partial F_2(1,1,0)}{\partial p(2)} \end{bmatrix} = \frac{1}{2} \begin{bmatrix} \dfrac{1}{6^2} & \dfrac{1}{3^2} \\ \dfrac{1}{3^2} & \dfrac{1}{6^2} \end{bmatrix}$$

and the rank condition is clearly satisfied. Again we can apply the Implicit Function Theorem Appendix A, to conclude that there is a neighborhood $\mathcal{N}_0$ of $\theta = 0$ and a unique function $P \in \mathbb{C}^1(\mathcal{N}_0, \mathbb{R}^2)$ where $\theta \longrightarrow (p(1), p(2))$ such that $P(0) = (1, 1)$ and $(\theta, (p(1), p(2))) \in \mathcal{N}_0 \times \mathbb{R}^2_{++}$ satisfies 5.4.C if and only if

$$(p(1), p(2)) = P(\theta)$$

Cf. Figure 5.4.A. Hence there is a continuum of Radner equilibria parametrized by $\theta \in \mathcal{N}_0$. Since for each $\theta \in \mathcal{N}_0$, the equilibrium consumption of consumer a at date 0 is $c^a(0) = e^a(0) - \theta$, the equilibrium allocations differ for each of these equilibria.

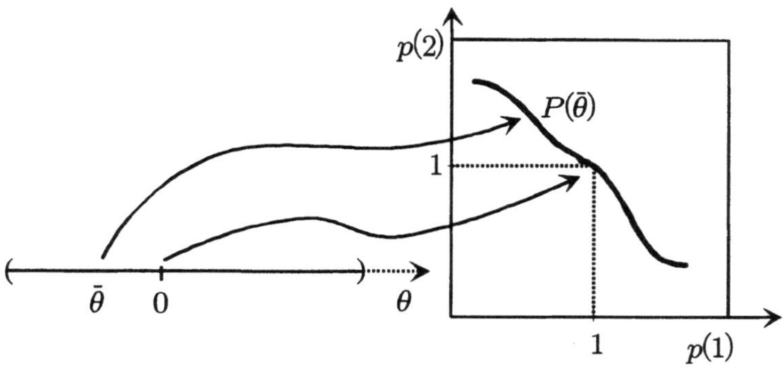

Figure 5.4.A: *Each value of $\theta \in \mathcal{N}_0$ gives a vector of equilibrium prices and the corresponding equilibrium allocations are distinct*

A General Theorem on Indeterminacy: Nominal Assets*

The situation encountered in the example of an economy with nominal assets and incomplete markets turns out to be typical. It is, perhaps, surprising that the degree of indeterminacy is unrelated to the degree of incompleteness of the asset markets. In relation to the theorem below we remark that a $S \times J$ matrix V, with $J < S$, is in *general position* if every $J \times J$ submatrix has rank J. The following theorem is due to Geanakoplos and Mas-Colell [1989]. There is a closely related result by Balasko and Cass [1989].

Theorem 5.4.A *Let $\mathbb{E} = \{(C^i, u^i, e^i)_{i \in \mathbb{I}}, V\}$ be a family of economies with nominal assets parametrized by the initial endowment, $e \in \mathbb{R}^{L(S+1)I}_{++}$. Let $\mathcal{E}(e)$ denote the economy with initial endowment e and assume*

(a) $0 < J < S$

(b) $I > J$

(c) *the matrix V is in general position.*

Then there is a generic set of endowments, $D \subset \mathbb{R}^{L(S+1)I}_{++}$, such that for each $e \in D$ there is a differentiable injective mapping defined on $\mathbb{R}^{S-1}$

whose image is contained in the equilibrium allocations of the economy $\mathcal{E}(e)$.

Intuitively, one might think of the equilibrium allocations as containing an open set of dimension $S-1$. In our example above $S-1=1$ and the mapping $\theta \longrightarrow (c^a, c^b)$ was injective since $c^a(0) = e^a(0) - \theta$.

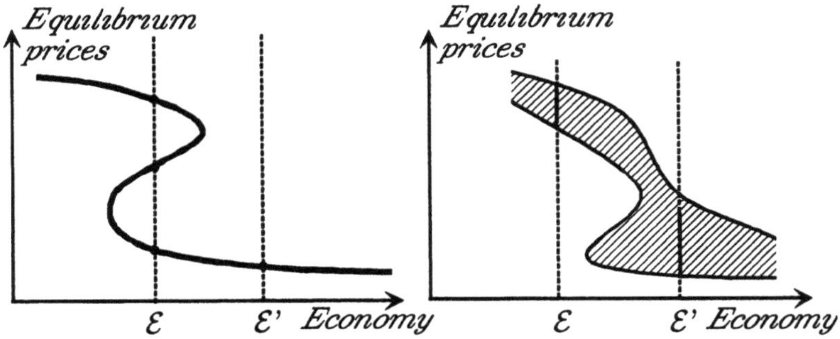

Figure 5.4.B: *In the left panel the economy $\mathcal{E}'$ has a single equilibrium price system, the economy $\mathcal{E}$ has three equilibrium price systems. In the right panel each economy has a continuum of equilibria*

Consequences of Indeterminacy

In an economy extending over time and, possibly, with uncertainty and no forward or contingency markets the agents have to act according to expectations if there is trade on the spot-markets.

Assume that the economy has a *single* equilibrium price system, made up of spot-prices at the current date and spot-prices, at future dates, see Figure 5.4.B, where this occurs for the economy $\mathcal{E}'$. In this case the assumption of common and correct expectations has some chance of being a reasonable approximation to real life. Consequently our models could serve to predict the outcome of the exchange process.

Assume now that the economy has *more than one* spot-market equilibrium price system, see Figure 5.4.B, where this occurs for the economy $\mathcal{E}$. Then it might be the case that even knowing spot-prices at the current date, there is more than one possible equilibrium spot-price system possible at future dates. Each of these future equilibrium spot-price system can be realized provided each consumer plans against this *same* future price system.

With several future spot-price systems possible there is nothing in the theory that forces all of the consumers to expect the same one. Assume that each consumer expects one of the equilibrium spot-price system, but that consumers expect different spot-price systems. Then, in general, the realized future spot-prices will not be among the equilibrium spot-price systems, at all. Hence there is a feed-back from the expectations to the realized outcome which is not modelled explicitly. In this case our model needs to specify in more detail how expectations are generated. The conditions for equilibrium may also need to be extended by conditions on expectations, which ensure that expectations are to some extent in accordance with realized prices. Models can be constructed where the current spot-prices reveal information about future spot-prices. This leads to a theory of rational expectations equilibria.

Example 5.4 shows that with nominal assets that there might be a *continuum of future spot-prices* which are consistent in the sense that if each consumer expects the same future spot-prices then these spot-prices may realize and give market balance in the future. This is illustrated in Figure 5.4.B. In our example the date 0 prices of the good and of the asset were normalized to be 1 and thus consumers could not get any information from the knowledge of current prices. Furthermore, there seems to be nothing else in the model which ensures that agents have common and correct expectations. If they act, at the current date, on different expectations about future spot-prices then the outcome may well be some future prices, which are not among the given (future) spot-market equilibrium price systems. Again there is a feed-back from the expectations to realized future spot-prices. Even if consumers have unlimited capacity to calculate and complete knowledge of the economy they can not predict the future spot-prices.

With incomplete markets the equilibrium allocation is in general not a Pareto optimal allocation. If the outcome of exchange was in some sense "better" when consumers had common and correct expectations than with differing expectations, then there would be a case for coordinating expectations. To study such problems we need other equilibrium concepts which would predict the outcome for a much larger class of expectations than perfect foresight or perfect foresight conditional on the state of the world realized.

Summary

In this chapter we have studied existence and determinacy of Radner equilibria in economies with real or nominal assets and incomplete asset markets.

Hart's example showed that, with real assets, one could not hope for existence of a Radner equilibrium in general. However, later results have shown that a Radner equilibrium will typically exist, so that if a given economy does not have an equilibrium, a small perturbation will give us an economy where a Radner equilibrium exists. Generic existence comes from the fact that the drop-in-rank of the value dividend matrix occurs for very few prices. Polemarchakis and Ku showed, by extending the concept of an asset somewhat, that non-existence could occur for a large set of parameters. With real assets an economy would typically have a finite number of equilibria.

For economies with nominal assets the existence problem was simpler. Under weak assumptions a Radner equilibrium has been shown to exist. We showed by an example that with nominal assets there could be a continuum of Radner equilibria. This turned out to be a general phenomena and the degree of indeterminacy was seen to equal the number of states of the world at date 1 less 1. The indeterminacy of equilibrium made the assumption of common and correct expectations particularly dubious.

Exercises

Exercise 5.A Consider an asset structure with real assets where $L = 1$. Show that the rank of the value dividend matrix is then constant with respect to (positive) spot-prices and thus the "drop-in-rank" problem does not occur.

(a) Consider an economy with an asset structure under uncertainty where $S = 3$. Assume that the only assets are Arrow-Debreu assets for states 1 and 2. Is the conclusion from Theorem 5.4.A valid?

(b) Consider the following matrices

$$A = \begin{bmatrix} 1 & 4 \\ 2 & 5 \\ 3 & 0 \end{bmatrix} \qquad B = \begin{bmatrix} 1 & 4 \\ 2 & 5 \\ 3 & 9 \end{bmatrix} \qquad C = \begin{bmatrix} 1 & 3 \\ 4 & 8 \\ 3 & 6 \end{bmatrix}$$

What is the rank of A, B and C? Are A, B and C in general position?

Exercise 5.B Exercise 5.C An application using a generic or typical property which is easy to describe, but whose proof is not trivial, is Borel's *normal number theorem.* A thorough discussion can be found in Billingsley [1986] .

Consider the interval $]0, 1]$ and the mapping taking the number $\alpha \in]0, 1]$ to an infinite sequence of 1's and 0's, $.d_1(\alpha)d_2(\alpha)d_3(\alpha)\ldots$, such that

$$\alpha = \sum_{n=1}^{\infty} \frac{d_n(\alpha)}{2^n}$$

Some numbers have two such dyadic expansions, like $\alpha = 1/2$, which has the expansions .10000 and .01111. Choose the expansion ending with 1's. Let D be the set of such dyadic expansions. A number $\alpha \in]0, 1]$ such that

$$\lim_{n\to\infty} \frac{1}{n} \sum_{i=1}^{n} d_i(\alpha) = \frac{1}{2}$$

is, by definition, a *normal number.* Let N be the set of normal numbers. Borel's normal number theorem asserts that typically a number is a normal number.

(a) The set N is uncountable. (*Hint:* Assume, in order to obtain a contradiction, that there is an enumeration of N. Then there is an enumeration

1	.0	δ_{12}	δ_{13}	δ_{14}	$\ldots$
2	$.\delta_{21}$	1	δ_{23}	δ_{24}	
3	$.\delta_{31}$	δ_{32}	0	δ_{34}	$\ldots$
4	$.\delta_{41}$	δ_{41}	δ_{41}	1	

such that the diagonal terms alternate between 0 and 1.)

(b) Consider the mapping

$$.d_1(\alpha)d_2(\alpha)d_3(\alpha)\ldots \longrightarrow .11d_1(\alpha)11d_2(\alpha)11d_3(\alpha)\ldots$$

taking a dyadic expansion to a new dyadic expansion where every third component is taken from the given dyadic expansion.

(c) Show that the mapping defined in (b) is injective and let D_3 denote its image. Show that D_3 is an uncountable set.

(d) Show that an element in D_3 is not a normal number.

(e) Let α be a normal number and let $\varepsilon > 0$. Show that there is a number α' which is not normal such that $|\alpha - \alpha'| < \varepsilon$.

(f) Let E be a countable subset of $]0,1]$. Given $\varepsilon > 0$, show that there is a countable set of intervals (not necessarily disjoint) $I_j =]a_j, b_j]$, for $j \in \mathbb{N}$, such that the sum of their lengths is smaller than ε, that is,

$$\textstyle\sum_{j\in\mathbb{N}}(b_j - a_j) \; < \; \varepsilon$$

and with the property $E \subset \cup_{j\in\mathbb{N}} I_j$.

One can show that, given $\varepsilon > 0$, there are intervals $I_j =]a_j, b_j]$, for $j \in \mathbb{N}$, of total length smaller than ε such that the complement of N is included in $\cup_{j\in\mathbb{N}} I_j$. Thus typically a dyadic expansion will have a frequency of 1's equal to 1/2. However, from (c), (d) and (e) above we know that there are uncountably many exceptions forming a dense subset of $]0,1]$.

Exercise 5.D Construct an economy over time, $\mathcal{E}$, with two consumers, $L = 2$ and $T = 1$ which has a unique Walras equilibrium (apart from price normalization) with equilibrium price system $(P_1(0), P_2(0), P_1(1), P_2(1)) = (1,1,1,1)$, where the equilibrium entails non-zero net income vectors for the consumers and which has a single pure spot market equilibrium with spot prices at date 1 such that $p_1(1) \neq p_2(1)$.

(a) Define a single real asset, to get an economy with an asset structure, so that the Walras equilibrium allocation can be realized as the equilibrium allocation for the economy with the asset structure.

(b) As in (a) define a single real asset, to get an economy with an asset structure, so that the Walras equilibrium allocation can not be realized as the equilibrium allocation for the economy with the given asset structure but such that a small perturbation of the initial endowments gives complete markets.

(c) Show that if the asset is chosen suitably the economy with an asset structure, from (b), will not have a Radner equilibrium.

(d) Modify the asset from (b) to an optionlike "asset" such that a small perturbation of the initial endowment(s) will not restore the equilibrium.

Exercise 5.E Consider the family of economies with an asset structure $\mathcal{E}_\pi = \{(C^i, u^i, e^i)_{i\in\{a,b\}}, V_\pi\}$ where $\pi > 0$. For $i = a, b$, the consumption set $C^i = \mathbb{R}^3_{++}$ and $u^i : C^i \longrightarrow \mathbb{R}$ is given by

$$u^i(c) = c(0) + \frac{1}{2}\ln c(1) + \frac{1}{2}\ln c(2)$$

We interpret the economy as an economy over time with $T = 2$ and $L = 1$. The initial endowments, which are independent of π, and the single nominal asset are given by

Date	End. a	End. b	Asset
0	4	4	$-q$
1	6	3	1
2	3	6	π

(a) State the conditions for a Radner equilibrium for $\mathcal{E}_\pi$.

(b) Show that the Consumer Problem can be reduced to the choice of a portfolio and derive the first order condition for optimal portfolio choice. (This presumes that the problem has a solution.)

(c) Show that if $\bar{\theta}^a$ and $\bar{\theta}^b$ are solutions to the consumer portfolio choice problem and $\bar{\theta}^a + \bar{\theta}^b = 0$ then $(\bar{\theta}^a, \bar{\theta}^b)$ induces a Radner equilibrium.

(d) Show that $\mathcal{E}_\pi$ with $\pi = 1$ has a Radner equilibrium

$$(\bar{p}(0), \bar{p}(1), \bar{p}(2), \bar{q}, (\bar{c}^i, \bar{\theta}^i)_{i \in \mathbb{I}})$$

with $((\bar{p}(0), \bar{p}(1), \bar{p}(2)) = (1, 1, 1)$ and $\bar{q} = 1/4$.

(e) Apply the Implicit Function Theorem to conclude that there is an interval, $A \subset \mathbb{R}_{++}$, with $1 \in A$, such that each economy $\mathcal{E}_\pi$, for $\pi \in A$, has a Radner equilibrium with $((\bar{p}(0), \bar{p}(1), \bar{p}(2)) = (1, 1, 1)$. Are the equilibrium allocations different for different values of π?

(f) Show that each Radner equilibrium, from (e), for $\mathcal{E}_\pi$ induces a Radner equilibrium for $\mathcal{E}_1$ with $\bar{p}(0) = \bar{p}(1) = 1$ and $\bar{p}(2) = 1/\pi$

(g) Conclude that $\mathcal{E}_1$ has a continuum of equilibria.

Date	State	Asset 1	Asset 2	Ext. value div. vector 1	Ext. value div. vector 2
0	0	–	–	$-q_1$	$-q_2$
1	1	$\begin{bmatrix} 1 \\ 0 \end{bmatrix}$	$\begin{bmatrix} 0 \\ 1 \end{bmatrix}$	$p_1(1)$	$p_2(1)$
	2	$\begin{bmatrix} 1 \\ 0 \end{bmatrix}$	$\begin{bmatrix} 0 \\ 1 \end{bmatrix}$	$p_1(2)$	$p_2(2)$

Exercise 5.F Hart [1975] considered an economy with an asset structure with two dates, two states $S = 2$, two goods, two real assets and two consumers; a and b. Neither consumer has any initial endowment at date 0. Hence the only possible trade at date 0 is in the assets. The first asset is a contract for the sure delivery of one unit of good 1 at date 2 and the second asset is a contract for the sure delivery of one unit of good 2 at date 2. Thus we have the situation depicted in the table above.

(a) How should a Radner equilibrium be defined in this setting (with no consumption at date 0)?

(b) Show that the rank of the extended value dividend matrix is 2 if and only if $(p_1(1), p_1(2))$ is not proportional to $(p_2(1), p_2(2))$.

(c) Assume that the rank of the extended value dividend matrix is 1. Show that the date 0 prices of the assets must be positive in an equilibrium.

(d) Assume that the rank of the extended value dividend matrix is 1. Show that the portfolios for both consumers must be 0 in an equilibrium.

(e) Let $\mathcal{E}$ be an economy having a unique Walras equilibrium and where $(P_1(1), P_1(2)) = (P_2(1), P_2(2))$ in the Walras equilibrium. Give conditions on $\mathcal{E}$ which either ensure that the Walras equilibrium for $\mathcal{E}$ induces a Radner equilibrium for $\mathcal{E}$, with the asset structure as given or prevents the Walras equilibrium from inducing a Radner equilibrium for $\mathcal{E}$.

(f) Assume that the economy from (e) is such that (i) the unique Walras equilibrium implies non-zero net income vectors for the consumers and (ii) with net income vectors equal to 0 the economy has a single spot-market equilibrium with spot price $(\bar{p}_1(1), \bar{p}_1(2))$ not proportional to $(\bar{p}_2(1), \bar{p}_2(2))$. Explain why there does not exist a Radner equilibrium for $\mathcal{E}$.

Chapter 6

OPTIMALITY AND INCOMPLETE MARKETS

Introduction

With incomplete asset markets the *conclusions of the First and Second Theorem* of Welfare economics are no longer true. An example by Hart [1975] shows that it is even possible to have Pareto dominance among the equilibrium allocations. The possible non-optimality of equilibrium allocations is related to the restricted possibilities of transferring income between dates or states of Nature. It presents a difficulty for the theory that Pareto optimality is linked to the allocation of commodities, while the restrictions in exchange are linked to the transfer of income. In Section 6.1 we give an example of Pareto dominance among the equilibrium allocations in the spirit of Hart.

We saw in Chapter 2 that financial decisions are irrelevant when asset markets are complete. This was not surprising since a Radner equilibrium in that case induced a Walras equilibrium where financial considerations are implicit. When asset markets are incomplete the agents will *evaluate net income vectors differently.* In Section 6.2 we use the example from Section 5.4 to show that even if the economy has a continuum of equilibrium allocations, each of these may fail to be a Pareto optimal allocation.

In Section 6.3 we show that Radner equilibrium allocations do have some restricted optimality properties and shortly discuss the concept of weak constrained efficiency introduced by Grossman [1977].

At the end of the chapter there is a summary of results for economies with complete and incomplete asset markets.

6.1 Hart's Example of Pareto Domination

Not only can Radner equilibrium allocations fail to be Pareto optimal allocations but it is possible to construct an example of an economy with, at least, two Radner equilibria such that one equilibrium allocation Pareto dominates another. The example is due to Hart [1975].

Hart's construction makes use of an exchange economy, with 2 consumers and 2 goods, which has, at least, two Walras equilibria. Mas-Colell et. al [1995, p.521] give an explicit example of such an economy which was part of Exercise 1.G. We will use that example as a starting point. Note however that it does not quite satisfy the Maintained Assumptions.

Example 6.1.A An economy with three Walras equilibria

Let $\mathcal{E}'$ be the economy from Exercise 1.G with commodity space $\mathbb{R}^2$, and two consumers, a and b. $\hat{C}^a = \hat{C}^b = \mathbb{R}^2_{++}$ and

$$\begin{aligned} \hat{u}^a(c_1^a, c_2^a) &= c_1^a - \frac{1}{8}(c_2^a)^{-8} \quad \text{and} \quad (\hat{e}_1^a, \hat{e}_1^a) = (2, \eta) \\ \hat{u}^b(c_1^b, c_2^b) &= -\frac{1}{8}(c_1^b)^{-8} + c_2^b \quad \text{and} \quad (\hat{e}_1^b, \hat{e}_1^b) = (\eta, 2) \end{aligned}$$

where $\eta = 2^{8/9} - 2^{1/9}$. These utility functions fail to satisfy (C3) of the Maintained Assumptions.

A calculation shows that the total demand for good 2 is

$$\left(\frac{p_2}{p_1}\right)^{-\frac{1}{9}} + 2 + \eta\left(\frac{p_1}{p_2}\right) - \left(\frac{p_1}{p_2}\right)^{\frac{8}{9}}$$

and total supply of good 2, is $2 + \eta$.

It is easy to check that demand, for good 2, equals supply if and only if $p_1/p_2 \in \{2, 1, 1/2\}$, which implies that the economy has precisely three

Walras equilibria. In fact, we need only two of these. We will restrict attention to the two Walras equilibria for $\mathcal{E}'$ given by

$$((\tilde{c}^a, \tilde{c}^b), (\tilde{p}_1, \tilde{p}_2) \quad \text{with} \quad (\tilde{p}_1, \tilde{p}_2) = (2, 1)$$
$$((\bar{c}^a, \bar{c}^b), (\bar{p}_1, \bar{p}_2)) \quad \text{with} \quad (\bar{p}_1, \bar{p}_2) = (\frac{1}{2}, 1)$$

Cf. Figure 6.1.A where the equilibrium allocations are marked with a thick dot and a square.

□

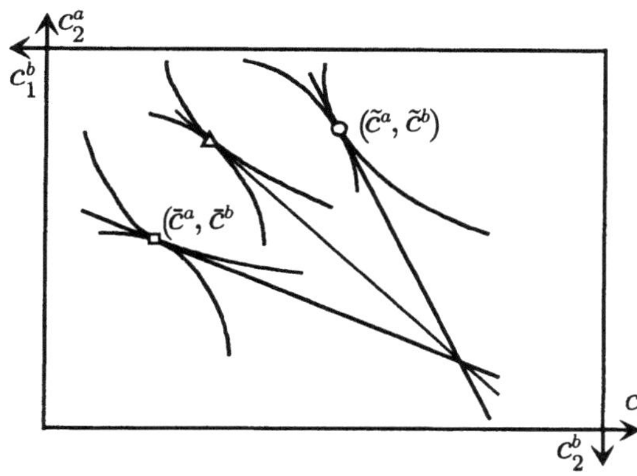

Figure 6.1.A: *The economy $\mathcal{E}'$ has three Walras equilibria. In the sequel we use only two of these*

Using the economy $\mathcal{E}'$ to construct the economy $\mathcal{E}$

We will use the economy $\mathcal{E}'$, from Example 6.1.A above to construct an economy, $\mathcal{E}$, with $T = 1$ and $L = 2$. The idea is essentially to "replicate" the economy in Example 6.1.A over two dates. Cf. Figure 6.1.B.

Date	End. a	End. b	Cons. i	Prices
0	$e^a(0) = \hat{e}^a$	$e^b(0) = \hat{e}^b$	$(c_1^i(0), c_2^i(0))$	$(p_1(0), p_2(0))$
1	$e^a(1) = \hat{e}^a$	$e^b(1) = \hat{e}^b$	$(c_1^i(1), c_2^i(1))$	$(p_1(1), p_2(1))$

Consumer a and b have consumption sets $C^a = C^b = \mathbb{R}^4_{++}$. We assume that there are no assets in the economy and study only pure spot market equilibria.

We now use the utility functions from the economy $\mathcal{E}'$ to define utility functions for the consumers in the economy $\mathcal{E}$

$$
\begin{aligned}
u^a(c^a(0), c^a(1)) &= \gamma^a \hat{u}^a(c^a(0)) + (1-\gamma^a)\hat{u}(c^a(1)) \\
u^b(c^b(0), c^b(1)) &= \gamma^b \hat{u}^b(c^b(0)) + (1-\gamma^b)\hat{u}^b(c^b(1))
\end{aligned}
$$

where $\gamma^a, \gamma^b \in \,]0,1[$.

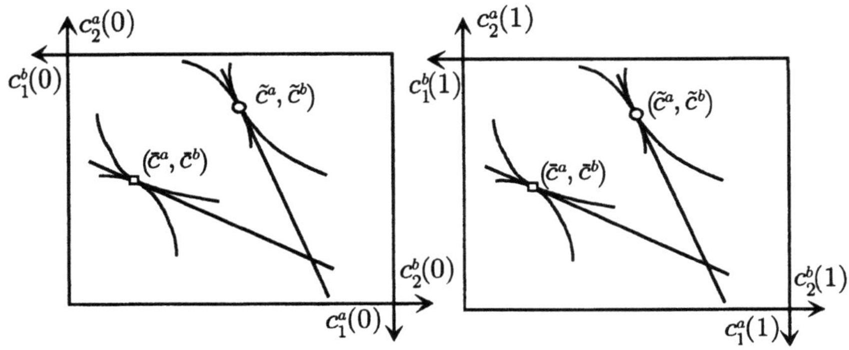

Figure 6.1.B: *Using two of the Walras equilibria from the economy $\mathcal{E}'$ in* Example 6.1.A *an economy $\mathcal{E}$ with four spot-market equilibria is constructed*

The pure spot market consumer problem

Consider the utility maximization problem of consumer a.

$$
\text{Max}_{(c^a(0),c^a(1))} \quad u^a(c^a(0), c^a(1)) \quad \text{s. to } c^a \in \mathbb{R}^4_{++} \text{ and}
$$
$$
\begin{cases}
p_1(0)c_1^a(0) + p_2(0)c_2^a(0) \leq p_1(0)e_1^a(0) + p_2(0)e_2^a(0) \\
p_1(1)c_1^a(1) + p_2(1)c_2^a(1) \leq p_1(1)e_1^a(1) + p_2(1)e_2^a(1)
\end{cases}
\tag{6.1.A}
$$

Since u^a is separable, $(\hat{c}^a(0), \hat{c}^a(1))$ is a solution if and only if $\hat{c}^a(t)$ is a solution to a's date t problem, for $t = 0, 1$.

$$
\text{Max}_{c^a(t)} \quad \hat{u}^a(c^a(t)) \quad \text{s. to } c^a(t) \in \mathbb{R}^2_{++} \text{ and}
$$
$$
p_1(t)c_1^a(t) + p_2(t)c_2^a(t) \leq p_1(t)e_1^a(t) + p_2(t)e_2^a(t)
\tag{6.1.B}
$$

A similar result is true for consumer b.

Two pure spot market equilibria for $\mathcal{E}$

Now we can use the two Walras equilibria from $\mathcal{E}'$ to construct four spot market equilibria for our economy but we will concentrate on two of them. Cf. Figure 6.1.B.

Using the relations 6.1.A and 6.1.B it follows that the consumptions

$$\begin{aligned}(c^a(0), c^a(1)) &= (\tilde{c}^a, \bar{c}^a)\\ (c^b(0), c^b(1)) &= (\tilde{c}^b, \bar{c}^b)\end{aligned}$$

form an equilibrium allocation relative to the prices

$$(p_1(0), p_2(0), p_1(1), p_2(1)) = (\tilde{p}_1, \tilde{p}_2, \bar{p}_1, \bar{p}_2)$$

and the consumptions

$$\begin{aligned}(c^a(0), c^a(1)) &= (\bar{c}^a, \tilde{c}^a)\\ (c^b(0), c^b(1)) &= (\bar{c}^b, \tilde{c}^b)\end{aligned}$$

form an equilibrium allocation relative to the prices

$$(p_1(0), p_2(0), p_1(1), p_2(1)) = (\bar{p}_1, \bar{p}_2, \tilde{p}_1, \tilde{p}_2,)$$

Pareto dominance among the equilibrium allocations

Assume, without loss of generality, that $\hat{u}^a(\tilde{c}^a) > \hat{u}^a(\bar{c}^a)$. If also $\hat{u}^b(\tilde{c}^b) \geq \hat{u}^b(\bar{c}^b)$ then $(\tilde{c}^a, \tilde{c}^b)$ would Pareto dominate $(\bar{c}^a, \bar{c}^b)$. However, this can not be true, since by the First Theorem on Welfare Economics $(\tilde{c}^a, \tilde{c}^b)$ and $(\bar{c}^a, \bar{c}^b)$ are both Pareto optimal allocations in $\mathcal{E}'$. It follows that $\hat{u}^b(\tilde{c}^b) < \hat{u}^b(\bar{c}^b)$. The same conclusion could be derived by direct computation in Example 6.1.A.

Despite the fact that our economy does not satisfy all the Maintained Assumptions we get the following

Proposition 6.1.A *Consider the economy $\mathcal{E}$ as defined above. For $\gamma^a \in \left]\frac{1}{2}, 1\right[$ and $\gamma^b \in \left]0, \frac{1}{2}\right[$ the pure spot market equilibrium allocation*

$$\begin{cases}(c_1^a(0), c_2^a(1)) = (\tilde{c}^a, \bar{c}^a)\\ (c_1^b(0), c_2^b(1)) = (\tilde{c}^b, \bar{c}^b)\end{cases}$$

Pareto dominates the pure spot market equilibrium allocation

$$\begin{cases} (c_1^a(0), c_2^a(1)) &= (\bar{c}^a, \tilde{c}^a) \\ (c_1^b(0), c_2^b(1)) &= (\bar{c}^b, \tilde{c}^b) \end{cases}$$

Proof: We have

$$\gamma^a \hat{u}^a(\tilde{c}^a) + (1-\gamma^a)\hat{u}^a(\bar{c}^a) > \gamma^a \hat{u}^a(\bar{c}^a) + (1-\gamma^a)\hat{u}^a(\tilde{c}^a)$$

if and only if

$$\gamma^a[\hat{u}^a(\tilde{c}^a) - \hat{u}^a(\bar{c}^a)] > (1-\gamma^a)[\hat{u}^a(\tilde{c}^a) - \hat{u}^a(\bar{c}^a)]$$

Since $\hat{u}^a(\tilde{c}^a) - \hat{u}^a(\bar{c}^a) > 0$ the last relation is obviously true, for $\gamma^a \in \left]\frac{1}{2}, 1\right[$.

We also have

$$\gamma^b \hat{u}^b(\tilde{c}^b) + (1-\gamma^b)\hat{u}^b(\bar{c}^b) > \gamma^b \hat{u}^b(\bar{c}^b) + (1-\gamma^b)\hat{u}^b(\tilde{c}^b)$$

if and only if

$$(1-\gamma^b)[\hat{u}^b(\bar{c}^b) - \hat{u}^b(\tilde{c}^b)] > \gamma^b[\hat{u}^b(\bar{c}^b) - \hat{u}^b(\tilde{c}^b)]$$

Since $\hat{u}^b(\bar{c}^b) - \hat{u}^b(\tilde{c}^b) > 0$ the last relation is obviously true, for $\gamma^b \in \left]0, \frac{1}{2}\right[$. It follows that for $\gamma^a \in \left]\frac{1}{2}, 1\right[$ and $\gamma^b \in \left]0, \frac{1}{2}\right[$

$$u^a(\tilde{c}^a, \bar{c}^a) > u^a(\bar{c}^a, \tilde{c}^a)$$
$$u^b(\tilde{c}^b, \bar{c}^b) > u^b(\bar{c}^b, \tilde{c}^b)$$

which is the desired conclusion

6.2 Abundance of Non-optimal Equilibrium Allocations

In Chapter 5 we gave an example of an economy with nominal assets which had a continuum of Radner equilibria. There the asset market was incomplete and we could not expect each equilibrium allocation to be a Pareto optimal allocation. Still it is perhaps a bit surprising that none of the equilibrium allocations is a Pareto optimal allocation. That this is indeed the case can be seen as follows.

Recall that for some neighborhood of $\theta = 0$, , we could find the equilibrium prices $(p(1), p(2))$ as functions of $\theta \in \mathcal{N}_0$. For any Radner equilibrium the equilibrium allocation, $(\bar{c}^a, \bar{c}^b)$, and the equilibrium portfolios $(\bar{\theta}^a, \bar{\theta}^b)$ were related through

$$\bar{c}^a = \begin{bmatrix} e^a(0) - \bar{\theta}^a \\ e^a(1) + \dfrac{\bar{\theta}^a}{p(1)} \\ e^a(2) + \dfrac{\bar{\theta}^a}{p(2)} \end{bmatrix} \quad \text{and} \quad \bar{c}^b = \begin{bmatrix} e^b(0) - \bar{\theta}^b \\ e^b(1) + \dfrac{\bar{\theta}^b}{p(1)} \\ e^b(2) + \dfrac{\bar{\theta}^b}{p(2)} \end{bmatrix} \qquad \textbf{6.2.A}$$

By Corollary 1.2.H an allocation, $(\bar{c}^a, \bar{c}^b)$, given by 6.2.A, is a Pareto optimal allocation if and only if the gradients, evaluated at $\bar{c}^a$ and $\bar{c}^b$, of the utility functions are proportional, that is, for some $\lambda \in \mathbb{R}_{++}$

$$\operatorname{grad} u^a(\bar{c}^a) = \begin{bmatrix} \dfrac{1}{4-\theta} \\ \dfrac{1}{2}\dfrac{1}{6p(1)+\theta} \\ \dfrac{1}{2}\dfrac{1}{3p(2)+\theta} \end{bmatrix} = \lambda \begin{bmatrix} \dfrac{1}{4+\theta} \\ \dfrac{1}{2}\dfrac{1}{3p(1)-\theta} \\ \dfrac{1}{2}\dfrac{1}{6p(2)-\theta} \end{bmatrix} = \lambda \operatorname{grad} u^a(\bar{c}^b)$$

Note that for θ sufficiently close to 0

$$\theta \le 0 \quad \text{implies} \quad \begin{cases} \dfrac{1}{4-\theta} \le \dfrac{1}{4+\theta} \\ \dfrac{1}{2}\dfrac{1}{3p(1)+\theta} > \dfrac{1}{2}\dfrac{1}{6p(1)-\theta} \end{cases}$$

and

$$\theta \ge 0 \quad \text{implies} \quad \begin{cases} \dfrac{1}{4-\theta} \ge \dfrac{1}{4+\theta} \\ \dfrac{1}{2}\dfrac{1}{6p(2)+\theta} < \dfrac{1}{2}\dfrac{1}{3p(2)-\theta} \end{cases}$$

Hence the gradients can not be proportional and there is no Pareto optimal allocation among the equilibrium allocations.

Thus there is a continuum of Radner equilibria but no equilibrium allocation is a Pareto optimal allocation. An intuitive way to understand this is through the evaluation of net income vectors. The trade in assets, when there are incomplete asset markets, will not result in an equalization at any of the equilibrium allocations although there is a continuum of equilibrium allocations.

6.3 Welfare Properties of Spot-Market Equilibria

We have argued in Chapter 2 that the basic reason for the occurrence of spot market equilibrium allocations, which fail to be Pareto optimal allocations, is the restrictions on the possibilities of income transfers. Proposition 6.3.A below shows that, at least, spot market equilibrium allocations are market-by-market Pareto optimal allocations.

Market-by-market optimality

Proposition 6.3.A *Let $((\bar{c}^i)_{i\in\mathbb{I}}, p)$ be a spot market equilibrium for $\mathcal{E}$ relative to $\mathbb{M}$. Choose $\tau \in \mathbb{T}$. There does not exist an allocation $(\hat{c}^i)_{i\in\mathbb{I}}$ which agrees with $(\bar{c}^i)_{i\in\mathbb{I}}$ for $t \in \mathbb{T}$ and $t \neq \tau$, such that $(\hat{c}^i)_{i\in\mathbb{I}}$ Pareto dominates $(\bar{c}^i)_{i\in\mathbb{I}}$.*

Proof: Assume, in order to obtain a contradiction, that there is an allocation $(\hat{c}^i)_{i\in\mathbb{I}}$, which Pareto dominates $(\bar{c}^i)_{i\in\mathbb{I}}$ and such that $(\hat{c}^i)_{i\in\mathbb{I}}$ agrees with $(\bar{c}^i)_{i\in\mathbb{I}}$ for $t \in \mathbb{T}$ and $t \neq \tau$. Then

$$
\begin{aligned}
u^i((\bar{c}^i(t))_{t\neq\tau}, \hat{c}^i(\tau)) &\geq u^i((\bar{c}^i(t))_{t\neq\tau}, \bar{c}^i(\tau)) \quad \text{for } i \in \mathbb{I} \quad \text{and} \\
u^i((\bar{c}^i(t))_{t\neq\tau}, \hat{c}^i(\tau)) &> u^i((\bar{c}^i(t))_{t\neq\tau}, \bar{c}^i(\tau)) \quad \text{for some } i \in \mathbb{I}
\end{aligned}
\qquad \textbf{6.3.A}
$$

Consider the exchange economy, $\mathcal{E}'$, induced by $\mathcal{E}$ and the equilibrium allocation, $(\bar{c}^i)_{i\in\mathbb{I}}$. The commodity space of $\mathcal{E}'$ is $\mathbb{R}^L$ and the consumers are defined by, for $i \in \mathbb{I}$,

$$
\begin{aligned}
\hat{C}^i &= \mathbb{R}^L_{++} \\
\hat{u}^i(c^i(\tau)) &= u^i(\bar{c}^i(0), \ldots, \bar{c}^i(\tau-1), c^i(\tau), \bar{c}^i(\tau+1), \ldots, \bar{c}^i(T)) \\
\hat{e}^i &= e^i(\tau)
\end{aligned}
$$

It is easy to see that $((\bar{c}^i(\tau))_{i\in\mathbb{I}}, p(\tau))$ is a Walras equilibrium for $\mathcal{E}'$. By the First Theorem of Welfare Economics the allocation, $(\bar{c}^i(\tau))_{i\in\mathbb{I}}$, is a Pareto optimal allocation in $\mathcal{E}'$. This is a contradiction to 6.3.A, which asserts that, for $\mathcal{E}'$, the allocation, $(\hat{c}^i(\tau))_{i\in\mathbb{I}}$, Pareto dominates the allocation, $(\bar{c}^i(\tau))_{i\in\mathbb{I}}$. We conclude that $(\bar{c}^i(\tau))_{i\in\mathbb{I}}$ is a Pareto optimal allocation for $\mathcal{E}'$.

Pareto optimality, net income vectors and portfolio choice

Hart's example illustrates the fairly weak conclusion of Proposition 6.3.A. Although we have market-by-market Pareto optimality, we can still have that one spot market equilibrium allocation Pareto dominates another.

With incomplete markets there are restrictions on the possibilities of transferring income. Given spot-prices, $(p(t))_{t\in\mathbb{T}}$, and a set of feasible income vectors, $\mathbb{M}$, define a set of net income vectors, $(r^i)_{i\in\mathbb{I}}$, to be a *net income allocation* in $\mathbb{M}$ if $\Sigma_{i\in\mathbb{I}} r^i = 0$ and $r^i \in E^i(p) \cap \mathbb{M}$, for $i \in \mathbb{I}$. A net income allocation, $(\bar{r}^i)_{i\in\mathbb{I}}$, is a *Pareto optimal net income allocation* in $\mathbb{M}$, if there is no net income allocation $(\hat{r}^i)_{i\in\mathbb{I}}$ in $\mathbb{M}$ such that

$$\begin{cases} v^i(p,\hat{r}^i) \geq v^i(p,\bar{r}^i) & \text{for } i \in \mathbb{I} \quad \text{and} \\ v^i(p,\hat{r}^i) > v^i(p,\bar{r}^i) & \text{for some } i \in \mathbb{I} \end{cases}$$

Proposition 6.3.B shows that, although income transfers may be restricted, the exchange of spot income results in a net income allocation, which is a Pareto optimal net income allocation relative to the set of income tranfers. Since the net income vectors in $\mathbb{M}$ can be freeely traded among the agents their subjective evaluations of these income vectors are equalized.

Proposition 6.3.B *Let $((\bar{c}^i)_{i\in\mathbb{I}}, p)$ be a spot market equilibrium for $\mathcal{E}$ relative to $\mathbb{M}$. If $\mathbb{M}$ is a linear subspace of $\mathbb{R}^{S+1}$ then the net income allocation, $r^i(s) = p(s)(\bar{c}^i(s) - e^i(s))$ for $s \in \mathbb{S}$ and $i \in \mathbb{I}$, is a Pareto optimal net income allocation in $\mathbb{M}$.*

Proof: Let $(\bar{r}^i)_{i\in\mathbb{I}}$ be the net income allocation induced by $((\bar{c}^i)_{i\in\mathbb{I}}, p)$. Since preferences are monotone, we have

$$\left(\bar{r}^i(s)\right)_{s\in\mathbb{S}} = \left(p(s)(\bar{c}^i(s) - e^i(s))\right)_{s\in\mathbb{S}} \in E^i(p) \cap \mathbb{M} \quad \text{for } i \in \mathbb{I}$$

Assume, in order to obtain a contradiction, that there is a net income allocation $(\hat{r}^i)_{i\in\mathbb{I}} = (\hat{r}^1, \hat{r}^2, ..., \hat{r}^I)$, which Pareto dominates the net income allocation, $(\bar{r}^i)_{i\in\mathbb{I}}$, from the equilibrium. We have $v^i(p, \hat{r}^i) \geq v^i(p, \bar{r}^i)$ for $i \in \mathbb{I}$, and without loss of generality we may assume that $v^1(p, \hat{r}^1) > v^1(p, \bar{r}^1)$.

Since $(\hat{r}^i)_{i\in\mathbb{I}}$ is feasible we have $\hat{r}^1 = -(\Sigma_{i=2}^I \hat{r}^i)$. The net income vectors $\hat{r}^i$, $i = 2, 3, ...I$, all belong to $\mathbb{M}$. Since $\mathbb{M}$ is a *linear* subspace $\hat{r}^1 = -(\Sigma_{i=2}^I \hat{r}^i) \in \mathbb{M}$. On the other hand, from the definition of a spot market equilibrium relative to $\mathbb{M}$ and the decomposition of the consumer problem, we have $v^1(p, \bar{r}^1) \geq v^1(p, \hat{r}^1)$, since $\hat{r}^1 \in \mathbb{M}$. This contradicts $v^1(p, \hat{r}^1) > v^1(p, \bar{r}^1)$. Hence $(\bar{r}^i)_{i\in\mathbb{I}}$ is a Pareto optimal net income allocation in $\mathbb{M}$.

□

Proposition 6.3.B carries over to the choice of portfolios.

Corollary 6.3.C *Let $((\bar{c}^i, \bar{\theta}^i)_{i\in\mathbb{I}}, p, q)$ be a Radner equilibrium for the economy with an asset structure $\mathcal{E} = ((C^i, u^i, e^i)_{i\in\mathbb{I}}, V)$. Then there does not exist $\hat{\theta}^i \in \mathbb{R}^J$, for $i \in \mathbb{I}$, with $\Sigma_{i\in\mathbb{I}}\, \hat{\theta}^i = 0$ such that*

$$\begin{cases} v^i(p, W\hat{\theta}^i) \geq v^i(p, W\bar{\theta}^i) & \textit{for } i \in \mathbb{I} \quad \textit{and} \\ v^i(p, W\hat{\theta}^i) > v^i(p, W\bar{\theta}^i) & \textit{for some } i \in \mathbb{I} \end{cases}$$

This follows immediately from Proposition 6.3.B, since the net income allocation $(\bar{r}^i)_{i\in\mathbb{I}}$, with $\bar{r}^i = W\bar{\theta}^i$ for $i \in \mathbb{I}$, is a Pareto optimal net income allocation in $\mathbb{M} = \langle W \rangle$.

Weak constrained Pareto optimality

Grossman [1977] has defined a constrained Pareto optimality concept, *weak constrained Pareto optimality*, where redistributions over states of Nature, can take place only using the (real) asset structure. With this optimality concept there is a relation between Radner equilibrium allocations and weakly constrained Pareto optimal allocations. A Radner equilibrium allocation is a weak constrained Pareto optimal allocation and a weak constrained Pareto optimal allocation is Radner equilibrium allocation, with the appropriate redistribution of initial resources and a suitable choice of portfolios.

The difficulty that occurs in defining optimality concepts is that the agents trade in the assets, not for final consumption, but as a means to transfer income between states or dates. If trade was restricted to the commodity bundles defined by the real assets and retrade on the spot-markets was not allowed, then we would expect an equilibrium allocation to be Pareto optimal allocation in a restricted sense. The relevant constraints would arise from the asset structure.

Summary

The First Theorem of Welfare Economics states that the use of prices as an institution to allocate commodities has at least one desirable property. The (Walras) equilibrium allocations are Pareto optimal allocations. Alternatively, we could say that goods are not wasted so that the utility levels attained by the agents in an equilibrium could not have been achieved with a smaller total initial endowment for the economy. Of course, the theorem is valid for economies over time and/or with uncertainty but careful examination of the required market structure indicates that the result may be less relevant once time and uncertainty are taken into account.

While the First Theorem is related to a particular set of institutions, although somewhat incompletely specified, and the related equilibrium concept the Second Theorem concerns Pareto optimal allocations which are thought of as free of "institutions".

In the theory of incomplete asset markets the distinction between equilibrium and the concept of Pareto optimality is to some extent muddled. This is related to the fact that the possible reallocations are done through the use of assets and this implies that there will be an interplay between the possible reallocations and the institution of "prices".

We saw that this interplay might result in equilibrium allocations that Pareto dominate each other. In this case we considered a pure spot market equilibrium which did not admit the agents to make any trades to equalize their evaluation of net income vectors for the two dates.

We also gave an example to show that even in the case of a continuum of equilibrium allocations it was possible that none of them were a Pareto optimal allocation. Again this was seen to be a consequence of the asset structure which for none of the equilibria allowed consumers to equalize their evaluation of net income vectors.

A Summary of Results for Economies with Complete or Incomplete Markets

The following table summarizes the results on incomplete markets in Chapter 5 to 6. Many of the properties are true only generically which we have indicated with a "*".

	Assets	Exist.	Determ.	Pareto opt.
Gen. compl.	Real	Yes*	Yes*	Yes*
Complete	Nominal	Yes	Yes*	Yes
Incomp.	Real	Yes*	Yes*	No*
	Nominal	Yes(*)	No*	No*

With generically complete asset markets, in the case of real assets, or complete, in the case of nominal assets the results from general equilibrium theory are essentially unchanged. Typically an economy has a finite number of equilibria and the equilibrium allocations are all Walras equilibrium allocations and hence Pareto optimal allocations. Mainly one has to be careful with the case of real assets where the completeness of the asset markets is itself a generic property.

With incomplete asset markets and real assets existence is ensured generically (in endowments and the asset structure). This is also the case for nominal assets but if frequent trading is considered then also in this case the result holds only generically (in endowments).

Examples of pure exchange economies with a continuum of Walras equilibria show that one can not hope to prove that economies, with real or nominal assets and (generically) complete markets, have a finite number of equilibria in general. Hence determinacy can not be more than a generic property.

However, under suitable assumptions, the number of equilibria in economies with incomplete asset markets and nominal assets turn out to be generically uncountable .

With incomplete asset markets equilibrium allocations are typically not Pareto optimal allocations.

Exercises

Exercise 6.A Consider an economy over two dates, satisfying the Maintained Assumptions, with L goods and with an asset structure with nominal assets where the dividend matrix, V, is such that $\mathbb{R}^k \times \{0\}^{S-k} \subset \langle V \rangle$ where $0 < k < S$. Let $((\bar{c}^i)_{i\in\mathbb{I}}, p, q)$ be a Radner equilibrium for $\mathcal{E}$ with net income vectors $(\bar{r}^i)_{i\in\mathbb{I}}$.

(a) Let $\beta = (1, \beta_{\mathbf{1}}, \beta_{\mathbf{2}})$ and $\beta' = (1, \beta'_{\mathbf{1}}, \beta'_{\mathbf{2}}) \in \mathbb{R}^{S+1}_{++}$ with $\beta_{\mathbf{1}}, \beta'_{\mathbf{1}} \in \mathbb{R}^k$ and $\beta_{\mathbf{2}}$, $\beta'_{\mathbf{2}} \in \mathbb{R}^{S-k}$ be such that

$$\beta \begin{bmatrix} -q \\ V \end{bmatrix} = \beta' \begin{bmatrix} -q \\ V \end{bmatrix} = 0$$

Show that $\beta_{\mathbf{1}} = \beta'_{\mathbf{1}}$.

(b) Let consumer 1's equilibrium consumption be $\bar{c}^1 = (\bar{c}^1_{\mathbf{0}}, \bar{c}^1_{\mathbf{1}}, \bar{c}^1_{\mathbf{2}})$ where $\bar{c}^1_{\mathbf{0}} \in \mathbb{R}^L, \bar{c}^1_{\mathbf{1}} \in \mathbb{R}^{kL}$ and $\bar{c}^1_{\mathbf{2}} \in \mathbb{R}^{(S-k)L}$. Show that $c^1 = (\bar{c}^1_{\mathbf{0}}, c^1_{\mathbf{1}}, \bar{c}^1_{\mathbf{2}})$ satisfies the budget restrictions for consumer 1 if and only if

$$\beta(1)p(1)(c^1(1) - e^1(1)) + \ldots + \beta(k)p(k)(c^1(k) - e^1(k) \leq$$
$$\beta(1)p(1)(\bar{c}^1(1) - e^1(1)) + \ldots + \beta(k)p(k)(\bar{c}^1(k) - e^1(k))$$

(c) Extend the result from Proposition 6.3.A and show that there does not exist an allocation, $(\hat{c}^i)_{i\in\mathbb{I}}$, which agrees with $(\bar{c}^i)_{i\in\mathbb{I}}$ for $s = 0$ and $s = k+1, \ldots, S$, such that $(\hat{c}^i)_{i\in\mathbb{I}}$ Pareto dominates $(\bar{c}^i)_{i\in\mathbb{I}}$

Exercise 6.B (continuation of Exercise 5.E) consider the economy defined in Exercise 5.E.

(a) Show that an allocation, (c^a, c^b) is a Pareto optimal allocation if and only if for some α such that $\alpha \in]0, 8[$,

$$c^a = (\alpha, \frac{9}{2}, \frac{9}{2})$$
$$c^b = (8 - \alpha, \frac{9}{2}, \frac{9}{2})$$

(b) Let $(\bar{c}^a, \bar{c}^b)$ be an equilibrium allocation for $\mathcal{E}_1$ for some $\pi \in A$. Show that $(\bar{c}^a, \bar{c}^b)$ is not a Pareto optimal allocation. (*Hint:* Consider the consumptions found in part (b) of Exercise 5.E.)

(c) It is easy to visualize the Pareto optimal allocations from (a) in a three dimensional Edgeworth's box; it is simply a line segment. Where is the image of the mapping $\pi \longrightarrow (c^a, c^b)$ taking $\pi \in A$ to an equilibrium allocation in $\mathcal{E}_1$, with an associated price system where $p(2) = 1/\pi$ in relation to this line segment?

Chapter 7

OVERLAPPING GENERATIONS ECONOMIES

Introduction

In the present chapter overlapping generations economies (OG economies), are introduced and the demographic structure, the market structure as well as perfect foresight, rational expectations and the notion of equilibrium are discussed.

Time extends from $-\infty$ to ∞ and at every date there is a single perishable consumption good. At every date a consumer is born who lives and consumes at that date and the succeeding one. Hence, at every date a young consumer; the consumer born at that date and an old consumer; the consumer born at the previous date, are present.

The young and the old consumer may exchange the good and money. The preferred interpretation is that there are spot-markets for the consumption good and for money at every date. A young consumer decides how much to consume at the first date and plans an amount to consume at the next date. In order to make that decision the consumer has to form expectations about prices at the next date. With the assumption of perfect foresight the preferred interpretation is seen to be equivalent to an interpretation with forward markets.

7.1 Structure and Assumptions

In this section we will introduce the elements making up an OG economy.

The set of dates

The assumption of a finite number of dates is crucial for Arrow-Debreu economies. This assumption is given up in the study of OG economies. Time is discrete and extends indefinitely into the future and thus the economy has no last date. Still there is a choice about having a first date or not. It turns out that for many problems it is convenient to let time extend indefinitely into the past as well. Since economies with a first date can be embedded in economies with no first date the assumption of no first date does not restrict the generality of the model and often simplifies statements of results since we avoid having an old consumer acting only at the first date. For example, any equilibrium in an OG economy with a first date can be regarded as a truncation of an equilibrium in a suitably chosen economy with no first date, cf. Exercise 7.D. Thus the set of dates is taken to be the set of integers $\mathbb{Z} = \{\ldots, -2, -1, 0, 1, 2, \ldots\}$.

The commodity space

We assume that at each date there is a single good. To avoid the introduction of production activities in the form of storage between dates the good is assumed to be perishable. A commodity is characterized by the delivery date and the single good induces at each date a specific commodity. Thus the set of commodities is also indexed by $\mathbb{Z}$ and the commodity space is $\mathbb{R}^\infty = \prod_{t \in \mathbb{Z}} \mathbb{R} = \ldots \mathbb{R} \times \mathbb{R} \times \mathbb{R} \times \mathbb{R} \ldots$. A *price system* assigns to each commodity a price and will be denoted $(p_t)_{t \in \mathbb{Z}}$. Prices are assumed to be positive, that is, $(p_t)_{t \in \mathbb{Z}} \in \prod_{t \in \mathbb{Z}} \mathbb{R}_{++}$. The interpretation will be discussed below.

Consumers

It is an essential assumption that an OG economy should have an infinite number of dates. Each consumer is assumed to live only at a finite number of consecutive dates and we consider the simple case where there is a single consumer born at each date who lives only at two consecutive dates; as *young* and as *old.*

The consumer born at date t is referred to as consumer t and is characterized by her consumption set, C_t, initial endowment, e_t, and utility function, $u_t : C_t \to \mathbb{R}$. Though a strict application of the conventions adhered to up till now would force us to consider the consumption set as a subset of the commodity space and the initial endowment as a point in the commodity space, the particular structure of OG economies allow us to focus on only the two dates when the consumer acts. Thus the consumption set is taken to be a subset of $\mathbb{R}^2$ and the initial endowment is a point in $\mathbb{R}^2$. Thus we assume implicitly that for consumer t, the components of a consumption and the initial endowment are 0 at all dates except dates t or $t+1$.

A consequence is that although two consumers may have the same consumption set, when this set is regarded as a subset of $\mathbb{R}^2$, their induced consumption sets in the commodity space $\mathbb{R}^\infty$ will differ.

Maintained assumptions on consumers

Under the interpretation of the previous section the Maintained Assumptions from Chapter 1, with $\mathbb{L}$ =2 makes sense, and we will in the sequel take it as part of the definition of a consumer that these assumptions are satisfied, unless otherwise stated. The five Maintained Assumptions are quite strong and could often be relaxed considerably. The advantage of working with a single set of assumptions has been deemed to outweigh the loss in generality.

Simple OG economies defined

We can now state the following

Definition 7.1.A *A* **simple OG economy,** $\mathcal{E}$, *is a sequence of consumers*

$$\mathcal{E} = (C_t, u_t, e_t)_{t\in\mathbb{Z}}$$

We will refer to such an economy as an "OG economy" or simply "economy". From the definition it is seen that we have an economy with private ownership and implicit is also the assumption that the consumers of the economy are related as indicated in Figure 7.1.A. Thus at any single date there is an old consumer who can trade with the young consumer and this young consumer will, at the next date, trade with the consumer born at that date.

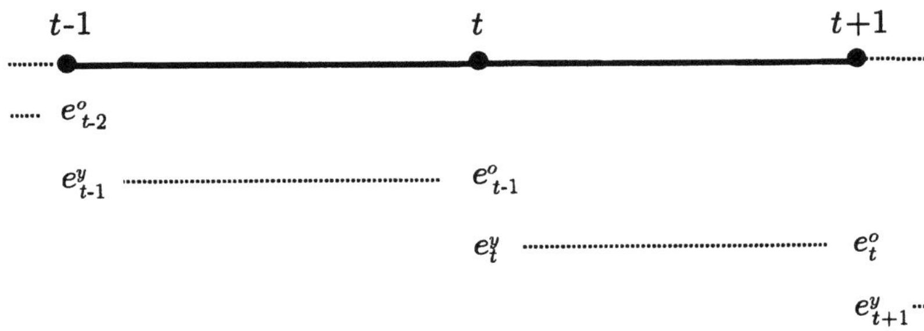

Figure 7.1.A: *At date t consumer t, who is at that date young, has initial endowment e^y_t and consumer $t-1$, who is at date t old, has initial endowment e^o_{t-1}*

Allocations and Reallocations

Allocations

For an OG economy $\mathcal{E}$ an *allocation* is a sequence of consumption bundles such that the consumption bundle of every consumer is in her consumption set and total consumption is equal to total endowment at every date. We thus continue to take it as part of the definition of an allocation that it is balanced.More formally we have

Definition 7.1.B $(c_t)_{t\in\mathbb{Z}} = (c^y_t, c^o_t)_{t\in\mathbb{Z}}$ *is an* **allocation** *for $\mathcal{E}$ if, for $t \in \mathbb{Z}$,*

(a) $c_t \in C_t$

(b) $c^o_{t-1} + c^y_t = e^o_{t-1} + e^y_t$

Reallocations

For simple OG economies there is a convenient way to relate two allocations using the notion of a *reallocation*. Let $(c_t)_{t\in\mathbb{Z}} = (c^y_t, c^o_t)_{t\in\mathbb{Z}}$ be an allocation and let $(a_t)_{t\in\mathbb{Z}}$ be a sequence of real numbers. The sequence

$$(\hat{c}_t^y, \hat{c}_t^o)_{t\in\mathbb{Z}} \;=\; (c_t^y - a_t, c_t^o + a_{t+1})$$

is market balanced and thus is an allocation provided that the inherent consumptions are individually feasible. Cf. Figure 7.1.B. We make the following

Definition 7.1.C *Let $(c_t)_{t\in\mathbb{Z}}$ be an allocation for $\mathcal{E}$. A sequence of real numbers, $(a_t)_{t\in\mathbb{Z}}$, is a* **reallocation** *for the allocation $(c_t)_{t\in\mathbb{Z}}$ if $(\hat{c}_t^y, \hat{c}_t^o)_{t\in\mathbb{Z}}$ $= (c_t^y - a_t, c_t^o + a_{t+1})_{t\in\mathbb{Z}}$ is an allocation for $\mathcal{E}$.*

Hence an allocation and a reallocation for that allocation results in another allocation.

On the other hand, let the allocation $(c_t)_{t\in\mathbb{Z}}$ be given and let $(\hat{c})_{t\in\mathbb{Z}}$ denote an arbitrary allocation. Define a sequence $(a_t)_{t\in\mathbb{Z}}$ through

$$\hat{c}_t^y \;=\; c_t^y - (c_t^y - \hat{c}_t^y) \;=\; c_t^y - a_t \qquad \text{or} \qquad a_t = c_t^y - \hat{c}_t^y$$

Note that $a_t = -(\hat{c}_t^y - c_t^y) = \hat{c}_{t-1}^o - c_{t-1}^o$ by the market balance condition so that

$$\hat{c}_{t-1}^o \;=\; c_{t-1}^o + (\hat{c}_{t-1}^o - c_{t-1}^o) \;=\; c_{t-1}^o + a_t$$

With this definition of the sequence $(a_t)_{t\in\mathbb{Z}}$ we have $(\hat{c}_t)_{t\in\mathbb{Z}} = (c_t^y - a_t, c_t^o + a_{t+1})_{t\in\mathbb{Z}}$ so that $(a_t)_{t\in\mathbb{Z}}$ is a reallocation for the allocation $(c_t)_{t\in\mathbb{Z}}$ resulting in the allocation $(\hat{c}_t)_{t\in\mathbb{Z}}$.

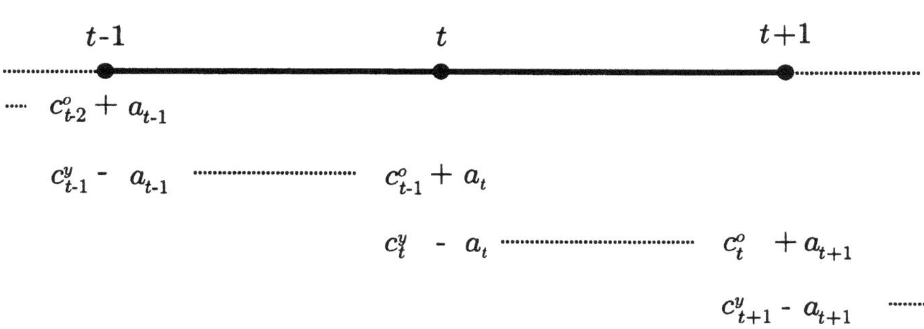

Figure 7.1.B: *The reallocation $(a_t)_{t\in\mathbb{Z}}$ for the allocation $(c_t^y, c_t^o)_{t\in\mathbb{Z}}$ induces the allocation $(c_t^y - a_t, c_t^o + a_{t+1})_{t\in\mathbb{Z}}$*

Improving reallocations

A reallocation $(a_t)_{t\in\mathbb{Z}}$ for $(c_t)_{t\in\mathbb{Z}}$ is an *improving reallocation* (for$(c_t)_{t\in\mathbb{Z}}$) if $(c_t^y - a_t, c_t^o + a_{t+1})_{t\in\mathbb{Z}}$ dominates $(c_t)_{t\in\mathbb{Z}}$ in the sense that no consumer gets lower utility and at least one consumer gets higher utility.

Definition 7.1.D *Let be a reallocation for $(c_t)_{t\in\mathbb{Z}}$ then $(a_t)_{t\in\mathbb{Z}}$ is an* **improving reallocation** *for $(c_t)_{t\in\mathbb{Z}}$ if*

$$u_t(c_t^y - a_t, c_t^o + a_{t+1}) \geq u_t(c_t^y, c_t^o) \quad \textit{for } t \in \mathbb{Z}$$

with strict inequality for at least one $t \in \mathbb{Z}$..

If the given allocation is the initial endowment allocation $(e_t)_{t\in\mathbb{Z}}$ a reallocation $(a_t)_{t\in\mathbb{Z}}$ induces the net trades $(-a_t, a_{t+1})_{t\in\mathbb{Z}}$ and the consumptions $\hat{c}_t = e_t + (-a_t, a_{t+1})$. Cf. Figure 7.1.C.

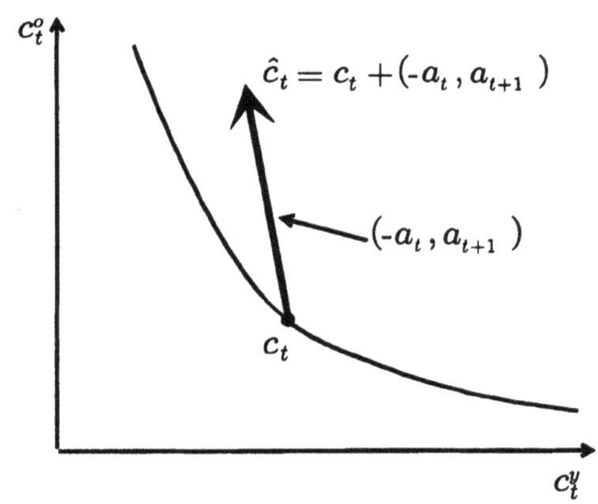

Figure 7.1.C: *A reallocation $(a_t)_{t\in\mathbb{Z}}$ for the allocation $(c_t^y, c_t^o)_{t\in\mathbb{Z}}$ induces the allocation $(c_t^y - a_t, c_t^o + a_{t+1})_{t\in\mathbb{Z}}$*

7.2 Expectations and Equilibrium

Spot-markets with money

The market structure is assumed to be spot-markets with money where young consumers decide how much to consume and how much to save and old consumers decide how much to consume.

At each date, the good can be exchanged for money. Money can be thought of as a durable good that *per se* yields no utility. It turns out to be convenient to allow also for a negative amount of money. When this is the case we can think of money at date t as a debt of consumer t which she has to transfer to consumer $t+1$, at date $t+1$, by refraining from consumption at that date.

Indeed the pattern of exchanges may be: At date t consumer $t-1$ buys some of good t from consumer t with her stock of money. At date $t+1$ consumer t buys some of good $t+1$ from consumer $t+1$ with her stock of money, which she got from consumer $t-1$, and so on.

Expectations

For spot-markets with money as market structure, young consumers know the prices at the first date but they do not know the prices at the next date. Therefore young consumers are assumed to form expectations about prices at the next date. The expectations of a young consumer is assumed to be a probability distribution on next day prices. Young consumers are assumed to maximize their expected utility with $\tilde{u}$ as state utility function.

We restrict attention to the case where the expectations can be described by a probability distribution with finite support. Assume that $p_t > 0$ is the price of good t and that $\tilde{p}_{t+1} = \{p^1_{t+1}, \dots, p^n_{t+1}\} \subset \mathbb{R}_{++}$ is the set of possible prices of good $t+1$. Let $\mu_{t+1} : \{p^1_{t+1}, \dots, p^n_{t+1}\} \to \mathbb{R}_+$ where $\mu_{t+1}(p^1_{t+1}) + \dots + \mu_{t+1}(p^n_{t+1}) = 1$ be the true probability distribution at date t on the price of good $t+1$ and let: $\{p^1_{t+1}, \dots, p^n_{t+1}\} \to \mathbb{R}_+$ where $\nu_{t+1}(p^n_{t+1}) + \dots + \nu_{t+1}(p^n_{t+1}) = 1$ be the expectations of consumer t regarding the next date price, at date t.

The consumer problem under perfect foresight

If there is no randomness in prices there exists $\bar{p}_{t+1} \in \{p^1_{t+1}, \dots, p^n_{t+1}\}$ such that $\mu_{t+1}(\bar{p}_{t+1}) = 1$. If expectations are correct so that $\nu_{t+1} = \mu_{t+1}$ then the expectations of consumer t are referred to as *perfect foresight*. In this case consumer t believes that only one price is possible at the next date and acts on this belief at the current date.

In this case the problem of consumer t at date t is

$$\text{Max}_{(c^y,c^o,m)} \quad u_t(c^y, c^o)$$

$$\text{subject to} \quad \begin{cases} p_t c^y + m \leq p_t e_t^y \\ \bar{p}_{t+1} c^o \leq \bar{p}_{t+1} e_t^o + m \\ c \in C_t \quad \text{and} \quad m \in \mathbb{R} \end{cases} \qquad \textbf{7.2.A}$$

where m is her nominal saving and $\nu_{t+1}(\bar{p}_{t+1}) = \mu_{t+1}(\bar{p}_{t+1}) = 1$. A solution to her problem is an amount to consume at date t, a (planned) amount to consume at date $t+1$ and nominal savings, m, at date t.

The consumer problem under rational expectations

If expected future spot-prices are truly random there does not exist $p_{t+1}^i \in \{p_{t+1}^1, \ldots, p_{t+1}^n\}$ such that $\mu_{t+1}(p_{t+1}^i) = 1$. If this is the case then expectations are correct if $\nu_{t+1} = \mu_{t+1}$. The expectations of consumer t are referred to as *rational expectations* and in this case the problem of consumer t at date t is

$$\text{Max}_{(c^y,(c_i^o)_{i=1}^n,m)} \quad \sum_{i=1}^{n} \mu_{t+1}(p_{t+1i}^i) u_t(c^y, c_i^o)$$

$$\text{subject to} \quad \begin{cases} p_t c^y + m \leq p_t e^y \\ p_{t+1}^i c_i^o \leq p_{t+1}^i e^o + m \quad \text{for } i \text{ s. t. } \mu_{t+1}(p_{t+1}^i) > 0 \\ (c^y, c_i^o) \in C_t \quad \text{for } i \in \{1, 2, ..., n\} \text{ and } m \in \mathbb{R} \end{cases}$$

A solution to her problem is an amount to consume at date t, an amount to consume at date $t+1$ and nominal savings at date t. The amount to consume at date $t+1$ is contingent on the realized spot-price of the commodity at date $t+1$. Clearly, perfect foresight coincides with rational expectations when there is no randomness in prices.

The problem of old consumers

Let c^y be the consumption of consumer t at date t and let m be her nominal savings. Assume that $\bar{p}_{t+1} > 0$ is the realized spot-price of good

$t+1$. The problem of the old consumer t at date $t+1$ is

$$\text{Max}_{c^o} \quad u_t(c^y, c^o)$$
$$\text{subject to} \quad \begin{cases} \bar{p}_{t+1}c^o \leq \bar{p}_{t+1}e^o + m \\ (c^o, c^y) \in C_t \end{cases}$$

Clearly, if c^y and m are such that a solution exists then $c^o = e^{\,o} + m/p_{t+1}$ by the monotonicity of preferences.

Assumptions on expectations in the sequel

It is assumed that there is no randomness in prices and that consumers have perfect foresight in all sections except in Section 12.2 on sunspot equilibria. There exposition there proceeds on the assumption that there is randomness in the prices and that consumers have rational expectations.

Equilibrium

Definition of an equilibrium

In the chapters to come we will often interpret the market structure to be spot-markets with money. Accordingly the equilibrium concept will be spot-market equilibrium but we will often refer to it simply as equilibrium. However, the price system associated with a spot-market equilibrium will also be the price system of an equilibrium with forward markets so that there is often no need to distinguish the two equilibrium concepts.

Consider a sequence of prices $(p_t)_{t\in\mathbb{Z}}$ an allocation and a sequence of money holdings $(\bar{c}_t, \bar{m}_t)_{t\in\mathbb{Z}}$ such that the consumers' budget restrictions from 7.2.A are satisfied with equality. Then, in particular, for $t \in \mathbb{Z}$

$$p_t\bar{c}_t^y + \bar{m}_t = p_t e_t^y$$
$$p_t\bar{c}_{t-1}^o = p_t e_{t-1}^o + \bar{m}_{t-1}$$

and the market balance condition for the good market at each date implies that (dis)savings are invariant over time so that $\bar{m}_t = \bar{m}_{t+1}$ for

$t \in \mathbb{Z}$. Let M denote the amount of savings, or equivalently, the stock of money which can be positive, negative or 0. Since a given economy may have several equilibria, the stock of money is determined by the equilibrium conditions and hence is not intrinsic to the economy and in view of this does M does not occur in the definition of the OG economy. We only consider equilibria where the price of money is normalized to unity. Exercise 7.F shows that we can do so without any loss of generality. We can now state the following

Definition 7.2.A *An* **equilibrium** $((p_t)_{t\in\mathbb{Z}}, (\bar{c}_t, \bar{m}_t)_{t\in\mathbb{Z}}, M)$ *is a sequence of prices, a sequence of consumption bundles and nominal savings and a stock of money such that, for* $t \in \mathbb{Z}$*:*

(a) $(\bar{c}_t, \bar{m}_t)$ *is a solution to the Consumer Problem*

$$\mathrm{Max}_{(c^y, c^o, m)} \quad u_t(c^y, c^o)$$

$$\text{subject to} \quad \begin{cases} p_t c^y + m \le p_t e^y \\ p_{t+1} c^o \le p_{t+1} e^o + m \\ c \in C_t \text{ and } m \in \mathbb{R} \end{cases}$$

(b) *the good market and the money market balance,*

$$\bar{c}_t^y + \bar{c}_{t-1}^o = e_t^y + e_{t-1}^o$$

$$\bar{m}_t = M$$

The sequence $(p_t)_{t\in\mathbb{Z}}$ *is the* **equilibrium price system** *and* $(\bar{c}_t)_{t\in\mathbb{Z}}$ *is the* **equilibrium allocation***.*

Clearly the price *level* is without importance for equilibrium allocations, that is,

$$((\bar{p}_t)_{t\in\mathbb{Z}}, (\bar{c}_t, \bar{m}_t)_{t\in\mathbb{Z}}, M)$$

is an equilibrium if and only if, or $\lambda > 0$,

$$((\lambda\bar{p}_t)_{t\in\mathbb{Z}}, (\bar{c}_t, \lambda\bar{m}_t)_{t\in\mathbb{Z}}, \lambda M)$$

is an equilibrium. Therefore equilibria with identical relative prices, $(\bar{p}_t/\bar{p}_{t+1})_{t\in\mathbb{Z}}$, and identical allocation, $(\bar{c}_t)_{t\in\mathbb{Z}}$, are considered to be identical equilibria and often an equilibrium will be identified merely by the price-consumption sequence $((\bar{p}_t)_{t\in\mathbb{Z}}, (\bar{c}_t)_{t\in\mathbb{Z}})$. The sequence $(\bar{m}_t)_{t\in\mathbb{Z}}$ and number M are then implicitly given by $M = \bar{m}_t = p_t(e^y - \bar{c}_t^y)$.

Equivalence to forward markets

Although the institution of spot-markets with money is natural in OG economies one can also imagine a benchmark interpretation where consumers are assumed to enter into contracts of deliveries "before" the economy unravels. With no first date for the economy this interpretation becomes a bit contrived but leads to the same possibilities as spot-markets and money.

Money makes it possible for consumer t to transfer purchasing power between date t and date $t+1$. The future dividend of holding one unit of money is 1 which implies that the discount factor is 1. Hence the spot-price at date $t+1$ is also the implicit date t price of one unit of the good to be delivered at date $t+1$. Since the consumptions available to consumer t are the same with spot-markets and money as when the consumer can, at date t, buy the good for delivery at date $t+1$.Thus

$$\{c \in C_t \mid \text{there exists } m \in \mathbb{R} \text{ such that } (c, m) \text{ satisfies the restrictions in 7.2.A}\}$$
$$= \left\{ c \in C_t \;\middle|\; p_t c^y + p_{t+1} c^o \le p_t e_t^y + p_{t+1} e_t^o \right\}$$

and there exists $\bar{m}_t$ such that is $(\bar{c}_t, \bar{m}_t)$ a solution to the Consumer Problem with spot-markets and money if and only if $\bar{c}_t$ is a solution to the Consumer Problem with a the single budget restriction defining the right-hand set above.

Invariance of savings; classification of equilibria

At an equilibrium, $((p_t)_{t\in\mathbb{Z}}, (\bar{c}_t, \bar{m}_t)_{t\in\mathbb{Z}}, M)$, nominal savings of consumers are constant across time, that is,

$$M = \ldots = \bar{m}_{t-1} = \bar{m}_t = \bar{m}_{t+1} = \ldots$$

Thus at an equilibrium savings can be 0 at each date, in which case the equilibrium is a *real equilibrium,* or savings can be non-zero at each date.

In the latter case the equilibrium is a *classical equilibrium* if $m_t < 0$ for $t \in \mathbb{Z}$ so that savings are negative at each date and the equilibrium is a *Samuelson equilibrium* if $m_t > 0$ for $t \in \mathbb{Z}$ so that savings are positive at each date. Cf. Figure 7.2.A.

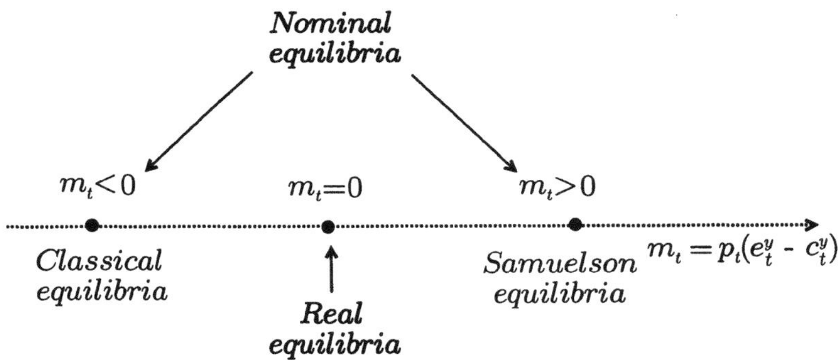

Figure 7.2.A: *The invariance of savings makes possible a classification of equilibria*

Existence of equilibria

The simplicity of simple OG economies together with our Maintained Assumptions imply that each allocation can be supported by prices. This in turn ensures that each economy has an equilibrium with the initial endowments as equilibrium allocation.

Theorem 7.2.B (Existence of Equilibrium). *Each OG economy has an equilibrium.*

Proof: First, note that if $(\bar{c}_t, \bar{m}_t)$ is a solution to consumer t's problem then $(\bar{c}_t, \bar{m}_t)$ satisfies $p_t\bar{c}^y + p_{t+1}\bar{c}^o = p_te^y_t + p_{t+1}e^o_t$ and $\bar{m}_t = p_t(e^y_t - \bar{c}^y_t)$.

By Proposition 1.1.B, $\bar{c}_t$ is a solution to the consumer problem if and only if there exists $\lambda > 0$ such that

$$\begin{cases} Du_t(\bar{c}_t) - \lambda \begin{pmatrix} p_t \\ p_{t+1} \end{pmatrix} = 0 \\ p_t\bar{c}^y_t + p_{t+1}\bar{c}^o_t - (p_te^y_t + p_{t+1}e^o_t) = 0 \end{cases} \qquad \textbf{7.2.B}$$

that is, $\bar{c}_t$ belongs to the budget hyperplane and the gradient of the utility function, evaluated at $\bar{c}_t$, is proportional to the prices.

We now show that $((p_t)_{t\in\mathbb{Z}}, (e_t, \bar{m}_t)_{t\in\mathbb{Z}}, M)$ with $0 = M = \bar{m}_t$ for $t \in \mathbb{Z}$ is an equilibrium for a suitably chosen price system. Since $(\bar{c}_t)_{t\in\mathbb{Z}}$ is an allocation for $\mathcal{E}$ we only have to prove that there is a price system $(p_t)_{t\in\mathbb{Z}}$ such that, for $t \in \mathbb{Z}$, e_t is a solution to consumer t's problem.

Let

$$Du_t(e_t) = (D_y u_t(e_t), D_o u_t(e_t))$$

be the gradient of u_t, evaluated at $e_t = (e_t^y, e_t^o)$. By the Maintained Assumption (C4) both components of $Du_t(e_t)$ are positive.

Put $q_0 = 1$ and, for $t \in \mathbb{Z}\backslash\{0\}$, define q_t by

$$\begin{cases} (1, q_{t+1}) = \left(1 , \dfrac{D_o u_t(e_t)}{D_y u_t(e_t)} \right) & \text{for } t \geq 0 \\ (q_t, 1) = \left(\dfrac{D_y u_t(e_t)}{D_o u_t(e_t)}, 1 \right) & \text{for } t < 0 \end{cases}$$

Obviously $(1, q_{t+1})$, for $t \geq 0$, and $(q_t, 1)$, for $t < 0$, are proportional to the gradient of u_t evaluated at e_t.

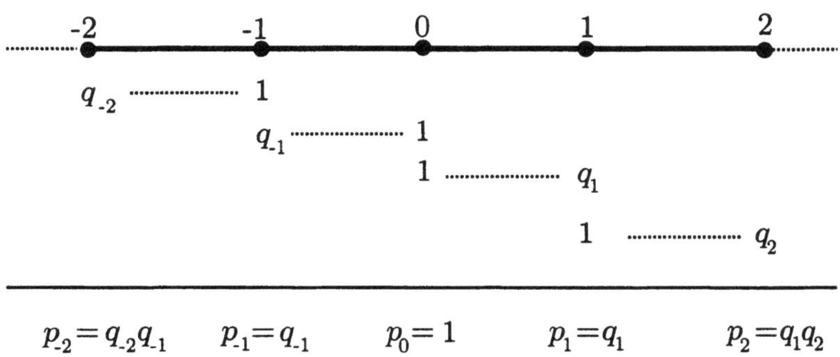

Figure 7.2.B: *The normalized gradients are used to construct an equilibrium price system*

We now use the sequence $(q_t)_{t\in\mathbb{Z}}$ to construct an equilibrium price system. Cf. Figure 7.2.B. Let $p_0 = q_0 = 1$ and

$$\begin{cases} p_t = q_t q_{t+1} \cdots q_{-1} q_0 = \prod_{\tau=t}^{0} q_\tau & \text{for } t < 0 \\ p_t = q_0 q_1 \cdots q_{t-1} q_t = \prod_{\tau=0}^{t} q_\tau & \text{for } t > 0 \end{cases}$$

Consider $t \geq 0$. Then

$$(p_t, p_{t+1}) = \left[\prod_{\tau=0}^{t} q_\tau\right] \cdot (1, q_{t+1})$$

and hence, by the definition of q_{t+1}, the first two relations of 7.2.B are satisfied with $\bar{c} = e_t$ and $\lambda = \left(1/\prod_{\tau=0}^{t} q_\tau\right) D_y u_t(e_t)$. An analogous result holds for $t < 0$.

The initial endowment e_t trivially satisfies the last relation in 7.2.B. This implies that $(e_t, 0)$ is a solution to consumer t's problem and hence that $((p_t)_{t\in\mathbb{Z}}, (e_t, 0)_{t\in\mathbb{Z}}, 0)$ is an equilibrium.

□

Only the existence of a real equilibrium, that is, an equilibrium with zero savings or, equivalently, an equilibrium where the intensity of trade as measured by $c_t - e_t$ is zero at all dates, is established – see Balasko, Cass and Shell [1980], Balasko and Shell [1980], Borglin and Keiding [1986] for further results on the existence of equilibrium in OG economies with many goods, many consumers and/or production.

Forward Markets

Definition of a forward markets equilibrium

On the one hand spot-markets with money as market structure does respect the demographic structure of OG economies since at every date only the old consumer and the young consumer trade, that is, only consumers alive at date t trade at date t. On the other hand forward markets as market structure, where all consumers trade before the "first" date of the OG economy start, that is, consumers trade at $t = -\infty$ – whatever that means – does not respect the demographic structure of OG economies. However, as for Arrow-Debreu economies, forward markets is a natural (but unrealistic) benchmark for OG economies. For OG economies there is also an additional interpretation with one date forward markets but in this interpretation the young consumer at date t has to make a contract with the yet unborn consumer $t+1$.

Let $P_t > 0$ be the forward price of good t and let $c_t \in C_t$ be consumer t's consumption.

Definition 7.2.C *A* **forward markets equilibrium**, $((P_t)_{t\in\mathbb{Z}}, (\bar{c}_t)_{t\in\mathbb{Z}})$, *is a sequence of prices and consumptions such that for* $t \in \mathbb{Z}$*:*

(a) $\bar{c}_t$ *is a solution to the Consumer Problem at prices* (P_t, P_{t+1})

$$\text{Max}_{(c^y,c^o)} \quad u_t(c^y, c^o)$$
$$\text{subject to} \quad \begin{cases} P_t c^y + P_{t+1} c^o \leq P_t e^y_t + P_{t+1} e^o_t \\ c \in C_t \end{cases}$$

(b) *the good market balances*

$$\bar{c}^y_t + \bar{c}^o_{t-1} = e^y_t + e^o_{t-1}$$

Equilibrium and forward markets equilibrium

In Section 7.2 we saw that the Consumer Problem with spot-markets and money was equivalent to the Consumer Problem with forward markets. This immediately gives the following

Theorem 7.2.D (Equilibria and Forward Markets Equilibria) *Let* $\mathcal{E} = (C_t, u_t, e_t)_{t\in\mathbb{Z}}$ *be an OG economy. Then* $((p_t)_{t\in\mathbb{Z}}, (\bar{c}_t)_{t\in\mathbb{Z}})$ *is a forward markets equilibrium if and only if* $((p_t)_{t\in\mathbb{Z}}, (\bar{c}_t, \bar{m}_t)_{t\in\mathbb{Z}}, M)$ *is an equilibrium with* $\bar{m}_t = p_t(e^y_t - \bar{c}^y_t) = M$, *for* $t \in \mathbb{Z}$.

Hence, an allocation is an equilibrium allocation for spot-markets with money as market structure if and only if it is an equilibrium allocation for forward markets as market structure according to Theorem 7.2.D. In particular, the equilibrium prices have a dual interpretation as spot-prices or prices for forward delivery. This will be true in the following except for the discussion of sun-spot equilibria in Chapter 11.

Example 7.2.A (Real and nominal equilibria)

Let $\mathcal{E} = (\mathbb{R}^2_{++}, \ln c^y + \ln c^o, (3,1))_{t\in\mathbb{Z}})$ be an OG economy and $(p_t)_{t\in\mathbb{Z}}$ a price system. Denote by w_t

$$p_t e^y_t + p_{t+1} e^o_{t+1} = 3p_t + p_{t+1} \qquad \textbf{7.2.C}$$

consumer t's wealth. The Walrasian demand function is here independent of $t \in \mathbb{Z}$ and consumer t's Walrasian demand is

$$\begin{bmatrix} f_t^y(p_t, p_{t+1}, w_t) \\ f_t^o(p_t, p_{t+1}, w_t) \end{bmatrix} = \begin{bmatrix} \frac{1}{2}\frac{w_t}{p_t} \\ \frac{1}{2}\frac{w_t}{p_{t+1}} \end{bmatrix} = \begin{bmatrix} \frac{1}{2}\left(\frac{3p_t + p_{t+1}}{p_t}\right) \\ \frac{1}{2}\left(\frac{3p_t + p_{t+1}}{p_{t+1}}\right) \end{bmatrix}$$

The sequence $(p_t)_{t\in\mathbb{Z}}$ is a (forward markets) equilibrium price system if and only if for $t \in \mathbb{Z}$ total demand equals total endowment

$$f_t^y(p_t, p_{t+1}, w_t) + f_{t-1}^o(p_{t-1}, p_t, w_t) = e_t^y + e_{t-1}^o = 4$$

which is equivalent to

$$p_{t+1} - 4p_t + 3p_{t-1} = 0 \quad \text{for } t \in \mathbb{Z}$$

This second order homogenous difference equation with constant coefficients has the solutions

$$\begin{cases} p_t = 3^t & \text{for } t \in \mathbb{Z} \text{ and} \\ p_t = 1 & \text{for } t \in \mathbb{Z} \end{cases}$$

and for $v_1, v_2 \in \mathbb{R}$ the sequence $p_t = v_1 3^t + v_2 \cdot 1$, $t \in \mathbb{Z}$, is a solution. Cf. Sydsaeter and Hammond [1995]. To make sense prices should be positive and hence we restrict v_1 and v_2 to be positive. Using the demand functions and budget restrictions to find the corresponding equilibrium allocations and money holdings, it is seen that $((3^t)_{t\in\mathbb{Z}}, ((3,1),0)_{t\in\mathbb{Z}}, 0)$ is a real equilibrium and $((1)_{t\in\mathbb{Z}}, ((2,2),1)_{t\in\mathbb{Z}}, 1)$ is a Samuelson equilibrium.

Alternatively, one could use the desired savings to generate the difference equation for an equilibrium price system. Indeed we have

$$\begin{aligned} m_{t-1} = p_{t-1}z_{t-1} &= p_{t-1}(e_{t-1}^y - f_{t-1}^y(p_{t-1}, p_t, w_{t-1}) \\ &= p_t(e_t^y - f_t^y(p_t, p_{t+1}, w_t) = p_t z_t = m_t \end{aligned}$$

if and only if

$$p_{t-1}\left(\frac{3}{2} + \frac{1}{2}\frac{p_t}{p_{t-1}}\right) = p_t\left(\frac{3}{2} + \frac{1}{2}\frac{p_{t+1}}{p_t}\right)$$

if and only if

$$p_{t+1} - 4p_t + 3p_{t-1} = 0 \quad \text{for } t \in \mathbb{Z}$$

Summary

In this chapter we have defined simple OG economies. The set of integers was used to index dates as well as commodities and consumers. We showed that for simple OG economies we could compare allocations using the notion of a reallocation.

It was noted that the exposition in the sequel is, with a minor exception, based on perfect foresight for the consumers. The institution of spot-markets with money lent itself to a simple interpretation and turned out to lead to the same equilibrium allocations as a more contrived benchmark interpretation, where consumers choose before the "start" of the economy.

The invariance of savings made possible a classification of equilibria into real and nominal equilibria. Each OG economy was shown to have a real equilibrium with the equilibrium allocation equal to the initial endowments.

Exercises

Exercise 7.A Let $\mathcal{E} = (\mathbb{R}^2_{++}, \ln c^y + \ln c^o, e)_{t\in\mathbb{Z}}$ be an OG economy with identical consumers where $e^y_t + e^o_{t-1} = 4$ for $t \in \mathbb{Z}$.

(a) Let $(e_t)_{t\in\mathbb{Z}}$ be the allocation with $e_t = (3, 1)$ for $t \in \mathbb{Z}$ and let $\tau \in \mathbb{Z}$. Check that $(a_t)_{t\in\mathbb{Z}}$ with

$$a_t = \begin{cases} 1 & \text{for } t \leq \tau \\ 0 & \text{for } t > \tau \end{cases}$$

is a reallocation for $(e_t)_{t\in\mathbb{Z}}$. What is the induced allocation? Is $(a_t)_{t\in\mathbb{Z}}$ an improving reallocation?

(b) Check that $(a_t)_{t\in\mathbb{Z}}$ with $a_t = -1$ for $t \in \mathbb{Z}$ is a reallocation for $(e_t)_{t\in\mathbb{Z}}$ with $e_t = (1, 3)$ for $t \in \mathbb{Z}$. What is the induced allocation? Is $(a_t)_{t\in\mathbb{Z}}$ an improving reallocation?

(c) Let $(e_t)_{t\in\mathbb{Z}}$ be the allocation with $e_t = (3, 1)$ for $t \in \mathbb{Z}$ and let $(\hat{c}_t)_{t\in\mathbb{Z}}$ be the allocation with $\hat{c}_t = (2, 2)$ for $t \in \mathbb{Z}$. Find a reallocation for

$(e_t)_{t\in\mathbb{Z}}$ which induces the allocation $(\hat{c}_t)_{t\in\mathbb{Z}}$ and a reallocation for $(\hat{c}_t)_{t\in\mathbb{Z}}$ which induces the allocation $(e_t)_{t\in\mathbb{Z}}$.

Exercise 7.B Let $\mathcal{E} = (\mathbb{R}^2_{++}, \ln c^y + \ln c^o, e)_{t\in\mathbb{Z}}$ be an OG economy with identical consumers.

(a) What are the equilibrium conditions?

(b) Find a real equilibrium.

(c) Find an equilibrium where $p_t = 1$ for $t \in \mathbb{Z}$ and characterize it with respect to savings.

Exercise 7.C Let $\mathcal{E} = (R^2_{++}, \ln c^y + \ln c^o, e_t)_{t\in\mathbb{Z}}$ where

$$e_t = \begin{cases} (1,2) \text{ for } t \text{ odd} \\ (2,1) \text{ for } t \text{ even} \end{cases}$$

(a) Find an equilibrium $((p_t)_{t\in\mathbb{Z}}, (\bar{c}_t, \bar{m}_t)_{t\in\mathbb{Z}}, M)$ for the OG economy.

(b) Consider the economy truncated from date 1. It is assumed that there is at date 1 an old consumer, born at date 0, who consumes only at date 1. Show that the equilibrium found in (a) induces, for the truncated economy, an equilibrium, with the same consumption bundles for consumer $t = 1, 2, \ldots,$ as in the equilibrium found in (a) and the consumption of consumer 0 at date 1 equal to $c_0^o = \bar{c}_0^o$. What should be the preferences and the budget restriction of the consumer born at date 0?

Exercise 7.D Let $((\bar{p}_t)_{t\in\mathbb{Z}}, (\bar{c}_t, \bar{m}_t)_{t\in\mathbb{Z}}, M)$ be an equilibrium for the economy $\mathcal{E} = (C_t, u_t, e_t)_{t\in\mathbb{Z}}$ with $\bar{p}_t = 1$. Show that there is an economy $\mathcal{E}'$ $= (\hat{C}_t, \hat{u}_t, \hat{e}_t)_{t\in\mathbb{Z}}$ such that $(C_t, u_t, e_t) = (\hat{C}_t, \hat{u}_t, \hat{e}_t)$ for $t \in N$, which has an equilibrium $((\hat{p}_t)_{t\in\mathbb{Z}}, (\hat{c}_t, \hat{m}_t)_{t\in\mathbb{Z}}, \hat{M})$ where $\hat{p}_t = 1$ for $t < 0$, and $\hat{c}_0^o = \bar{c}_0^o$, $(\hat{c}_t, \hat{m}_t) = (\bar{c}_t, \bar{m}_t)$, $p_t = \hat{p}_t$ for $t \in \mathbb{N}$ and $\hat{M} = M$.

Exercise 7.E Show that every OG economy has a unique real equilibrium .

Exercise 7.F In Section 7.2 the price of money is normalized to be 1 at all dates. Assume that we allow for the price of money to be $(v_t)_{t\in\mathbb{Z}}$ where $v_t > 0$ for $t \in \mathbb{Z}$.

(a) What is the problem for consumer t, for $t \in \mathbb{Z}$?

(b) Suggest a definition of an equilibrium where the price of money is not normalized to be 1 at all dates.

(c) Let $((p_t)_{t\in\mathbb{Z}}, (\bar{c}_t, \bar{m}_t)_{t\in\mathbb{Z}}, M)$ be an equilibrium (where the price of money is normalized to be 1 at all dates) and let $(v_t)_{t\in\mathbb{Z}}$ where $v_t > 0$ for $t \in \mathbb{Z}$ be given. Find an equilibrium where the price of money is $(v_t)_{t\in\mathbb{Z}}$ and $(\bar{c}_t)_{t\in\mathbb{Z}}$ is the associated equilibrium allocation.

(d) Let $((p_t)_{t\in\mathbb{Z}}, (\bar{c}_t, \bar{m}_t)_{t\in\mathbb{Z}}, M)$ be an equilibrium (where the price of money is normalized to be 1 at all dates) and let $(q_t)_{t\in\mathbb{Z}}$ where $q_t > 0$ for $t \in \mathbb{Z}$ be given. Find an equilibrium where $(v_t)_{t\in\mathbb{Z}}$ with $v_t > 0$ for $t \in \mathbb{Z}$ are the equilibrium prices for money, $(q_t)_{t\in\mathbb{Z}}$ is the equilibrium price system for goods and $(\bar{c}_t)_{t\in\mathbb{Z}}$ is the equilibrium allocation.

(e) Let $((p_t, v_t)_{t\in\mathbb{Z}}, (\bar{c}_t, \bar{m}_t)_{t\in\mathbb{Z}}, M)$ be an equilibrium where the price of money is $(v_t)_{t\in\mathbb{Z}}$. Find an equilibrium where the price of money is 1 at all dates and $(\bar{c}_t)_{t\in\mathbb{Z}}$ is the associated equilibrium allocation.

(f) Does it restrict the set of equilibrium allocations that we require the price of money to be normalized to 1 at all dates?

Chapter 8

OPTIMAL ALLOCATIONS IN OG ECONOMIES

Introduction

We have seen in Chapter 1 that in an Arrow-Debreu economy an allocation is a Pareto optimal allocation if and only if the subjective evaluations are equalized among consumers. Thus the existence of a common price system is for such economies a necessary and sufficient condition for an allocation to be a Pareto optimal allocation.

For the simple OG economies considered here every allocation is a an equilibrium allocation and so trivially consumers' subjective evaluations are equalized. Of course, this is a consequence of the fact that there is a single good at each date. The definition of a Pareto optimal allocation can be formulated in the same way as for Arrow-Debreu economies; an allocation is a Pareto optimal allocation if and only if there is no alternative allocation giving at least the same utility to each consumer and higher utility to at least one consumer. With an infinite time horizon the alternative allocation may differ from the given one for dates arbitrarily far in the future and/or in the past. It might occur that any intervention aimed at improving the allocation should have been carried out "before the start of the economy". While the verbatim translation of the definition for a Pareto optimal allocation fromArrow-Debreu economies to

OG economies makes sense, it is overly strong in that it demands that the present allocation should compare favorably with "too many" alternative allocations. An allocation which is a Pareto optimal allocation in the sense described is a *strongly optimal allocation* (we will drop the reference to Pareto to save some space). Although this optimality concept appears too demanding for policy decisions strongly optimal allocations serve as a benchmark in the investigation of equilibria of OG economies.

A less demanding concept of optimality arises if we allow the alternative allocations to differ from the given one only from some date and onwards. Intuitively we think of the situation where we observe the economy at some date and where the past is given. Intervention to improve the allocation at some date would have to take the past of the economy as given. An intervention to improve would then have a first date where improved allocation does not agree with the given allocation, say date t. The old consumer, born at date $t-1$, can not at date t give up some amount of the good to the young consumer since then consumer $t-1$ would be worse off. Hence the alternative allocation must involve consumer t giving up some amount of the good at date t. In order for consumer t not be worse off she has to be compensated at date $t+1$, which can only be achieved if consumer $t+1$ gives up some amount of the good. Clearly, the alternative allocation will then involve infinitely many dates and consumers. This "chain-letter" phenomena might appear dubious, but with an infinity of agents and dates such a "chain-letter" can be perfectly sound. Closely related to this is the possibility of an equilibrium allocation where each consumer has an income larger than the value of her initial endowment, evaluated at the equilibrium prices. An allocation that compares favorably with each allocation which agrees with it from some date and backwards is referred to as an *forwards optimal allocation.* This optimality concept appears to be the relevant one for policy considerations.

If an intervention can affect only a finite number of dates and the given allocation compares favorably with each allocation in this set, then the allocation is referred to as a *finitely optimal allocation.* For this concept of optimality the two theorems of welfare economics carry over to OG economies. This is the subject of Section 8.2 below while Section 8.1 discusses the different notions of optimality.

Some intuition for the concept of optimality in OG economies can be had by introducing *reduced models*. These are essentially the "no-worse-than sets" of a specific allocation in an OG economy. In Section 8.3 we introduce such models and define a composition of the sets of a reduced model. It is then possible to formulate a *general efficiency criterion* for reduced models which provides some insight into the conditions for an allocation to be a forwards optimal allocation or a strongly optimal allocation.

The general efficiency criterion may be difficult to apply to an equilibrium allocation of an OG economy. Therefore we study reduced models where each set belongs to a family that is totally ordered by set inclusion and which is described by a single parameter measuring "curvature". In Section 8.4 we show that there is a family of such sets; *the family of hyperbola*, where the composition of the sets corresponds to the addition of the parameters of the inherent sets. Approximation of a reduced model by sets from the family of hyperbola results in a criterion for optimality based on the sum of parameters in the approximation. In this way we get *parametric efficiency criteria*.

In Section 8.5 it is shown that in case the "no-worse-than-sets" can be approximated by hyperbola one can determine whether an equilibrium allocation is a forwards optimal allocation or a strongly optimal allocation from knowledge merely of the sequence of equilibrium prices.

8.1 Notions of Optimality

In this section we will define and discuss three different notions of optimality for allocations in an OG economy.

Strongly optimal allocations

In general, an allocation is not Pareto optimal if there exists an improving reallocation or equivalently another allocation such that no consumer is worse off and at least one consumer is better off. For Arrow-Debreu economies with complete markets any redistribution of commodities consistent with the total initial endowment is allowed. Also for OG economies it may seem natural to consider arbitrary reallocations. However, to allow for all reallocations in OG economies is very demanding since then reallocations without a first date and extending into the indefinite past are admitted. If consumer t gives some of good $t+1$ to consumer $t+1$, then in order to compensate consumer t consumer $t-1$ must give some of good t to consumer t, and so on. Cf. Figure 8.1.A where the sequence $(a_t)_{t \in \mathbb{Z}}$ could be nonpositive for $t \in \mathbb{Z}$.

Definition 8.1.A *An allocation,* $(\hat{c}_t)_{t \in \mathbb{Z}}$*, is* **a strongly optimal allocation** *if there is no allocation,* $(c_t)_{t \in \mathbb{Z}}$*, such that:*

$$u_t(c_t) \geq u_t(\hat{c}_t) \quad \text{for } t \in \mathbb{Z} \text{ with at least one strict inequality}$$

If $(\hat{c}_t)_{t \in \mathbb{Z}}$ is not a strongly optimal allocation and $(c_t)_{t \in \mathbb{Z}}$ satisfies the condition in the Definition 8.1.A, then the sequence defined by $a_t = \hat{c}_t^y - c_t^y$ is an improving reallocation for$(\hat{c}_t)_{t \in \mathbb{Z}}$.

Now restrict attention to reallocations which are not 0 at each date. The case where $a_t \geq 0$ corresponds to allocations where savings are too low (from some date and forwards) in the sense that there is an improving reallocation which leads to less consumption for consumers as young and more consumption for consumers as old. The case $a_t \leq 0$, on the other hand, corresponds to allocations where savings are too high (at some date and backwards) in the sense that there is an improving reallocation which leads to more consumption for consumers as young and less consumption for consumers as old.

Hence, if an allocation is not a strongly optimal allocation then saving is either too high from some date and forwards or too low from some date and backwards.

Forwards optimal allocations

Obviously, it is quite hard to imagine how reallocations without a first date, where the term of the reallocation is not 0, are arranged. In view of this it may be argued that strong optimality is too strong a notion and that only reallocations with a first date should be allowed. With such a reallocation no consumer gives anything to anybody before a first date T_L, consumer T_L gives some of good T_L to consumer $T_L - 1$, consumer

$T_L + 1$ has to give some of good $T_L + 1$ to consumer T_L in order to compensate her and so on. Cf. Figure 8.1.A which depicts an improving reallocation. If $(a_t)_{t\in\mathbb{Z}}$ is an improving reallocation then either $a_t \geq 0$ for $t \in Z$ or $a_t \leq 0$ for $t \in Z$. In the first case, we might have a date T_L such that $a_t = 0$ if and only if $t < T_L$ and in the second case a date T^U such that $a_t = 0$ if and only if $t > T^U$.

Definition 8.1.B *An allocation,* $(\hat{c}_t)_{t\in\mathbb{Z}}$, *is* **a forwards optimal allocation** *if there is no allocation,* $(c_t)_{t\in\mathbb{Z}}$, *such that:*

(a) $u_t(c_t) \geq u_t(\hat{c}_t)$ *for* $t \in Z$ *with at least one strict inequality*

(b) *there exists* $T_L \in Z$ *such that* $c_t = \hat{c}_t$ *for each* $t \leq T_L - 1$.

Thus if an allocation, $(\hat{c}_t)_{t\in\mathbb{Z}}$, is not a forwards optimal allocation and $(c_t)_{t\in\mathbb{Z}}$ satisfies the condition in Definition 8.1.B then there exists an improving reallocation for $(\hat{c}_t)_{t\in\mathbb{Z}}$ defined by $a_t = \hat{c}^y_t - c^y_t$ which is such that $a_t = 0$ for each $t < T_L$.

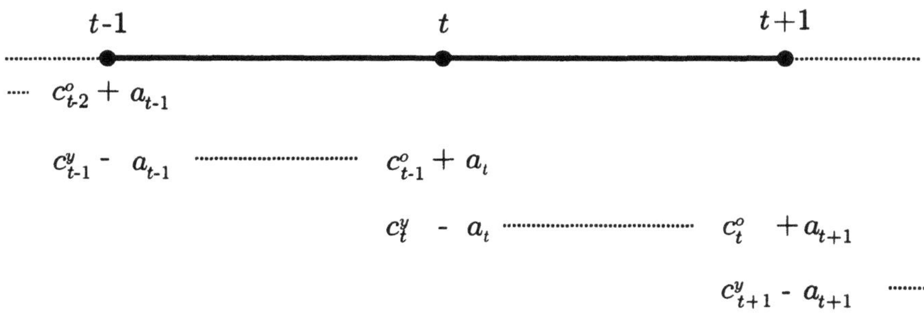

Figure 8.1.A: *If* $(a_t)_{t\in\mathbb{Z}}$ *is an improving reallocation then either* $a_t \geq 0$ *for* $t \in Z$ *or* $a_t \leq 0$ *for* $t \in Z$.

Hence, allocations which are not forwards optimal allocations correspond to allocations where savings are too low, or equivalently, first date consumptions are too high.

Example 8.1.A Let $\mathcal{E} = (\mathbb{R}^2_{++}, \ln c^y + \ln c^o, e)_{t\in\mathbb{Z}}$. Suppose that $e = (1, 3)$. Then the initial endowments do not form a strongly optimal allocation since the allocation $(2, 2)_{t\in\mathbb{Z}}$, which is obtained by the improving reallocation $(a_t)_{t\in\mathbb{Z}}$, with $a_t = -1$ for $t \in \mathbb{Z}$, makes each consumer better off.

Assume instead that $e = (3,1)$. Then the initial endowments do not form a forwards optimal allocation since the allocation $(c_t)_{t\in\mathbf{Z}}$ defined by

$$c_t = \begin{cases} (3,1) & \text{for } t \le -1 \\ (3,2) & \text{for } t = 0 \\ (2,2) & \text{for } t \ge 1 \end{cases}$$

which is obtained by the improving reallocation $(a_t)_{t\in\mathbf{Z}}$ defined by

$$a_t = \begin{cases} 0 & \text{for } t \le 0 \\ 1 & \text{for } t \ge 1 \end{cases}$$

makes no consumer worse off and each consumer, from date 0 and onward, better off.

Finitely optimal allocations

Obviously, a reallocation without a final date involves an infinite number of consumers. Thus it may be argued that forward optimality is too demanding and only reallocations with a first and a final date should be allowed. Accordingly only reallocations which are 0 except for a finite number of dates should be considered.

Definition 8.1.C *An allocation, $(\hat{c}_t)_{t\in\mathbf{Z}}$, is* **a finitely optimal allocation** *if there is no allocation, $(c_t)_{t\in\mathbf{Z}}$, such that:*

(a) $u_t(c_t) \ge u_t(\hat{c}_t)$ *for $t \in Z$ with at least one strict inequality*

(b) *there exists $T_L \in Z$ such that $c_t = \hat{c}_t$ for $t < T_L - 1$ and*

(c) *there exists $T^U \in Z$ such that $c_t = \hat{c}_t$ for $t \ge T^U + 1$.*

If an allocation $(c_t)_{t\in\mathbf{Z}}$ is not a finitely optimal allocation then there exists an improving reallocation, $(a_t)_{t\in\mathbf{Z}}$, and $T_L, T^U \in Z$ such that $a_t = 0$ for each $t < T_L$ and each $t > T^U$. In the simple OG models considered here each allocation is a finitely optimal allocation. Such an allocation can be realized as an equilibrium allocation. This property will be true also for OG models with long lived consumers and many goods.

Applications of the notions of optimality

Finite optimality is the relevant notion of optimality with regard to the two theorems of welfare economics as will be seen in Section 8.2. Strong optimality may seem contrived, but it has a bearing on the study of the set of equilibria which will be pursued for stationary economies in the next chapter. Forwards optimality is of interest with regard to economic policy. If it is realized at some date that an equilibrium allocation is not a forwards optimal allocation then there exists an improving reallocation which starts at that date and so it may be possible to transform the equilibrium allocation into a forwards optimal allocation.

Note that for an Arrow-Debreu economy the three notions of optimality are indistinguishable since there is by assumption only a finite number of dates.

A note on terminology*

Most of the literature on OG economies consider economies with a first date so that the notion of a forwards optimal allocation is equivalent to a Pareto optimal allocation whereas a strongly optimal allocation has no counterpart. We have preferred "finitely optimal" to the often used "weakly Pareto optimal". Thus

Here	**Common usage**
Strongly optimal allocation	
Forwards optimal allocation	Pareto optimal allocation
Finitely optimal allocation	Weakly Pareto optimal allocation

8.2 The Theorems of Welfare Economics

The two theorems of welfare economics for Arrow-Debreu economies state that an equilibrium allocation is a Pareto optimal allocation and that a Pareto optimal allocation is an equilibrium allocation with a suitable redistribution of the initial endowment. The welfare theorems carry over to OG economies with "finitely optimal allocation" taking the place of "Pareto optimal allocation". The First Theorem of Welfare Economics is an immediate consequence of the properties of an improving reallocation proved below.

Improving reallocations

In Chapter 7 we defined a reallocation, $(a_t)_{t\in\mathbb{Z}}$, for an allocation $(\hat{c}_t)_{t\in\mathbb{Z}}$as a sequence of numbers, $(a_t)_{t\in\mathbb{Z}}$, such that $c_t = (\hat{c}_t^y - a_t, \hat{c}_t^o + a_t)$ was an allocation. As defined the consumptions $(c_t)_{t\in\mathbb{Z}}$ are always market balanced so the condition on the reallocation simply ensures that the consumptions are individually feasible. The sign conventions imply that a_t, if *positive*, is a delivery *from* the young consumer t *to* the old consumer $t-1$, at date t and a_t, if *negative*, is a delivery *from* the old consumer $t-1$ *to* the young consumer t, at date t.

Recall that a reallocation $(a_t)_{t\in\mathbb{Z}}$ is an improving reallocation if

$$u_t(c_t^y - a_t, c_t^o + a_{t+1}) \geq u_t(c_t^y, c_t^o) \quad \text{for } t \in Z \qquad \textbf{8.2.A}$$

with at least one strict inequality. In Lemma 8.2.A we have gathered some results on the properties of an improving reallocation.

Lemma 8.2.A *Let $(\hat{c}_t)_{t\in\mathbb{Z}}$ be an allocation and $(a_t)_{t\in\mathbb{Z}}$ an improving reallocation for $(\hat{c}_t)_{t\in\mathbb{Z}}$. Then one and only one of the following alternatives is true*

(a) $a_t > 0$ *for* $t \in \mathbb{Z}$

(b) *there is a* T_L *such that* $a_t > 0$ *for* $t \geq T_L$ *and* $a_t = 0$ *for* $t < T_L$

(c) $a_t < 0$ *for* $t \in Z$

(d) *there is* T^U *such that* $a_t < 0$ *for* $t \leq T^U$ *and* $a_t = 0$ *for* $t > T^U$

Proof: Let $(a_t)_{t\in\mathbb{Z}}$ be an improving reallocation. From relation 8.2.A follows that there is a $\tau \in \mathbb{Z}$ such that $a_\tau \neq 0$. Assume that $a_\tau > 0$. By the monotonicity of preferences and relation 8.2.A; if $a_t > 0$ then $a_{t+1} > 0$ and if $a_t = 0$ then $a_{t-1} = 0$. Hence the improving reallocation satisfies precisely one of the alternatives (i) and (ii) below

(i) $\quad \ldots, a_{T_L-2}, a_{T_L-1}, a_{T_L}, a_{T_L+1}, a_{T_L+2}, \ldots \quad$ with $a_t > 0$ for $t \in \mathbb{Z}$

(ii) $\quad \ldots, 0, 0, a_{T_L}, a_{T_L+1}, a_{T_L+2}, \ldots \quad$ with $a_t > 0$ for $t \geq T_L$

and one of the cases **(a)** and **(b)** occurs.

Assume that $a_\tau < 0$. Again by the monotonicity of preferences and relation 8.2.A; if $a_t < 0$ then $a_{t-1} < 0$ and if $a_t = 0$ then $a_{t+1} = 0$. Hence the improving reallocation satisfies precisely one of the alternatives (iii) and (iv) below

(iii) $\ldots, a_{T^U-2}, a_{T^U-1}, a_{T^U}, a_{T^U+1}, a_{T^U+2}, \ldots$ with $a_t < 0$ for $t \in \mathbb{Z}$

(iv) $\ldots, a_{T^U-2}, a_{T^U-1}, a_{T^U}, 0, 0, \ldots$ with $a_t < 0$ for $t \leq T^U$

and one of the cases **(c)** and **(d)** occurs.

□

If $(a_t)_{t\in\mathbb{Z}}$ is a reallocation with $a_t > 0$ for $t \in \mathbb{Z}$ then, since relation 8.2.A is satisfied with strict inequality for some $t \in \mathbb{Z}$, we can truncate below and get an improving reallocation which is 0 from some $T_L - 1$ and backwards. On the other hand, if $(a_t)_{t\in\mathbb{Z}}$ is a reallocation with $a_t < 0$ for $t \in \mathbb{Z}$ then we can truncate above and get an improving reallocation which is 0 from some date $T^U + 1$ and onwards.

The first theorem of welfare economics

We can now prove the First Theorem of Welfare Economics.

Theorem 8.2.B *Let $((\bar{p}_t)_{t\in\mathbb{Z}}, (\bar{c}_t, \bar{m}_t)_{t\in\mathbb{Z}}, M)$ be an equilibrium for $\mathcal{E} = (C_t, u_t, e_t)_{t\in\mathbb{Z}}$. Then $(\bar{c}_t)_{t\in\mathbb{Z}}$ is a finitely optimal allocation.*

Proof: Let $(c_t)_{t\in\mathbb{Z}}$ be an arbitrary allocation for $\mathcal{E}$. Assume, in order to obtain a contradiction, that $(c_t)_{t\in\mathbb{Z}}$ is not a finitely optimal allocation. Then there is an improving reallocation $(a_t)_{t\in\mathbb{Z}}$ for $(c_t)_{t\in\mathbb{Z}}$, which is non-zero only for finitely many dates. By Lemma 8.2.A there can be no such improving reallocation. Hence $(c_t)_{t\in\mathbb{Z}}$ is a finitely optimal allocation.

□

In the proof of Theorem 8.2.B above that $(\bar{c}_t)_{t\in\mathbb{Z}}$ is an equilibrium allocation it is not used. Hence, each allocation is a finitely optimal allocation. Clearly this is a feature of the simple OG economies considered here which is not shared by OG economies with many goods at every date or many consumers in every generation.

Theorem 8.2.B is quite weak for OG economies since an equilibrium allocation is only shown to be a finitely optimal allocation rather than a forwards optimal or a strongly optimal allocation. The following example reveals that there is no hope for a stronger version of the first welfare theorem.

Example 8.2.A Let $\mathcal{E} = (\mathbb{R}^2_{++}, \ln c^y + \ln c^o, (3,1)_{t\in\mathbb{Z}})$ Then

$$((3^t)_{t\in\mathbb{Z}}, ((3,1),0)_{t\in\mathbb{Z}}, 0)$$

is an equilibrium as explained in the example on page 235 and the equilibrium allocation is not a forwards optimal allocation as explained in the example on page 245.

The second theorem of welfare economics

Theorem 8.2.B shows that the market outcome - the equilibrium allocations - are at least finitely optimal allocations, while Example 8.2.A shows that in general the market outcome can not be expected to result in a forwards optimal allocation and hence not in a strongly optimal allocation. We would expect the Second Theorem of Welfare Economics to show that an allocation, where the subjective evaluations of the consumers are equalized, can be realized as an equilibrium allocation. For the simple OG economies considered here the subjective evaluations of the consumers are indeed equalized at each allocation and hence each allocation can be realized as an equilibrium allocation.

Theorem 8.2.C *Let* $(\hat{c}_t)_{t\in\mathbb{Z}}$ *be a finitely optimal allocation for* $\mathcal{E} = (C_t, u_t, e_t)_{t\in\mathbb{Z}}$. *Then there exist* $(\bar{p}_t)_{t\in\mathbb{Z}}$, $(\bar{m}_t)_{t\in\mathbb{Z}}$ *and* M *such that*

$$((\bar{p}_t)_{t\in\mathbb{Z}}, (\hat{c}_t, \bar{m}_t)_{t\in\mathbb{Z}}, M)$$

is an equilibrium for $\hat{\mathcal{E}} = (C_t, \hat{c}_t, u_t)_{t\in\mathbb{Z}}$.

Proof: Define $(\bar{p}_t)_{t\in\mathbb{Z}}$ as in the proof of Theorem 7.2.B for the economy $\hat{\mathcal{E}} = (C_t, u_t, \hat{c}_t)_{t\in\mathbb{Z}}$. Put $\bar{m}_t = 0$ for $t \in \mathbb{Z}$ and $M = 0$. Then $((\bar{p}_t)_{t\in\mathbb{Z}}, (\hat{c}_t, \bar{m}_t)_{t\in\mathbb{Z}}, M)$ is an equilibrium for $\hat{\mathcal{E}}$.

□

In the proof of Theorem 8.2.C it is not used that $(\hat{c}_t)_{t\in\mathbf{Z}}$ is a finitely optimal allocation and it follows that each allocation is an equilibrium allocation after a suitable redistribution of initial endowments. Again this is a feature of the simple OG economies considered here which is, of course, not shared by OG economies with many goods at every date or many consumers in every generation.

Theorem 8.2.C, which is the Second Theorem of Welfare Economics, is quite strong since every finitely optimal allocation is shown to be an equilibrium allocation.

The welfare theorems, relating finitely optimal allocations and equilibrium allocations, generalize to OG economies with many goods at every date and many consumers in every generation, cf. Balasko and Shell [1980].

8.3 Reduced Models

An equilibrium allocation for an OG economy is a finitely optimal allocation, but it need not be a forwards optimal or a strongly optimal allocation as shown in the previous section. In Chapter 1 we saw that it was a necessary and sufficient condition for Pareto optimality in an Arrow-Debreu economy that the subjective evaluations of the consumers were equalized. This is no longer true for OG economies. The infinite number of dates implies that allocations may fail to be Pareto optimal also in case the consumers act against a common price system. The new phenomena occurring in OG economies can be studied using reduced models where the idea is to abstract from all the properties of preferences except those of importance for optimality considerations.

Allocations and Reduced Models

In this section we first show how an allocation induces a reduced model. We then define the concept of a reduced model, disregarding its origin, and introduce the notion of efficiency for such models .

Upper contour sets

The question of optimality reduces to whether there exists a suitable improving reallocation, where "suitable" depends on whether finitely optimal, forwards optimal or strongly optimal allocations are considered. For an allocation $(\bar{c}_t)_{t\in\mathbb{Z}}$ the existence of an improving reallocation $(a_t)_{t\in\mathbb{Z}}$ is related to the properties of the upper contour sets or the translations of these sets, by the consumptions, to the origin, while all other aspects of utility functions are irrelevant. For $t \in \mathbb{Z}$, define the sets

$$\hat{R}_t(\bar{c}_t) = \{c \in \mathbb{R}^2 \mid u_t(c^y, c^o) \geq u_t(\bar{c}_t^y, \bar{c}_t^o)\}$$

$$R_t(\bar{c}_t) = \{(z_1, z_2) \in \mathbb{R}^2 \mid u_t(\bar{c}_t^y + z_1, \bar{c}_t^o + z_2) \geq u_t(\bar{c}_t^y, \bar{c}_t^o)\}$$

Then $\hat{R}_t(\bar{c}_t) - \{\bar{c}_t\} = R_t(\bar{c}_t)$. For $t \in \mathbb{Z}$, the sequence of sets $(R_t(\bar{c}_t))_{t\in\mathbb{Z}}$ associates to t the set $R_t(\bar{c}_t)$. Cf. Figure 8.3.A.

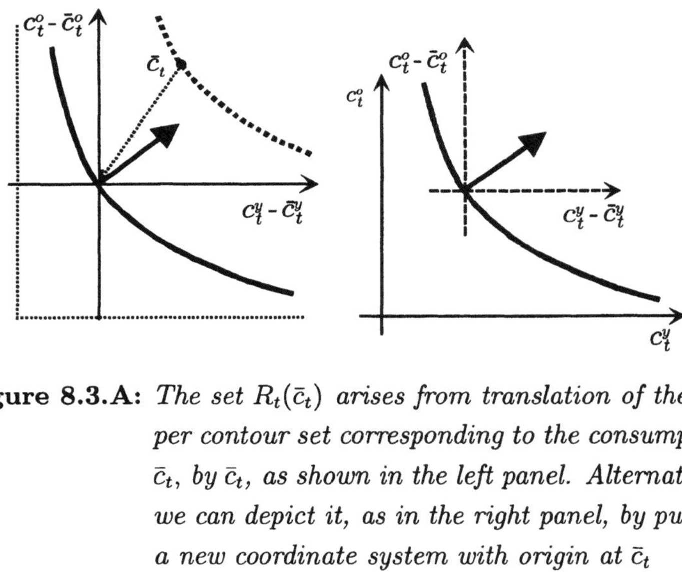

Figure 8.3.A: *The set $R_t(\bar{c}_t)$ arises from translation of the upper contour set corresponding to the consumption $\bar{c}_t$, by $\bar{c}_t$, as shown in the left panel. Alternatively we can depict it, as in the right panel, by putting a new coordinate system with origin at $\bar{c}_t$*

Definition of a reduced model

The study of forwards optimal allocations and strongly optimal allocations can be carried out through the study of the sequence of sets $(R_t(\bar{c}_t))_{t\in\mathbb{Z}}$. Such a sequence of sets is a particular instance of a *reduced model.*

Definition 8.3.A *A* **reduced model** *is a sequence of subsets $(S_t)_{t\in\mathbb{Z}}$ with $S_t \subset \mathbb{R}^2$, such that, for $t \in \mathbb{Z}$,*

(R1) *S_t is a closed set*

(R2) $S_t + \mathbb{R}^2_+ \subset S_t$

(R3) $S_t \cap -\mathbb{R}^2_+ = \{0\}$

Note that if $(S_t)_{t\in\mathbb{Z}}$ is a reduced model then, for $t \in \mathbb{Z}$, the point $0 \in S_t$ since $S_t \cap -\mathbb{R}^2_+ = \{0\}$. Reduced models were introduced and studied in Borglin and Keiding [1986].

Let $(c_t)_{t\in\mathbb{Z}}$ be an allocation. Then it is straightforward to show that $(R_t(c_t))_{t\in\mathbb{Z}}$ is a reduced model. Indeed, the Maintained Assumptions imply that (R1)-(R3) are satisfied and that $R_t(c_t)$, for $t \in \mathbb{Z}$, is a convex set which is bounded below.

Let $(S_t)_{t\in\mathbb{Z}}$ be a reduced model. Then a sequence, $(p_t)_{t\in\mathbb{Z}}$, of non-negative numbers defines a *support* to $(S_t)_{t\in\mathbb{Z}}$ or *supporting prices* if, for $t \in \mathbb{Z}$,

$$p_t z_t + p_{t+1} z_{t+1} \geq 0 \quad \text{for } (z_t, z_{t+1}) \in S_t$$

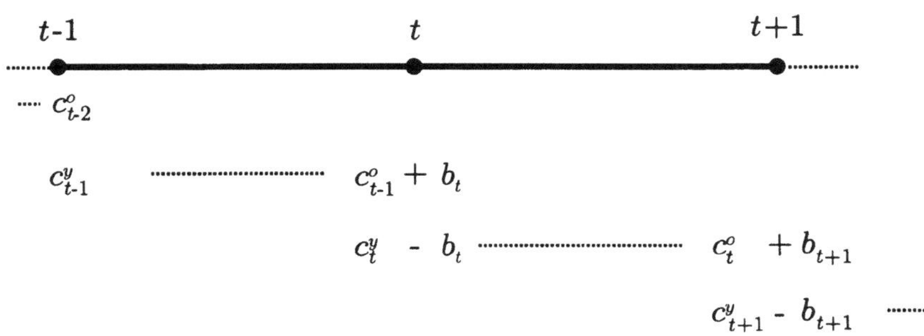

Figure 8.3.B: *A forward improvement $(b_t)_{t\in\mathbb{Z}}$ for a reduced model induced by an allocation in an OG economy defines an improving reallocation with $a_t = b_t$*

Improvements and efficient reduced models

Consider an allocation which is not a forwards optimal allocation. Then there exists an improving reallocation, which is 0 from some date and backwards. For the associated reduced model this improving reallocation induces a *forward improvement* as defined below.

For an allocation that is not a strongly optimal allocation there exists an improving reallocation. The improving reallocation can be used to define a *forward improvement* or a *backward improvement* for the associated reduced model. Cf. Figure 8.3.B where the case where $b_\tau = 0$ for $\tau < t$ and $b_\tau > 0$ for $\tau \geq t$ is depicted.

Definition 8.3.B *A* **forward improvement** *for a reduced model* $(S_t)_{t\in\mathbb{Z}}$ *is a sequence of nonnegative numbers* $(b_t)_{t\in\mathbb{Z}}$ *such that*

(a) $(-b_t, b_{t+1}) \in S_t$ *for* $t \in \mathbb{Z}$

(b) *there exists* $T_L \in Z$ *such that* $b_t = 0$ *for* $t < T_L$ *and* $b_t > 0$ *for* $t \geq T_L$

A **backward improvement** *for a reduced model* $(S_t)_{t\in\mathbb{Z}}$ *is a sequence of nonnegative numbers,* $(b_t)_{t\in\mathbb{Z}}$*, such that*

(a') $(b_t, -b_{t+1}) \in S_t$ *for* $t \in \mathbb{Z}$

(b') *there exists* $T^U \in Z$ *such that* $b_t = 0$ *for* $t > T^U$ *and* $b_t < 0$ *for* $t \leq T^U$

A reduced model $(S_t)_{t\in\mathbb{Z}}$ *is* **efficient** *if there exists no forward improvement and it is* **strongly efficient** *if there is no forward and no backward improvement.*

Note that for reallocations and improving reallocations we put no restriction on the sign of the terms in the sequence. It is however convenient to do so for forward and backward improvements.

Let $(c_t)_{t\in\mathbb{Z}}$ be an allocation and $(R_t(c_t))_{t\in\mathbb{Z}}$ the induced reduced model. If there is a forward improvement then it is possible to take the amount b_{T_L} from consumer T_L at date T_L and compensate her at date $T_L + 1$ with the amount b_{T_L+1}. This compensation can only come from consumer $T_L + 1$, who then has to be compensated at date $T_L + 2$ and so on. The existence of a forward improvement ensures that it is possible to continue to compensate the consumers in the indefinite future.

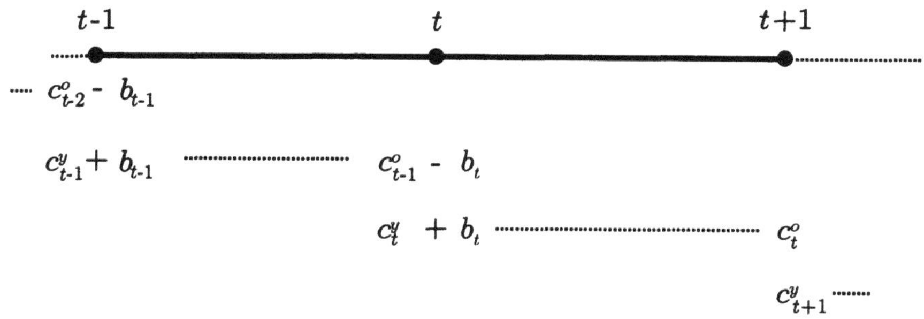

Figure 8.3.C: *A backward improvement $(b_t)_{t\in\mathbb{Z}}$ for a reduced model induced by an allocation in an OG economy defines an improving reallocation with $a_t = -b_t$. The case where $b_\tau = 0$ for $\tau \geq t+1$ and $b_\tau > 0$ for $\tau \leq t$ is depicted*

The interpretation of a backward improvement is analogous. Consumer T^U gives up an amount of b_{T^U+1} at date $T^U + 1$ and is compensated at the preceding date by the amount b_{T^U}. This compensation can only be made by consumer $T^U - 1$ who, in turn, has to be compensated at date $T^U - 1$. The existence of a backward improvement ensures that the compensation can go on into the indefinite past.

In order to ascertain that an allocation is a strongly optimal allocation it is sufficient to consider forward improvements and backward improvements, corresponding to improving reallocations which are 0 from some date and onward or which are 0 from some date and backward. Thus the improvement does not have to extend indefinitely into both the past *and* the future, in a non-trivial way.

Optimality and Efficiency

The raison d'être for reduced models is that the notions of optimality for OG economies (as well as other economies) are closely related to the efficiency or strong efficiency of the associated reduced model as shown by the following lemma.

Lemma 8.3.C *Let $(c_t)_{t\in\mathbb{Z}}$ be an allocation and $R_t(c_t))_{t\in\mathbb{Z}}$ the associated reduced model. Then*

(a) *$(c_t)_{t\in\mathbb{Z}}$ is forwards optimal allocation if and only if $R_t(c_t))_{t\in\mathbb{Z}}$ is an efficient reduced model.*

(b) *$(c_t)_{t\in\mathbb{Z}}$ is a strongly optimal allocation if and only if $R_t(c_t))_{t\in\mathbb{Z}}$ is a strongly efficient reduced model.*

Proof: To prove **(a)** assume that $R_t(c_t))_{t\in\mathbb{Z}}$ is not an efficient reduced model. Then there is a forward improvement, $(b_t)_{t\in\mathbb{Z}}$, for $R_t(c_t))_{t\in\mathbb{Z}}$. The forward improvement is also an improving reallocation for $(c_t)_{t\in\mathbb{Z}}$ since there is a gain for consumer $t = T_L - 1$, where T_L is the smallest t such that $b_t > 0$. On the other hand, if $(c_t)_{t\in\mathbb{Z}}$ is not a forwards optimal allocation then there is an improving reallocation, $(a_t)_{t\in\mathbb{Z}}$, for $(c_t)_{t\in\mathbb{Z}}$ and a T_L such that $a_t = 0$ for $t < T_L$. By Lemma 8.2.A, on the properties of an improving reallocation, the numbers a_t are nonnegative and so they define a forward improvement for $R_t(c_t))_{t\in\mathbb{Z}}$.

(b) If $R_t(c_t))_{t\in\mathbb{Z}}$ is a reduced model which is not strongly efficient then there exist a forward or backward improvement. If $(b_t)_{t\in\mathbb{Z}}$ is a forward improvement then $a_t = b_t$ for $t \in \mathbb{Z}$ is an improving reallocation for $(c_t)_{t\in\mathbb{Z}}$ and if $(b_t)_{t\in\mathbb{Z}}$ is a backward improvement then $a_t = -b_t$ for $t \in \mathbb{Z}$ is an improving reallocation for $(c_t)_{t\in\mathbb{Z}}$.

Characterization of Efficient Reduced Models

We have seen in the preceding section that different notions of optimality for an allocation are equivalent to the efficiency and strong efficiency of the corresponding reduced model.

In this section we give a general criterion for efficiency and strong efficiency of a reduced model. The general criterion has the advantage of providing good intuition but the drawback that it can be difficult to apply to an equilibrium allocation.

The composition of sets

Let S and S' be sets in $\mathbb{R}^2$. Define the composition, $\circ$, of S and S' by

$$S \circ S' = \{(a,b) \in \mathbb{R}^2 \mid \exists c \in \mathbb{R} : (a,c) \in S \text{ and } (-c,b) \in S'\}$$

This composition has a nice interpretation when applied to the reduced model $(R_t(c_t))_{t\in\mathbb{Z}}$ arising from an allocation $(c_t)_{t\in\mathbb{Z}}$. Cf. Figure 8.3.D While $R_t(c_t)$ depicts the willingness of consumer t to substitute between consumption at date t and date $t+1$, the set $R_t(c_t) \circ R_{t+1}(c_{t+1})$ is the set of redistributions between the consumers t, $t+1$ and other consumers such that consumers t and $t+1$ will not be worse off. Hence this set depicts the willingness of consumers t and $t+1$ to substitute between consumption at date t and $t+2$.

Applying the composition repeatedly we get the set

$$R_t(c_t) \circ \cdots \circ R_{t'}(c_{t'}) \;=\; [R_t(c_t) \circ \cdots \circ R_{t'-1}(c_{t'-1})] \circ R_{t'}(c_{t'})$$

which depicts the willingness of consumers $t, t+1, \ldots, t'$ to substitute between date t and $t'+1$. Thus if $(a, b) \in R_t(c_t) \circ \cdots \circ R_{t'}(c_{t'})$ and $a < 0$ then consumer t is willing to give up $|a|$ units of good t provided that consumer $t+1$ compensates her, consumer $t+1$ is willing to compensate consumer t provided that consumer $t+2$ compensates her, ... and consumer t' is willing to compensate consumer $t'-1$ provided that she gets b units of good $t'+1$. Thus the composition depicts how the group of consumers $t, \ldots, t'$ are willing to substitute, directly and indirectly, between date t and date $t'+1$ consumption.

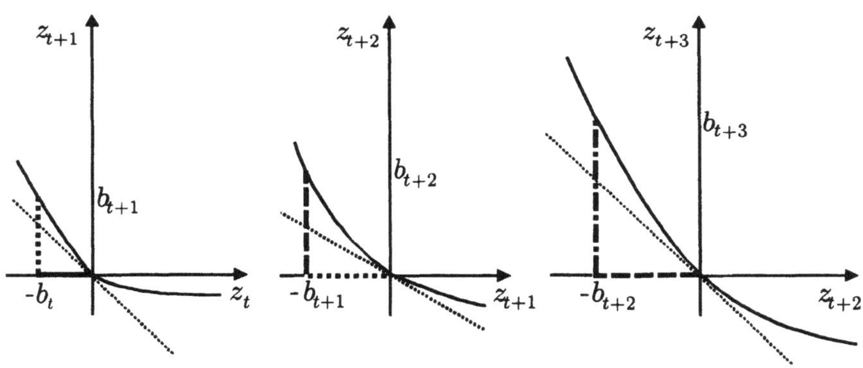

Figure 8.3.D: *The forward improvement $(b_t)_{t\in\mathbb{Z}}$ induces vectors $(-b_t, b_{t+1}) \in S_t$, $(-b_{t+1}, b_{t+2}) \in S_{t+1}$ and $(-b_{t+2}, b_{t+3}) \in S_{t+2}$. The vector $(-b_t, b_{t+3})$ belongs to $S_t \circ S_{t+1} \circ S_{t+2}$*

Similarly if $b < 0$ then we can compose the sets from t and "backwards" to t'. Thus if $b < 0$ and

$$(a, b) \in R_{t'}(c_{t'}) \circ \cdots \circ R_t(c_t)$$

then consumer t would be willing to give up $|b|$ units at date $t+1$ provided she was compensated at date t by consumer $t-1$ who in turn would have to be compensated at date $t-1$. Cf. Figure 8.3.E. In this case the composition will depict the necessary compensation for consumer t', at date t', if the group of consumers $t', \dots, t$ are, directly and indirectly, to deliver $|b|$ units to consumer $t+1$ at date $t+1$.

This suggests that *efficiency* and *strong efficiency* of reduced models are related to properties of

$$[(S_t \circ \cdots \circ S_{t'})_{t' \geq t}]_{t \in \mathbb{Z}} \quad \text{and} \quad [(S_{t'} \circ \cdots \circ S_t)_{t' \leq t}]_{t \in \mathbb{Z}}$$

For an Arrow-Debreu economy the consumers' willingness to substitute, given an allocation, would be given by the sum of their upper contour sets . In Exercise 8.H we indicate how the composition of sets is related to the sum of upper contour sets for the consumers. The pleasant feature of the composition is that it allows us to remain in $\mathbb{R}^2$.

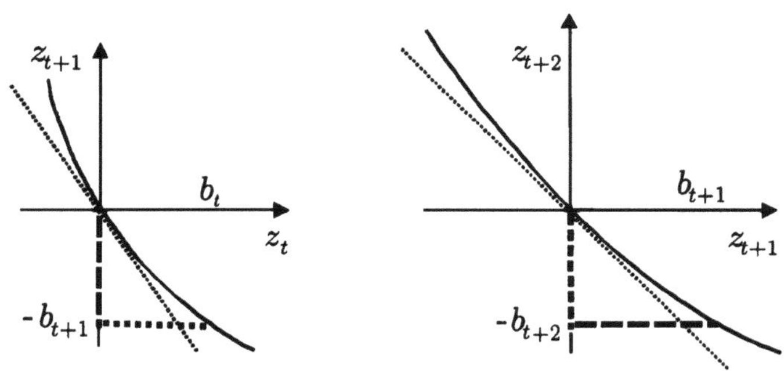

Figure 8.3.E: *Consumer $t+1$, whose upper contour set is depicted in the right panel, receives a positive amount as young from consumer t, as old. Consumer t is in turn compensated by consumer $t-1$*

For A a non-empty subset of $\mathbb{R}^2$, let $\mathrm{pr}_1 : A \longrightarrow \mathbb{R}$ be the projection on the first component so that pr_1 takes the point (c_1, c_2) to c_1 and similarly let pr_2 be the projection on the second component. We can now state a general efficiency criterion.

Theorem 8.3.D *Let $(S_t)_{t\in\mathbb{Z}}$ be a reduced model. Then*

(a) $(S_t)_{t\in\mathbb{Z}}$ *is efficient if and only if*

$$[\cup_{t\in\mathbb{Z}} \cap_{t'\geq t} \operatorname{pr}_1 (S_t \circ \cdots \circ S_{t'})] \cap -\mathbb{R}_+ = \{0\} \qquad \textbf{8.3.A}$$

(b) $(S_t)_{t\in\mathbb{Z}}$ *is strongly efficient if and only if*

$$\text{(i)} \quad [\cup_{t\in\mathbb{Z}} \cap_{t'\geq t} \operatorname{pr}_1 (S_t \circ \cdots \circ S_{t'})] \cap -\mathbb{R}_+ = \{0\} \quad \textit{and}$$
$$\text{(ii)} \quad [\cup_{t\in\mathbb{Z}} \cap_{t'\leq t} \operatorname{pr}_2 (S_{t'} \circ \cdots \circ S_t)] \cap -\mathbb{R}_+ = \{0\} \qquad \textbf{8.3.B}$$

Proof: To prove the "if part" of **(a)** assume that the reduced model, $(S_t)_{t\in\mathbb{Z}}$, is not efficient. Then there exists a forward improvement, $(b_t)_{t\in\mathbb{Z}}$, for $(S_t)_{t\in\mathbb{Z}}$, and a T_L so that $b_t = 0$ for each $t < T_L$ and $b_t > 0$ for each $t \geq T_L$

By the definition of a forward improvement we have

$$\begin{aligned} (-b_{T_L}, b_{T_L+1}) &\in S_{T_L} \\ (-b_{T_L+1}, b_{T_L+2}) &\in S_{T_L+1} \\ &\vdots \\ (-b_t, b_{t+1}) &\in S_t \end{aligned}$$

and applying the composition repeatedly we get $(-b_{T_L}, b_t) \in S_{T_L} \circ \cdots \circ S_t$ for $t \geq T_L$. Hence

$$-b_{T_L} \in \cap_{t\geq T_L} \operatorname{pr}_1(S_{T_L} \circ \cdots \circ S_t)$$

and

$$[\cup_{t\in\mathbb{Z}} \cap_{t'\geq t} \operatorname{pr}_1(S_t \circ \cdots \circ S_{t'})] \cap -\mathbb{R}_+ \neq \{0\}.$$

To prove the "only if" part of (a). Assume that

$$[\cup_{t\in\mathbb{Z}} \cap_{t'\geq t} \operatorname{pr}_1(S_t \circ \cdots \circ S_{t'})] \cap -\mathbb{R}_+ \neq \{0\}.$$

Then there exist $T_L \in \mathbb{Z}$ and $b_{T_L} > 0$ such that

$$-b_{T_L} \in \cap_{t'\geq T_L} \operatorname{pr}_1(S_{T_L} \circ \cdots \circ S_{t'}) \quad \text{for } t' \geq T_L$$

By definition of the composition there exists, for each $t' \geq T_L$, a sequence $(b_\tau^{t'})_{\tau=T_L}^{t'+1}$ such that $b_{T_L}^{t'} = -b_{T_L}$ and $(-b_t^{t'}, b_{t+1}^{t'}) \in S_t$ for $t \in \{T_L, \ldots, t'\}$. Let

$$b_t = \begin{cases} 0 & \text{for} \quad t < T_L \\ \inf_{t' \geq t}\{b_t^{t'}\} & \text{for} \quad t \geq T_L \end{cases}$$

Consider a fixed $t \geq T_L$ and $t' \geq t$. Since $b_t \leq b_t^{t'}$ we get trivially $(-b_t, b_{t+1}^{t'}) \geq (-b_t^{t'}, b_{t+1}^{t'})$. Since $(-b_t^{t'}, b_{t+1}^{t'}) \in S_t$ and, by assumption $S_t + \mathbb{R}_+^2 \subset S_t$, we conclude that $(-b_t, b_{t+1}^{t'}) \in S_t$.Thus $(-b_t, b_{t+1}^{t'}) \in S_t$ for $t' \geq t$. From the definition of b_t and the assumption that S_t is a closed set it follows that $(-b_t, b_{t+1}) = (-b_t, \inf_{t' \geq t}\{b_{t+1}^{t'}\}) \in S_t$.

(b) To prove the "if part" part of (b) assume that the reduced model, $(S_t)_{t\in\mathbb{Z}}$ is not strongly efficient. Then there exists a forward improvement; a case treated above, or a backward improvement. If $(b_t)_{t\in\mathbb{Z}}$ is a backward improvement then, for some T^U and $b_{T^U+1} > 0$

$$-b_{T^U+1} \in \cap_{t' \leq T^U} \operatorname{pr}_2(S_{t'} \circ \cdots \circ S_{T^U}) \quad \text{for } t' \leq T^U$$

and hence it is not true that

$$[\cup_{t\in\mathbb{Z}} \cap_{t' \leq t} \operatorname{pr}_2(S_{t'} \circ \cdots \circ S_t)] \cap -\mathbb{R}_+ = \{0\}$$

To prove the "only if" part of (b) note that we have shown in (a) how to construct a forward improvement in case 8.3.A is not true. The construction of a backward improvement in case (b) is not true is analogous and is left to the reader.

A monotonicity property

The assumptions $S_t \cap -\mathbb{R}_+^2 = \{0\}$ and $S_t + \mathbb{R}_+^2 \subset S_t$ imply $(a, 0) \in S_t$ for $a \geq 0$. It is easy to show that $(a, 0) \in S_t \circ \cdots \circ S_{t'}$ for $t' \geq t$. Hence $\mathbb{R}_+ \subset \operatorname{pr}_1(S_t \circ \cdots \circ S_{t'})$. Furthermore if $t' \geq t$ and $(a, b) \in S_t \circ \cdots \circ S_{t'+1}$ then, by the definition of the composition, we have for some $c \in \mathbb{R}$, that $(a, c) \in S_t \circ \cdots \circ S_{t'}$ and $(-c, b) \in S_{t'+1}$. Hence

$$\operatorname{pr}_1(S_t \circ \cdots \circ S_{t'+1}) \subset \operatorname{pr}_1(S_t \circ \cdots \circ S_{t'}) \quad \text{for } t' \geq t$$

so that the sequence of projections is decreasing with t'. Intuitively, as the number of consumers in the group, who are to be compensated in the future, increases the group is willing to give up less at date t.

Approximation of a reduced model

Trivially, if $S \subset S'$ then $\mathrm{pr}_1 S \subset \mathrm{pr}_1 S'$ and $\mathrm{pr}_2 S \subset \mathrm{pr}_2 S'$. An application of this principle gives the following useful corollary to the general efficiency criterion; Theorem 8.3.D.

Corollary 8.3.E *Let $(S_t)_{t\in\mathbb{Z}}$ and $(S'_t)_{t\in\mathbb{Z}}$ be reduced models. Then*

(a) *if $(S'_t)_{t\in\mathbb{Z}}$ is an efficient reduced model and $S_t \subset S'_t$ for $t \in Z$ then $(S_t)_{t\in\mathbb{Z}}$ is an efficient reduced model*

(b) *if $(S'_t)_{t\in\mathbb{Z}}$ is a strongly efficient reduced model and $S_t \subset S'_t$ for $t \in Z$ then $(S_t)_{t\in\mathbb{Z}}$ is a strongly efficient reduced model*

(c) *if the reduced model $(S'_t)_{t\in\mathbb{Z}}$ is not an efficient reduced model and $S'_t \subset S_t$ then $(S_t)_{t\in\mathbb{Z}}$ is not an efficient reduced model*

(d) *if the reduced model $(S'_t)_{t\in\mathbb{Z}}$ is not a strongly efficient reduced model and $S'_t \subset S_t$ then $(S_t)_{t\in\mathbb{Z}}$ is not a strongly efficient reduced model.*

It may be difficult to ascertain the efficiency of a given reduced model. The corollary suggests that it might at times be enough to consider an "inner" or "outer" approximation of the given reduced model. This is useful if the approximating sets in the reduced model can be chosen from a class of sets which behave nicely under the composition operation.

8.4 Parametric Reduced Models

Theorem 8.3.D characterizes efficient reduced models but will often be difficult to apply. In this section we introduce a family of sets referred to as hyperbola and study reduced models made up of sets from this family. It is easy to decide if a reduced model made up of sets which are hyperbola is efficient or not. This is due to the fact that applying the composition, which turned out to be crucial in the general efficiency theorem Theorem 8.3.D, to hyperbola results in another hyperbola.

When the reduced model consists of sets defined by hyperbola our general efficiency criterion translates directly into a criterion based on the parameters of the reduced model.

For a reduced model consisting of hyperbola an efficiency criterion is derived based on the sequence of prices defining supporting homogenous hyperplanes to the sets inherent in the reduced model. This criterion induces a corresponding criterion for forwards optimal or strongly optimal equilibrium allocations.

Reduced Model with Hyperbola

In this section we will define the family of sets given by hyperbola and state necessary and sufficient conditions for a reduced model made up of such sets to be efficient or strongly efficient.

The family of hyperbola

For $a \geq 0$ and $q = (q_1, q_2) \in \mathbb{R}^2_{++}$ let

$$S_H(a, q) \;=\; \{x \in \mathbb{R}^2 \mid q_1x_1+q_2x_2 \geq 0 \text{ and } q_1x_1+q_2x_2+a(q_1x_1)(q_2x_2) \geq 0\}$$

and let $S_H(\infty, q) = \mathbb{R}^2_+$. For $a \in \mathbb{R}_{++}$ the set $S_H(a, q)$ is a strictly convex set which is bounded below and satisfies conditions (R1)-(R3). The parameter a is, speaking loosely, a measure of "curvature" as suggested by the relation $2S_H(a, q) = S_H(a/2, q)$. Cf. Figure 8.4.A. We will refer to sets from this family as hyperbola. It is easy to verify that

- $\text{pr}_1 S_H(a, q) \;=\; \left]-\frac{1}{aq_1}, \infty\right[$
- $\text{pr}_2 S_H(a, q) \;=\; \left]-\frac{1}{aq_2}, \infty\right[$
- $S_H(a, q_1, q_2) \circ S_H(a', q_2, q_3) \;=\; S_H\left(a + a', q_1, q_3\right)$

The last relation shows that the composition of hyperbola results in a new hyperbola and that the parameter of the composition is simply the sum of the parameters of the sets entering the composition. It is this property that makes hyperbola attractive for approximation purposes.

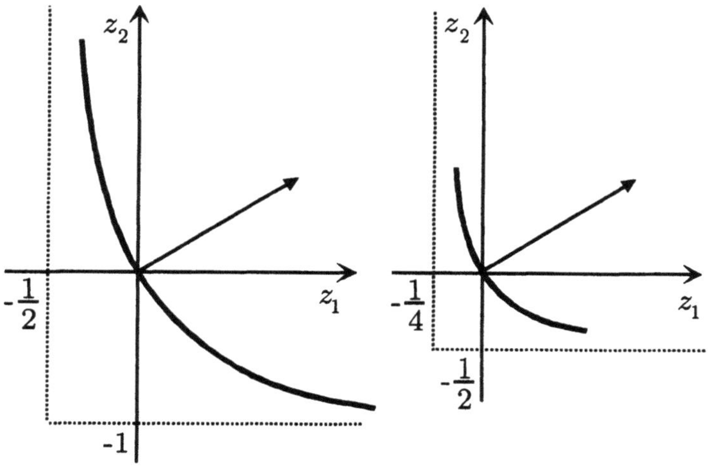

Figure 8.4.A: *On the left the hyperbola defined by $a = 2, q_1 = 1$ and $q_2 = 1/2$, On the right the hyperbola defined by the same q and $a = 4$*

Efficiency of a reduced model of hyperbola

As a corollary to Theorem 8.3.D, the general efficiency criterion, the following characterization of efficiency is obtained for a reduced model made up of hyperbola. Note that the "inf" in the sums of the corollary below occurs since we allow the parameter for the hyperbola to take the value ∞. Let $p^t = (p_t, p_{t+1})$.

Corollary 8.4.A *Let $(p_t)_{t\in\mathbb{Z}}$ be a price system and $(S_t)_{t\in\mathbb{Z}}$ a reduced model with $S_t = S_H(a_t, p^t)$ and $a_t \in \mathbb{R}_+ \cup \{\infty\}$ for $t \in \mathbb{Z}$. Then*

(a) *$(S_t)_{t\in\mathbb{Z}}$ is efficient if and only if*

$$\inf_{t\in\mathbb{Z}} \sum_{\tau=t}^{\infty} a_\tau = \infty$$

(b) *$(S_t)_{t\in\mathbb{Z}}$ is strongly efficient if and only if*

$$\inf_{t\in\mathbb{Z}} \sum_{\tau=t}^{\infty} a_\tau = \infty \quad \text{and} \quad \inf_{t\in\mathbb{Z}} \sum_{\tau=-\infty}^{t} a_\tau = \infty$$

Proof: To prove **(a)** note that

$$S_H(a_t, p_t, p_{t+1}) \circ \ldots \circ S_H(a_{t'}, p_{t'}, p_{t'+1}) = $$
$$= S_H\left(\sum_{\tau=t}^{t'} a_\tau, p_t, p_{t'+1}\right)$$

where the right hand set is thus the set of $(x_t, x_{t'+1})$ satisfying the inequality $p_t x_t + p_{t'+1} x_{t'+1} \geq 0$ and

$$p_t x_t + p_{t'+1} x_{t'+1} + \left[\sum_{\tau=t}^{t'} a_\tau\right] (p_t x_t)(p_{t'+1} x_{t'+1}) \geq 0$$

Since

$$\mathrm{pr}_1\, S_H\left(\sum_{\tau=t}^{t'} a_\tau, p_t, p_{t'+1}\right) = \left] -\frac{1}{\left(\sum_{\tau=t}^{t'} a_\tau\right) p_t}, \infty \right[$$

we have

$$\left[\cap_{t'\geq t} \mathrm{pr}_1 \left(S_H(a_t, p^t) \circ \cdots \circ S_H(a_{t'}, p^{t'})\right)\right] \cap -\mathbb{R}_+ = \{0\}$$

if and only if

$$\sum_{\tau=t}^{\infty} a_\tau = \infty$$

Thus

$$\left[\cup_{t\in\mathbb{Z}} \cap_{t'\geq t} \mathrm{pr}_1 \left(S_H(a_t, p^t) \circ \cdots \circ S_H(a_{t'}, p^{t'})\right)\right] \cap -\mathbb{R}_+ = \{0\}$$

if and only if

$$\inf_{t\in\mathbb{Z}} \sum_{\tau=t}^{\infty} a_\tau = \infty$$

(b) As in **(a)** the conditions

$$[\cup_{t\in\mathbb{Z}} \cap_{t'\geq t} \mathrm{pr}_1 (S_t \circ \cdots \circ S_{t'})] \cap -\mathbb{R}_+ = \{0\} \quad \text{and}$$
$$[\cup_{t\in\mathbb{Z}} \cap_{t'\leq t} \mathrm{pr}_2 (S_{t'} \circ \cdots \circ S_t)] \cap -\mathbb{R}_+ = \{0\}$$

are seen to be equivalent to

$$\inf_{t\in\mathbb{Z}} \sum_{\tau=t}^{\infty} a_\tau = \infty \quad \text{and} \quad \inf_{t\in\mathbb{Z}} \sum_{\tau=-\infty}^{t} a_\tau = \infty$$

We leave the details to the reader.

8.5 Characterization of Optimal Allocations

As shown in Lemma 8.3.C, an allocation is a strongly optimal allocation or a forwards optimal allocation for an OG economy if and only if the associated reduced model is an efficient or strongly efficient reduced model. Thus, by approximation the simple conditions derived in Corollary 8.4.A for reduced models made up of hyperbola carry over to conditions for strongly optimal allocations and forwards optimal allocations. Let $q(c)$ denote the normalized gradient of the utility function u_t evaluated at c so that $q(c) = Du(c)/\|Du(c)\|$ and let $p^t = (p_t, p_{t+1})$. In the next chapter we will see that an equilibrium allocation for a stationary economy satisfies the assumptions of the corollary.

Corollary 8.5.A *Let $((p_t)_{t\in\mathbb{Z}}, (\bar{c}_t, \bar{m}_t)_{t\in\mathbb{Z}}, M)$ be an equilibrium and assume that there exists $a, A \in \mathbb{R}_{++}$ such that*

$$a \leq \frac{p_{t+1}}{p_t} \leq A \quad \textit{and} \quad S_H(A, q(\bar{c}_t)) \subset R_t(\bar{c}_t) \subset S_H(a, q(\bar{c}_t))$$

Then

(a) *the equilibrium allocation is a forwards optimal allocation if and only if*

$$\sum_{t\in\mathbb{N}} \frac{1}{p_t} = \infty$$

(b) *the equilibrium allocation is a strongly optimal allocation if and only if*

$$\sum_{t\in\mathbb{N}} \frac{1}{p_t} = \infty \quad \textit{and} \quad \sum_{t\in-\mathbb{N}} \frac{1}{p_t} = \infty$$

Proof: Since $q(\bar{c}_t)$ is proportional to $p^t = (p_t, p_{t+1})$ we have

$$\begin{aligned} S_H\left(\frac{A}{\|p^t\|}, p^t\right) &= S_H\left(A, \frac{p^t}{\|p^t\|}\right) \subset R_t(\bar{c}_t) \\ &\subset S_H\left(\frac{a}{\|p^t\|}, p^t\right) = S_H\left(a, \frac{p^t}{\|p^t\|}\right) \end{aligned}$$

To prove **(a)** note that the reduced model $(S_H(A/\|p^t\|, p^t))_{t\in\mathbb{Z}}$ is efficient if and only, with $A_t = A/\|p^t\|$

$$\sum_{t\in\mathbb{N}} A_t = \sum_{t\in\mathbb{N}} \frac{A}{\|p^t\|} = A \sum_{t\in\mathbb{N}} \frac{1}{\|p^t\|} = \infty$$

Since

$$(1+a^2)^{\frac{1}{2}} p_t \leq \|p^t\| \leq (1+A^2)^{\frac{1}{2}} p_t$$

we get

$$\sum_{t\in\mathbb{N}} A_t = \infty \quad \text{if and only if} \quad \sum_{t\in\mathbb{N}} \frac{1}{p^t} = \infty$$

We can apply Lemma 8.3.C, on the relation between forwards optimal, strongly optimal allocations and reduced models, Corollary 8.3.E, on the "inner" and "outer" approximation of a reduced model and Corollary 8.4.A on the efficiency of a reduced model made up of hyperbola to conclude that $(\bar{c}_t)_{t\in\mathbb{Z}}$ is a forwards optimal allocation if and only if the reduced model $(R_t(\bar{c}_t))_{t\in\mathbb{Z}}$ is efficient.

The proof of **(b)** is analogous and is left to the reader.

□

The characterization of forwards optimal allocations in Corollary 8.5.A was first given by Cass [1972]. In view of this we will refer to the criterion as Cass' criterion. Cass studied production efficiency for an economy with an infinite number of dates. In such an economy, with finitely-lived producers who maximize profits, there is in an equilibrium the possibility of inefficiency due to an excessive amount of "savings" in the form of "overaccumulation" of capital.

Corollary 8.5.A generalizes to OG economies with many goods at every date and many consumers in every generation – Cf. Balasko and Shell [1980], Borglin and Keiding [1986)] and Burke [1987].

In Chapter 9 we will use Corollary 8.5.A to study optimality of equilibrium allocations for stationary economies.

In the literature the upper contour sets, $R_t(\bar{c}_t)$, are not explicitly approximated by hyperbola but rather assumptions on endowments and on the relation between first-order derivatives and second-order derivatives of utility functions are imposed. These assumptions imply that the sets $R_t(\bar{c}_t)$,$t \in \mathbb{Z}$, can be approximated by hyperbola as in Corollary 8.5.A. We will see in the next chapter that the approximation needed in Corollary 8.5.A can, under the Maintained Assumptions be carried out for an equilibrium allocation of a stationary economy. It turns out that the method of approximation can be carried out also for more general models. We postpone the discussion of this to the end of the next chapter.

Example 8.5.A Let $\mathcal{E} = (\mathbb{R}^2_{++}, \ln c^y + \ln c^o, e)_{t\in\mathbb{Z}}$. We will see in the next chapter that the assumptions of Corollary 8.5.A are satisfied. Assume that $e = (3, 1)$. Then $(p_t)_{t\in\mathbb{Z}}$ is an equilibrium price system if and only if there exist $(v_1, v_2) \in \mathbb{R}^2_+ \setminus \{0\}$ such that $p_t = v_1 3^t + v_2$ for $t \in \mathbb{Z}$ as explained in the example on page 235. Hence, the equilibrium allocation for the equilibrium with $v_1 = 0$ and $v_2 > 0$ is a strongly optimal allocation since

$$\sum_{t\in\mathbb{N}} \frac{1}{p_t} = \sum_{t\in-\mathbb{N}} \frac{1}{p_t} = \sum_{t\in\mathbb{N}} \frac{1}{v_2} = \infty$$

The equilibrium allocations for all other equilibria, where $v_1 > 0$, are finitely optimal allocations according to Theorem 8.2.B but not forwards optimal allocations since

$$\sum_{t\in\mathbb{N}} \frac{1}{p_t} = \sum_{t\in\mathbb{N}} \frac{1}{v_1 3^t + v_2} \leq \sum_{t\in\mathbb{N}} \frac{1}{v_1 3^t} = \frac{1}{v_1}\frac{1}{2} < \infty$$

Assume instead that $e = (1, 3)$. Then $(p_t)_{t\in\mathbb{Z}}$ is an equilibrium price system if and only if there exist $(v_1, v_2) \in \mathbb{R}^2_+ \setminus \{0\}$ such that $p_t = v_1 3^{-t} + v_2$ for $t \in \mathbb{Z}$ as the example on page 235 reveals. Hence, the equilibrium allocation for the equilibrium with $v_1 = 0$ and $v_2 > 0$ is a strongly optimal allocation since

$$\sum_{t\in\mathbb{N}} \frac{1}{p_t} = \sum_{t\in-\mathbb{N}} \frac{1}{p_t} = \sum_{t\in\mathbb{N}} \frac{1}{v_2} = \infty$$

The equilibrium allocations for all other equilibria where $v_1 > 0$ are forwards optimal allocations since

$$\sum_{t\in\mathbb{N}} \frac{1}{p_t} = \sum_{t\in\mathbb{N}} \frac{1}{v_1 3^{-t} + v_2} \geq \sum_{t\in\mathbb{N}} \frac{1}{v_1 + v_2} = \infty$$

Equilibria and Strongly Optimal Allocations

The following theorem shows that for an economy where the initial endowments form a strongly optimal allocation the set of equilibria is easy to describe; it is simply the allocation of the initial endowments and the induced prices. Thus at an equilibrium each consumer consumes her endowment and there is no trade.

Theorem 8.5.B *Assume that $(e_t)_{t\in\mathbb{Z}}$ is a strongly optimal allocation for the economy $\mathcal{E} = (C_t, u_t, e_t)_{t\in\mathbb{Z}}$. Then $(e_t)_{t\in\mathbb{Z}}$ is the unique equilibrium allocation for $\mathcal{E}$.*

Proof: From the Second Theorem of Welfare Economics follows that $(e_t)_{t\in\mathbb{Z}}$ is an equilibrium allocation. Assume, in order to arrive at a contradiction, that $(\bar{c}_t)_{t\in\mathbb{Z}}$ is another equilibrium allocation. Since trade is voluntary $u(\bar{c}_t) \geq u(e_t)$ for $t \in Z$ and since $\bar{c}_{t'} \neq e_{t'}$ for some t' we have $u(\bar{c}_{t'}) > u(e_{t'})$. From $\bar{c}_t^y = e_t^y + (\bar{c}_t^y - e_t^y)$ we get that $(\bar{c}_t^y - e_t^y)_{t\in\mathbb{Z}}$ is an improving reallocation for $(e_t)_{t\in\mathbb{Z}}$. This contradicts that $(e_t)_{t\in\mathbb{Z}}$ is a strongly optimal allocation. Hence there can be no equilibrium allocation apart from $(e_t)_{t\in\mathbb{Z}}$.

Summary

In the present chapter optimality of allocations was studied. Three different notions of optimality were introduced. The First and Second Welfare Theorem were seen to be related to the weakest notion of optimality, based on variations of an allocation at only a finite number of dates. For policy issues the relevant notion was seen to be forwards optimal allocations, based on variations in the allocation from some date and forward. Since time is assumed to extend into the indefinite past it was also possible to consider allocations which could not be improved upon by a variation of the allocation from some date and backwards.

The open-ended time structure of the model implied possible reasons for lack of optimality which were not present in an economy with a finite number of dates. In order to study these problems we introduced reduced models. We defined a composition operation between sets in a reduced model which depicted the willingness of (a group of) consumers to substitute consumption between dates. Using the composition a general criterion for efficiency and strong efficiency for reduced models was stated.

For the applications we considered parametric efficiency criterion based on hyperbola. The family of hyperbola had the advantage of having a simple relation between the parameters and the composition among such sets.

For OG economies and equilibrium allocations where the approximation by hyperbola was possible we derived a criterion for forwards and strong optimality based on the equilibrium prices. Finally, we demonstrated the relevance of the notion of strong optimality by showing that for an economy where the initial endowment allocation was a strongly optimal allocation it was also the unique equilibrium allocation.

Exercises

Exercise 8.A Show that the hyperbola sets have the properties stated in the text.

Exercise 8.B Let $(\bar{c}_t)_{t\in\mathbb{Z}}$ be an allocation. Show that $(R_t(\bar{c}_t))_{t\in\mathbb{Z}}$ where

$$R_t(\bar{c}_t) \;=\; \{(z_1, z_2) \in \mathbb{R}^2 \mid u_t(\bar{c}_t^y + z_1, \bar{c}_t^o + z_2) \;\geq\; u_t(\bar{c}_t^y, \bar{c}_t^o)\}$$

is a reduced model. What properties do the sets $R_t(\bar{c}_t)$ have besides those asserted in the definition of a reduced model?

Exercise 8.C Let $((p_t), (\bar{c}_t))_{t\in\mathbb{Z}}$ be an equilibrium for an economy $\mathcal{E} = (C_t, u_t, e_t)_{t\in\mathbb{Z}}$ satisfying the Maintained Assumptions such that

$$\sum_{t=0}^{\infty} p_t(e_{t-1}^o + e_t^y) \;<\; \infty$$

Show that $(\bar{c}_t)_{t\in\mathbb{Z}}$ is a forwards optimal allocation. (*Hint:* Check if the idea from the proof of the First Theorem of Welfare Economics for Arrow-Debreu economies can be applied.)

Exercise 8.D Let $\mathcal{E} = (C_t, u_t, e_t)_{t\in\mathbb{Z}}$ be an economy where

$$u_t(c^y, c^o) = c^y c^o, \ C_t = \mathbb{R}^2_{++} \text{ and } e_t = \lambda_t(1,1))$$

for some sequence of positive numbers λ_t, $t \in \mathbb{Z}$. Consider the sequence $(a_t)_{t\in\mathbb{Z}}$ with $a_t = 0$ for $t < 0$ and $a_t = 2^t$ for $t \geq 0$. Find a condition on the sequence $(\lambda_t)_{t\in\mathbb{Z}}$ which ensures that $(a_t)_{t\in\mathbb{Z}}$ is an improving reallocation for the allocation of initial endowments.

Exercise 8.E Let $\mathcal{E} = (C_t, u_t, e_t)_{t\in\mathbb{Z}}$ be an economy satisfying the Maintained Assumptions and $((p_t), (\bar{c}_t))_{t\in\mathbb{Z}}$ an equilibrium for $\mathcal{E}$. Assume that there is a forward improvement for the reduced model induced by $(\bar{c}_t)_{t\in\mathbb{Z}}$. Show that there is a forward improvement $(\hat{b}_t)_{t\in\mathbb{Z}}$ with $\hat{b}_t = 0$ for $t < 0$ and $\hat{b}_t \neq 0$ for $t \geq 0$

Exercise 8.F Show that the composition $\circ$ is associative, that is, if S, S', S'' are three subsets of $\mathbb{R}^2$ then

$$(S \circ S') \circ S'' = S \circ (S' \circ S'')$$

Is the composition commutative so that $S \circ S' = S' \circ S$?

Exercise 8.G Let S and S' be convex sets and let
$H^+ = \{z \in \mathbb{R}^2 \mid z_1 + z_2 \geq 0\}$
Assume that $S \subset H^+$ and $S' \subset H^+$

(a) Show that $S \circ H^+ = S$ and $H^+ \circ S = S$.

(b) Show that $S \circ S' \subset H^+ \circ S'$ and $S \circ S' \subset S \circ H^+$

(c) Conclude from (a) and (b) that $S \circ S' \subset S$ and $S \circ S' \subset S'$.

Exercise 8.H Let $u, u' : \mathbb{R}^2_{++} \longrightarrow \mathbb{R}$ be utility functions satisfying the Maintained Assumptions and let $\bar{z}, \hat{z} \in \mathbb{R}^2_{++}$. Define

$$\begin{cases} R &= \{z \in \mathbb{R}^3 \mid u(z_1, z_2) \geq u(\bar{z}) \text{ and } z_3 = 0\} \;-\; \{(\bar{z}, 0)\} \\ S &= \{z \in \mathbb{R}^2 \mid u(z_1, z_2) \geq u(\bar{z})\} \;-\; \{\bar{z}\} \end{cases}$$

$$\begin{cases} R' &= \{z \in \mathbb{R}^3 \mid u'(z_2, z_3) \geq u'(\hat{z}) \text{ and } z_1 = 0\} \;-\; \{(0, \hat{z})\} \\ S' &= \{z \in \mathbb{R}^2 \mid u'(z_1, z_2) \geq u'(\hat{z})\} \;-\; \{\hat{z}\} \end{cases}$$

Show that

$$\left\{(y_1, y_3) \in \mathbb{R}^2 \mid (y_1, 0, y_3) \in R + R'\right\} \;=\; S \circ S'$$

Exercise 8.I Let $((p_t), (\bar{c}_t))_{t \in \mathbb{Z}}$ be an equilibrium for an OG economy

$$\mathcal{E} \;=\; (C_t, u_t, e_t)_{t \in \mathbb{Z}}$$

satisfying the Maintained Assumptions and let $(b_t)_{t \in \mathbb{Z}}$ be a forward improvement, for the induced reduced model, with $b_t = 0$ for $t < 0$ and $b_0 \neq 0$. Show that $p_0 b_0 < p_1 b_1 < p_2 b_2 < \ldots$.

Exercise 8.J Let $((p_t), (\bar{c}_t))_{t \in \mathbb{Z}}$ be an equilibrium for $\mathcal{E} = (C_t, u_t, e_t)_{t \in \mathbb{Z}}$ and assume that $\liminf_{t \to \infty} p_t \bar{c}_t^y = 0$. Show that the equilibrium allocation is a forwards optimal allocation. (*Hint:* Use the result from Exercise 8.I. This is an efficiency criterion first given by Malinvaud [1953] for production models.)

Exercise 8.K Let $(a_t)_{t \in \mathbb{Z}}$ be a sequence of positive numbers and $(S_t)_{t \in \mathbb{Z}}$ a reduced model with

$$S_t \;=\; \left\{z \in \mathbb{R}^2 \mid z_1 + z_2 \geq 0,\;\; z_1 \geq -a_t\right\}$$

(a) Show that

$$S_t \circ S_{t+1} \;=\; \left\{z \in \mathbb{R}^2 \mid z_1 + z_2 \geq 0,\;\; z_1 \geq -\min(a_t, a_{t+1})\right\}$$

(b) Show that $\mathrm{pr}_1\, S_t \circ S_{t+1} = [-\min(a_t, a_{t+1}), +\infty[$

(c) Show that $(S_t)_{t\in\mathbb{Z}}$ is an efficient reduced model if and only if $\liminf_{t\to\infty} a_t = 0$.

Exercise 8.L Let $(p_t)_{t\in\mathbb{Z}}$ with $p_t > 0, t \in \mathbb{Z}$, be given and let

$$S_t = \left\{ z \in \mathbb{R}^2 \mid p_t z_1 + p_{t+1} z_2 \geq 0 \right\}$$

(a) Is $(S_t)_{t\in\mathbb{Z}}$ a reduced model?

(b) Is $(S_t)_{t\in\mathbb{Z}}$ an efficient reduced model?

(c) Is $(S_t)_{t\in\mathbb{Z}}$ a strongly efficient reduced model?

(d) Relate your answers in (b) and (c) to the Cass criterion and the assumptions needed to apply that criterion.

Exercise 8.M Let $(S_t)_{t\in\mathbb{Z}}$ be a reduced model with each S_t, for $t \in \mathbb{Z}$, a set from the family of hyperbola. Assume that for infinitely many $t \in \mathbb{Z}$, the set $S_t = S_H(\infty, q) = \mathbb{R}^2_+$.

(a) Can the reduced model be efficient?

(b) Can the reduced model be strongly efficient?

Exercise 8.N Show that if an OG economy has a Samuelson equilibrium then its real equilibrium allocation is not a forwards optimal allocation.

Exercise 8.O Show that if an OG economy has a classical equilibrium then its real equilibrium allocation is not a strongly optimal allocation.

Exercise 8.P Let $\mathcal{E} = (\mathbb{R}^2_{++}, \min(c^y, c^o), e)_{t\in\mathbb{Z}}$ where $e \in \mathbb{R}^2_{++}$.

(a) Does the economy satisfy the Maintained Assumptions?

(b) What are the equilibrium conditions?

(c) Find the demand function

(d) Does the economy have a real equilibrium?

(e) Find an equilibrium.

(f) Show that allocations, $(c_t)_{t\in\mathbb{Z}}$, where $c_t^y = c_t^o$ for $t \in \mathbb{Z}$ are strongly optimal allocations.

Chapter 9

STATIONARY OG ECONOMIES

Introduction

Already from the outset of our studies of OG economies the discussion was limited to OG economies which are simple in that there is only one good at each date and only one consumer born at each date, who acts at two consecutive dates. In Chapter 9 to 11 we will study an even more restricted class of OG economies referred to as stationary OG economies.

In a stationary OG economy consumer characteristics are stationary over time. Hence each consumer has the same consumption set, initial endowment and utility function and the economy is described by the characteristics of this representative consumer.

An Arrow-Debreu economy where all consumers are identical do not allow for interesting equilibria. But in an OG economy the young consumer trades with the old consumer at the same date and also a stationary economy opens up for interesting equilibrium phenomena like cycles or sunspot equilibria. The assumption of identical preferences implies that there will be a single demand function. At each date there is a young consumer, acting according to the value of the demand function when young, evaluated at the price today and the expected price the next date. There is also an old consumer acting according to the demand function as old, evaluated at the price today and the realized price at the date before. Thus the fact that, at each date, markets have to balance the total endowment with the total demand of the young and the old consumer, who furthermore act against different prices, provides an intuitive explanation of the richness of phenomena occurring also in stationary OG economies.

In Section 9.2 prices are normalized so that demand is given in relative prices and each consumer faces the price 1 as old. In that section we also take note of some properties of demand, which follow from properties of the Walrasian demand function given in Chapter 1. Finding an equilibrium price system (of relative prices) for the economy amounts to finding a solution to the market balance equation, or *equilibrium equation.* Once an equilibrium price system is found the consumption bundles are given by the demand function. The equilibrium equation is a difference equation. The study of this equation is complicated by two features. The equation, it is non-linear and it is implicit, so that one can not in general solve explicitly for the successor (or predecessor) of a given price.

In the literature prices are often given in terms of real interest rates and equilibrium conditions are stated using the savings function. At the end of Section 9.3 it is shown how that approach is related to the one adopted here.

The simplest type of equilibria for a stationary OG economy are referred to as *steady states,* which is the subject of Section 9.4. In a steady state each consumer gets the same commodity bundle and faces the same (relative) prices. From Chapter 7 we know that the initial endowments is the equilibrium allocation for a real equilibrium. If the economy is a stationary OG economy then the allocation made up of the initial endowments is also the equilibrium allocation for a steady state. This steady state equilibrium is referred to as the *real steady state (equilibrium),* and since each consumer demands her initial endowment the *intensity of trade* is 0 in such a steady state.

The relative prices at a real steady state are given by the (common) gradient since consumption equals the initial endowment. Unless this relative price is 1 there will, besides the real steady state, also be a *nominal steady state (equilibrium)* where prices are constant over time, so that the relative price faced by each consumer is 1. At a nominal steady state the demanded commodity bundle may differ markedly from the initial endowment so that the intensity of trade is high compared to the real steady state.

In the sequel we consider stationary economies and equilibria for these economies as the initial endowment varies. The economies where the real steady state and the nominal steady state coincide, so that each consumer demands his initial endowment and equilibrium prices are constant over time play a particular role. Such an economy has a unique equilibrium and the equilibrium allocation is the allocation formed by the initial endowment which is a strongly optimal allocation. The stationary economies where the only equilibrium is that formed by the initial endowment are referred to as *no-trade economies* since the equilibrium involves no trade.

In Section 9.5 we study optimality for a stationary economy and show that, under the Maintained Assumptions, knowledge of the equilibrium price system is sufficient to determine, whether an allocation is a forwards optimal allocation or a strongly optimal allocation.

9.1 Definition of a Stationary OG Economy

Let us begin by defining a stationary OG economy.

Definition 9.1.A *Let* $\mathcal{E} = (C_t, u_t, e_t)_{t\in\mathbb{Z}}$ *be an OG economy.* $\mathcal{E}$ *is a* **stationary OG economy** *if there is a consumer* (C, u, e) *such that* $(C_t, u_t, e_t) = (C, u, e)$ *for* $t \in \mathbb{Z}$.

As mentioned in Chapter 1 we will assume that the consumer defining the stationary OG economy satisfies the Maintained Assumptions and a stationary OG economy will be denoted $\mathcal{E}^s = (C, u, e)$. In order to study the relation between endowments and equilibria, the consumption set C and the utility function u are considered to be fixed while endowment e is taken to be a parameter. Hence the economy $\mathcal{E}^s = (C, u, e)$ will often be referred to simply as the economy e.

9.2 Demand

In this section demand as a function of relative price is defined and related to Walrasian demand, from Section 1.1. The income and substitution effects of a variation in relative price are studied and the derivative of demand is given in terms of derivatives and second derivatives of the utility function. Demand as a function of the relative price will be useful since it allows us to state the equilibrium conditions in relative prices.

Relative Price and Demand

The purpose of this section is to introduce demand as a function of the relative price faced by a consumer in an OG economy.

Walrasian demand in a stationary OG economy

In a stationary OG economy $\mathcal{E}^s = (C, u, e)$, the utility function u can be used to derive the Walrasian demand function $\hat{f}$ as defined in Section 1.1. For the case of two commodities $\hat{f} = (\hat{f}^y, \hat{f}^o) : \mathbb{R}^2_{++} \times \mathbb{R}_{++} \longrightarrow C$ where $\hat{f}$ takes prices and wealth to the demanded commodity bundle

$$(p, p', w) \longrightarrow (\hat{f}^y(p, p', w), \hat{f}^o(p, p', w))$$

Recall that p and p' can be interpreted as the prices at the current date for delivery at that date and the next date or as spot-prices at the current date and the next date. The demanded quantity at the current date is $\hat{f}^y(p, p', w)$ and $\hat{f}^o(p, p', w)$ is the demanded quantity the next date.

With wealth given by the value of the initial endowment we get

$$(p, p') \longrightarrow (p, p', pe^y + p'e^o) \longrightarrow (\hat{f}^y(p, p', pe^y + p'e^o), \hat{f}^o(p, p', pe^y + p'e^o))$$

Since the prices can be interpreted as spot prices we will refer to them as spot prices or nominal prices to distinguish them from relative prices to be introduced below.

Demand as a function of relative price

The Walrasian demand function with wealth given by the value of the initial endowment is homogenous of degree 0 in spot prices so that

$$\begin{aligned}&(\hat{f}^y(p,p',pe^y+p'e^o),\hat{f}^o(p,p',pe^y+p'e^o))\\&\qquad=(\hat{f}^y(\frac{p}{p'},1,\frac{p}{p'}e^y+e^o),\hat{f}^o(\frac{p}{p'},1,\frac{p}{p'}e^y+e^o))\end{aligned}$$

Let $r\in\mathbb{R}_{++}$ denote the relative price p/p'. We can now define demand as a function of relative price and initial endowment.

Definition 9.2.A *The function* $f:\mathbb{R}_{++}\times\mathbb{R}^2_{++}\longrightarrow C$ *defined by*

$$f(r,e)=(f^y(r,e),f^o(r,e)) = (\hat{f}^y(r,1,re^y+e^o),\hat{f}^o(r,1,re^y+e^o))$$

is the **demand function** *giving quantity demanded as a function of the relative price and the initial endowment.*

We will refer to f simply as the demand function, which is the same for all consumers when the economy is stationary.

Properties of the Demand Function

In this section we relate demand as a function of relative price to the Walrasian demand and derive some properties of demand as a function of relative price.

Income and substitution effects

A variation in the relative price at the current date affects the relative price but also the wealth of the consumer. Denoting the partial derivatives with respect to the relative price r by D_r and the partial derivatives of $\hat{f}^y$ and $\hat{f}^o$ by D_1,D_2 and D_3 we have the following relations

$$\begin{aligned}D_rf^y(r,e)&=D_r\hat{f}^y(r,1,re^y+e^o)\\&=D_1\hat{f}^y(r,1,re^y+e^o)+D_3\hat{f}^y(r,1,re^y+e^o)\cdot e^y\\D_rf^o(r,e)&=D_r\hat{f}^o(r,1,re^y+e^o)\\&=D_1\hat{f}^o(r,1,re^y+e^o)+D_3\hat{f}^o(r,1,re^y+e^o)\cdot e^y\end{aligned}$$

A variation in r; the relative price of the good to the consumer, generates substitution effects as well as an income effect. The size of the income effect is seen to depend on the initial endowment as young. Thus for $e = (e^y, e^o)$ such that $e \in \mathbb{R}^2_{++}$ and $re^y + e^o$ is constant the income effect is stronger the larger the endowment as young. For demand as old the effect of a variation in r on the demand as old is related to the demand as young, $\hat{f}^y$, by Walras' law, as is seen in Lemma 9.2.C below. An increase in r corresponds to a decrease in the price of the good for the consumer as old.

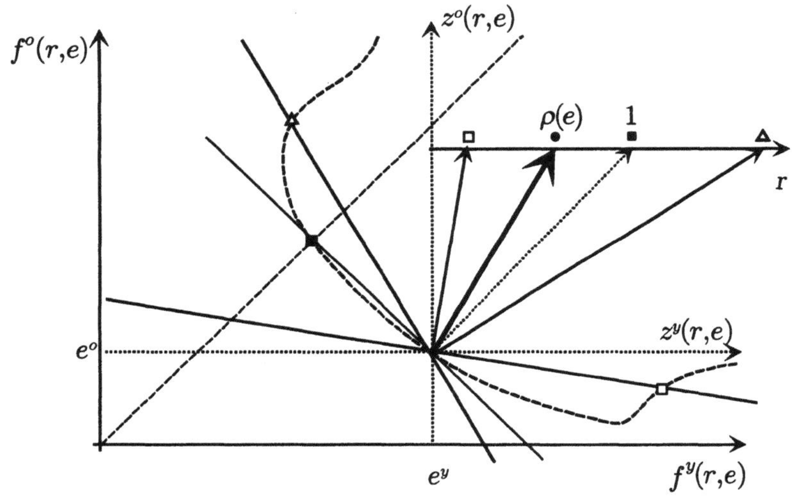

Figure 9.2.A: *For $r \in \mathbb{R}_{++}$ we can depict the consumer's demand and corresponding net trade. The budget restriction for $r = \rho(e)$ is not shown*

Using that demand is homogenous of degree 0 we have

$$\begin{aligned} D_r f^o(r,e) &= D_r \hat{f}^o(r, 1, re^y + e^o) \\ &= D_{\frac{1}{r}} \hat{f}^o(1, \frac{1}{r}, e^y + \frac{1}{r}e^o) \cdot \frac{\partial(1/r)}{\partial r} \\ &= \left[D_2 \hat{f}^o(1, \frac{1}{r}, e^y + \frac{1}{r}e^o) + D_3 \hat{f}^o(1, \frac{1}{r}, e^y + \frac{1}{r}e^o) \cdot e^o \right] \cdot -\frac{1}{r^2} \end{aligned} \quad \mathbf{9.2.A}$$

which shows that the sign of $D_r f^o(r,e)$ is opposite to that of the derivative with respect to the own-price of the good as old.

Basic properties of demand

Lemma 9.2.B below gives useful properties of demand as a function of relative price and Figure 9.2.A summarizes the intuition behind these results. The figure suggests how the curve connecting the consumer choices at different relative prices; the offer curve, is constructed. Note that, for each $r \in \mathbb{R}_{++}$, the consumer always has the opportunity to choose her initial endowment and hence $u(f(r,e)) \geq u(e)$. The strict quasi-concavity of the utility function implies that equality holds only if $r = \rho(e)$ where $(\rho(e), 1)$ is the normalized gradient of u evaluated at e.

Lemma 9.2.B *Let* $f : \mathbb{R}_{++} \times \mathbb{R}^2_{++} \longrightarrow C$ *be the demand function. Then*

(a) the *conditions* (i)-(iii) below *are equivalent*

(i) $r \geq \rho(e)$ (ii) $f^y(r,e) \leq e^y$ (iii) $f^o(r,e) \geq e^o$

(b) $\lim_{r \to 0} f^y(r,e) = \lim_{r \to \infty} f^o(r,e) = \infty$.

Proof: To prove **(a)** note that by Walras' law $r(f^y(r,e) - e^y) + (f^o(r,e) - e^o) = 0$ for $r > 0$ which immediately gives that (ii) and (iii) are equivalent.

We will now prove that (i) implies (ii). Let $r \geq \rho(e)$ and assume that $f^y(r,e) - e^y > 0$, in order to obtain a contradiction. Then

$$\rho(e)(f^y(r,e) - e^y) + (f^o(r,e) - e^0) \leq 0$$

which shows that the consumer could have chosen $f(r,e)$ at prices $(\rho(e), 1)$ Since e was chosen at these prices we have $u(e) = u(f(\rho(e), e)) > u(f^y(r,e)$ which is a contradiction. Hence $r \geq \rho(e)$ implies $f^y(r,e) - e^y \leq 0$. Cf. Figure 9.2.A.

To prove that (ii) implies (i), let $f^y(r,e) - e^y \leq 0$. Assume, in order to obtain a contradiction, that $r < \rho(e)$. Then $f(r,e) \neq e$ and $u(f(r,e)) > u(e)$. But

$$\rho(e)(f^y(r,e) - e^y) + (f^o(r,e) - e^0) \leq 0$$

implies that $u(f(r,e)) \leq u(e)$. From this contradiction follows $r \geq \rho(e)$ as was to be proved.

(b) Choose $k \geq \|e\|$ and consider the set

$$K = \{c \in C \mid u(c) \geq u(e), \|c\| \leq k\}$$

K is a non-empty, compact set. The mapping $\rho : \mathbb{R}^2_{++} \longrightarrow \mathbb{R}_{++}$ is continuous and hence the image of K is a compact set. Thus $\rho(K)$ is contained in an interval $[a, b]$ with $a > 0$.

If $f(r, e) \in K$ then, since $\rho(f(r, e)) = r$, we get $r \in [a, b]$. Conversely, if $r \notin [a, b]$ then $f(r, e) \notin K$, which implies $\|f(r, e)\| > k$.

Thus if $r \to \infty$ then $\|f(r, e)\| \to \infty$. But from (a) above, $0 < f^y(r, e) \leq e^y$ for $r \geq \rho(e)$, and thus $f^o(r, e) \to \infty$. If $r \to 0$ then an analogous reasoning shows that $f^y(r, e) \to \infty$. Cf. Figure 9.2.A.

Differentiability properties of demand

In the sequel it will be useful to know that demand as a function of the relative price is differentiable. It turns out that we can in fact calculate the partial derivatives and relate them to the derivatives of the utility function.

Let $\Delta(c, (1, -r))$ denote the quadratic form occurring in Assumption (C5) evaluted at $(1, -r)$. Thus

$$\begin{aligned} \Delta(c, (1, -r)) &= D^2_{yy}u(c) - rD^2_{yo}u(c) - rD^2_{oy}u(c) + r^2 D^2_{oo}u(c) \\ &= (1, -r)^T D^2 u(c) \begin{pmatrix} 1 \\ -r \end{pmatrix} \end{aligned}$$

The following lemma relates the derivatives of the demand function to the derivatives of the utility function.

Lemma 9.2.C *Let $f : \mathbb{R}_{++} \times \mathbb{R}^2_{++} \longrightarrow C$ be the demand function mapping (r, e) to $f(r, e) = (f^y(r, e), f^o(r, e))$. Then*

(a) $f \in \mathbb{C}^1(\mathbb{R}_{++} \times \mathbb{R}^2_{++}, C)$

(b) $rf^y(r, e) + f^o(r, e) = re^y + e^o$ *(Walras' law)*

(c) *the derivatives of the demand function with respect to the relative price are, with $c = f(r, e)$,*

$$\begin{cases} D_r f^y(r,e) &= -\dfrac{(e^y - c^y)(D^2_{yo}u(c) - rD^2_{oo}u(c)) - D_o u(c)}{\Delta(c,(1,-r))} \\ D_r f^o(r,e) &= e^y - f^y(r,e) - rD_r f^y(r,e) \\ &= \dfrac{(e^y - c^y)(D^2_{yy}u(c) - rD^2_{oy}u(c)) - D_y u(c)}{\Delta(c,(1,-r))} \end{cases}$$

Proof: To prove **(a)** note that by Propositon 1.1.B the Walrasian demand function $\hat{f}$ with values $\hat{f}(p,p',w)$ is a differentiable function of (p,p',w). Since the demand function $f(r,e)$ is a composition of differentiable functions

$$(r,e) \longrightarrow (r,1,re^y + e^o) \longrightarrow \hat{f}(r,1,re^y + e^o) = f(r,e)$$

it follows that $f \in \mathbb{C}^1(\mathbb{R}_{++} \times \mathbb{R}^2_{++}, C)$.

(b) Follows from the corresponding property of the Walrasian demand function.

(c) Since $f(r,e)$ is the unique solution to the Consumer Problem at prices $(r,1)$ and wealth $w = re^y + e^o$ it follows that $c^y = f^y(r,e)$ is the unique solution to

$$\text{Max}_{c^y}\ u(c^y, re^y + e^o - rc^y) \text{ where } 0 < c^y < e^y + \frac{e^o}{r}$$

and hence

$$\begin{aligned} &D_{c^y}u(c^y, re^y + e^o - rc^y) \\ &\quad = D_y u(c^y, re^y + e^o - rc^y) - rD_o u(c^y, re^y + e^o - rc^y) = 0 \end{aligned}$$

if and only if $c^y = f^y(r,e)$.

By (C5) $\Delta(c,(1,-r)) < 0$ and a calculation shows that

$$D^2_{c^y c^y}u(c^y, re^y + e^o - rc^y) = \Delta(c,(1,-r)) < 0$$

with the derivative evaluated at $c^y = f^y(r,e)$ and $c^o = re^y + e^o - rf^y(r,e) = f^o(r,e)$.

From the first order condition follows that for $r \in \mathbb{R}_{++}$

$$D_y u(f^y(r,e), re^y + e^o - rf^y(r,e)) - rD_o u(f^y(r,e), re^y + e^o - rf^y(r,e)) = 0$$

and differentiation with respect to r gives

$$D_r f^y(r,e) \;=\; -\frac{(e^y - c^y)(D^2_{yo}u(c) - rD^2_{oo}u(c)) - D_o u(c)}{\Delta(c,(1,-r))}$$

when evaluated at $c = f(r,e)$.

By Walras' law

$$rf^y(r,e) + f^o(r,e) \;=\; re^y + e^o$$

which implies, with $c = f(r,e)$,

$$\begin{aligned} D_r f^o(r,e) &= e^y - c^y - rD_r f^y(r,e) \\ &= \frac{(e^y - c^y)(D^2_{yy}u(c) - rD^2_{oy}u(c)) - D_y u(c)}{\Delta(c,(1,-r))} \end{aligned}$$

□

From the (c) part it is seen how the derivative of the demand function is related to the first order and second order derivatives of the utility function and the intensity of trade, $f(r,e) - e$.

Net trade and relative price

Using the demand function we can define the function $z : \mathbb{R}_{++} \times \mathbb{R}^2_{++} \longrightarrow C$ by

$$z(r,e) \;=\; (z^y(r,e), z^o(r,e)) \;=\; (f^y(r,e) - e^y, f^o(r,e) - e^o)$$

The function z thus gives the desired net trade of a consumer with initial endowment e at the relative price r. For a given e the function $(z^y(\cdot,e), z^o(\cdot,e))$ takes values in the set

$$\left\{x \in \mathbb{R}^2 \mid x_1 \;>\; -e^y, x_2 > -e^o\right\}$$

which is bounded below by the vector $(-e^y, -e^o)$. The interpretation is that the consumer can only supply a quantity less than her initial endowment.

Relative price and normalized gradient

Let $\rho \in \mathbb{C}^1(C, \mathbb{R}_{++})$ be the first component of the normalized gradient evaluated at c so that

$$\frac{1}{D_o u(c)} Du(c) = \left(\frac{D_y u(c)}{D_o u(c)}, 1\right) = (\rho(c), 1) \quad \text{or} \quad \rho(c) = \frac{D_y u(c)}{D_o u(c)}$$

Then $\rho(c)$ is *marginal rate of substitution* or *normalized gradient*, at $c \in C$. For $r > 0$ we have $\rho(f(r, e)) = r$ and in particular $f(\rho(e), e) = e$. Thus $\rho(e)$ is the unique relative price inducing the consumer to choose his initial endowment as consumption, resulting in the net trade 0. Note that above we also used "normalized gradient" to denote the vector $Du(c)/\|Du(c)\|$. The context will determine the normalization referred to.

9.3 Equilibrium and Relative Prices

The purpose of this section is to give alternative formulations of the equilibrium conditions using spot prices, relative prices or real interest rates and to relate the market balance conditions to savings.

Demand and equilibrium conditions

Let $((p_t)_{t\in\mathbb{Z}}, (\bar{c}_t)_{t\in\mathbb{Z}})$ be an equilibrium where p_t, for $t \in \mathbb{Z}$, could be interpreted as spot prices or forward prices. Then the consumptions solve the Consumer Problem so that for $t \in \mathbb{Z}$

$$\bar{c}_t = f\left(\frac{p_t}{p_{t+1}}, e\right)$$

and markets balance at each date $t \in \mathbb{Z}$

$$\bar{c}^y_{t+1} + \bar{c}^o_t = e^y + e^o$$

Substituting the demand into the market balance condition we have the equilibrium equation for date $t+1$ in nominal prices

$$f^y\left(\frac{p_{t+1}}{p_{t+2}}, e\right) + f^o\left(\frac{p_t}{p_{t+1}}, e\right) = e^y + e^o \qquad \textbf{9.3.A}$$

Therefore, $(p_t)_{t\in\mathbb{Z}}$ is an equilibrium price system if and only it is a solution to the market balance equation 9.3.A for each $t \in \mathbb{Z}$.

Given an equilibrium price system the demand function can be used to find the corresponding consumptions. To recover the amount of money demanded we set $\bar{m}_t = p_t(e^y - \bar{c}_t^y)$ for $t \in \mathbb{Z}$. Then, with $M = \bar{m}_t$ we recover the equilibrium under the market structure of spot-markets and money $((p_t)_{t\in\mathbb{Z}}, (\bar{c}_t, \bar{m}_t)_{t\in\mathbb{Z}}, M)$.

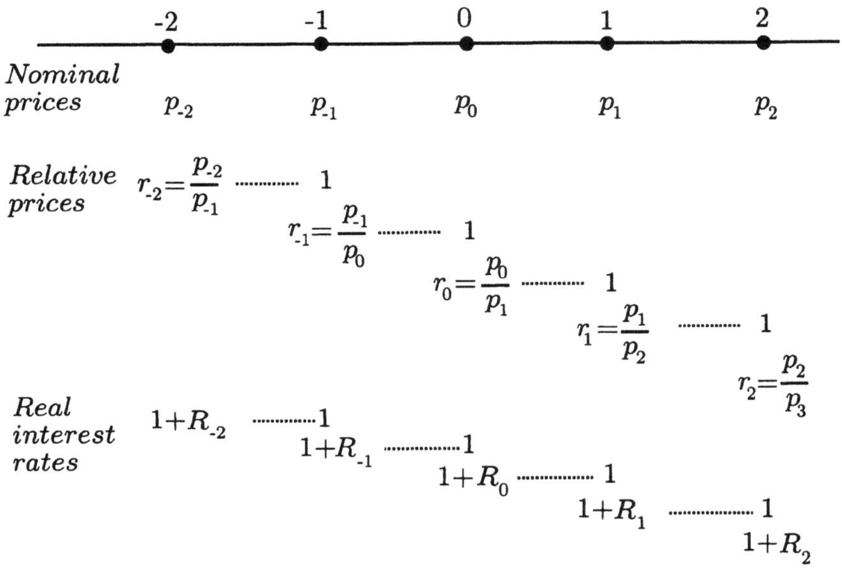

Figure 9.3.A: *Relation between nominal spot prices, relative prices and real interest rates*

Equilibrium price systems and relative prices

Obviously, a sequence of spot prices $(p_t)_{t\in\mathbb{Z}}$ induces a unique sequence of relative prices $(p_t/p_{t+1})_{t\in\mathbb{Z}}$; cf. Figure 9.3.A and a sequence of relative prices induces a sequence of spot prices, which is unique up to the choice of price level. Thus equilibrium price systems may be given either as spot prices, that is, p_t's, or as relative prices, that is, p_t/p_{t+1}'s. Indeed, on the one hand, $(p_t)_{t\in\mathbb{Z}}$ is a solution to the difference equation 9.3.A if only if $(r_t)_{t\in\mathbb{Z}}$, with

$$r_t = \frac{p_t}{p_{t+1}} \quad \text{for } t \in \mathbb{Z}$$

satisfies the date $t+1$ market balance condition for $t \in \mathbb{Z}$,

$$f^y(r_{t+1}, e) \;+\; f^o(r_t, e) \;=\; e^y + e^o \qquad \textbf{9.3.B}$$

for $t \in \mathbb{Z}$. This equation will be referred to as *the equilibrium equation in relative prices* for date $t+1$,where this date can be thought of as the current date. On the other hand, $(r_t)_{t\in\mathbb{Z}}$ satisfies the difference equation 9.3.B at each date if and only if $(p_t)_{t\in\mathbb{Z}}$, where $p_0 > 0$ is arbitrary, and

$$p_t \;=\; \begin{cases} r_t \ldots r_{-3} r_{-2} r_{-1} p_0 \;=\; p_0 \prod_{\tau=t}^{0} r_\tau & \text{for} \quad t \leq 0 \\[2ex] p_0 \dfrac{1}{r_0}\dfrac{1}{r_1} \cdots \dfrac{1}{r_{t-1}} \;=\; p_0 \prod_{\tau=0}^{t-1} \dfrac{1}{r_\tau} & \text{for} \quad t > 0 \end{cases}$$

satisfies the difference equation 9.3.A in nominal prices at each date.

The problem of finding an equilibrium price system for a stationary OG economy amounts to finding a solution to the difference equation 9.3.B. The difference equation is a first-order and one-dimensional difference equation since it involves only two consecutive values of the real variable r; it is stationary since the function f does not depend on time. It is implicit since we can not in general "solve" for r_{t+1} as a function of r_t and non-linear since the implicit relation between r_{t+1} and r_t is not given by a linear function.

Example 9.3.A Let $\mathcal{E}^s = (\mathbb{R}^2_{++}, \ln c^y + \ln c^o, (3,1))_{t\in\mathbb{Z}}$. Then the demand function $f : \mathbb{R}_{++} \times \mathbb{R}^2_{++} \to C$ is then given by

$$f(r,e) \;=\; \begin{pmatrix} \dfrac{1}{2}\left(e^y + \dfrac{e^o}{r}\right) \\[2ex] \dfrac{1}{2}\left(re^y + e^o\right) \end{pmatrix} \;=\; \begin{pmatrix} \dfrac{1}{2}\left(3 + \dfrac{1}{r}\right) \\[2ex] \dfrac{1}{2}(3r+1) \end{pmatrix}$$

according to the example on page 235. Hence, $(r_t)_{t\in\mathbb{Z}}$ is an equilibrium price system in relative prices if and only if

$$(f^y(r_{t+1}, e) - e^y) \;+\; (f^o(r_t, e) - e^o) \;=\; \frac{1}{r_{t+1}} + 3r_t - 4 \;=\; 0$$

for $t \in \mathbb{Z}$. Hence, $(r_t)_{t\in\mathbb{Z}}$ is an equilibrium price system if and only if there exist $(v_1, v_2) \in \mathbb{R}^2_+ \setminus \{0\}$ such that $r_t = (v_1 3^t + v_2)/(v_1 3^{t+1} + v_2)$ for $t \in \mathbb{Z}$ as explained in the example on page 235.

Equilibrium equation in net trades

Using the net trades the equilibrium equation at date $t+1$ becomes

$$z^y(r_{t+1}, e) + z^o(r_t, e) \;=\; 0$$

Total excess demand at date $t+1$ stems from the young consumer, acting given the prices $(r_{t+1}, 1)$, and the old consumer, acting at date $t+1$, who planned her consumption at date t given the prices $(r_t, 1)$.

Equilibrium, Real Savings and the Real Rate of Interest*

We saw above that the equilibrium price system may, disregarding the price level, equivalently be given as a sequence of relative prices. The purpose of this section is to show that the market balance conditions can be stated as conditions for savings to balance at each date and that an equilibrium price system may also be given as a set of real interest rates. The aim is to relate the terminology used here to that in other expositions, where equilibrium conditions are often stated using real interest rates and savings.

The real savings function

Let $\hat{g} \in \mathbb{C}^1(\mathbb{R}_{++} \times \mathbb{R}^2_{++}, \mathbb{R})$ be the real savings function defined by,

$$\hat{g}(r, e) \;=\; e^y - f^y(r, e)$$

Thus the real savings are simply the net trade as young with the sign adjusted. By Walras' law $rf^y(r,e) + f^o(r,e) = e^y + e^o$ and hence the real savings $\hat{g}(r, e)$ equal $(1/r)(f^o(r,e) - e^o)$.

The equilibrium conditions may be stated in terms of real savings: $(r_t)_{t\in\mathbb{Z}}$ is a equilibrium price system in relative prices if and only if the market for savings balances at each date so that $(r_t)_{t\in\mathbb{Z}}$ satisfies

$$-\,(e^y - f^y(r_{t+1}, e)) \;+\; r_t \left[\frac{1}{r_t}(f^o(r_t, e) - e^o)\right]$$
$$= -\hat{g}(r_{t+1}, e) + r_t\hat{g}(r_t, e) \;=\; 0$$

or $\hat{g}(r_{t+1}, e) = r_t \hat{g}(r_t, e)$, so that the real savings of the young consumer at date $t+1$ equals the real dissavings of the old consumer at the same date.

Part (a) of Lemma 9.2.B, on the properties of demand, implies

$$r \geq \rho(e) \quad \text{if and only if} \quad \hat{g}(r, e) \geq 0$$

Nominal values of savings are constant over time at an equilibrium which implies that real savings have constant sign. Thus if $(r_t)_{t \in \mathbb{Z}}$ is an equilibrium price system then precisely one of the following alternatives is true: $r_t < \rho(e)$ for $t \in \mathbb{Z}$, or $r_t = \rho(e)$ for $t \in \mathbb{Z}$, or $r_t > \rho(e)$ for $t \in \mathbb{Z}$.

Real interest rates and the equilibrium conditions

If consumer t consumes one unit less of commodity t then she may consume r_t units more of commodity $t+1$. Thus the real rate of interest between date t and date $t+1$ is

$$R_t = r_t - 1$$

The equilibrium conditions may also be stated using real rates of interest: $(R_t)_{t \in \mathbb{Z}}$ is an equilibrium system of real interest rates if and only if the markets for savings balance at each date so that $(R_t)_{t \in \mathbb{Z}}$ is a solution to

$$\hat{g}(1 + R_{t+1}, e) = (1 + R_t)\hat{g}(1 + R_t, e).$$

Equilibrium prices can thus be given in terms of relative prices clearing the goods markets as well as relative prices or real rates of interest such that the market for savings balances at each date.

9.4 Steady State Equilibria

In this section we will define a steady state equilibrium and show that stationary economy has precisely two steady state equilibria if $\rho(e) \neq 1$ and a unique steady state equilibrium if $\rho(e) = 1$.

Definition of a steady state equilibrium

The simplest type of equilibrium is an equilibrium where each consumer faces the same relative price and gets the same consumption bundle. Such an equilibrium is referred to as a steady state equilibrium or a steady state for short.

Definition 9.4.A *A* **steady state equilibrium** *for a stationary OG economy* $\mathcal{E}^s = (C, u, e)$ *is an equilibrium* $(p_t)_{t\in\mathbb{Z}}, (\bar{c}_t)_{t\in\mathbb{Z}})$ *for which there exists a relative price,* r*, and a consumption bundle,* $\bar{c}$*, such that, for* $t \in \mathbb{Z}$

$$\begin{aligned} \bar{c}_t &= \bar{c} \\ p_t &= \frac{1}{r^t} p_0 \end{aligned}$$

To a steady state corresponds a whole class of equilibria with spot-markets and money but since these differ only in the price level and the amount of money we disregard the difference and use $(r, \bar{c})$ to denote a steady state.

Existence of steady state equilibria

The conditions for $(r, \bar{c})$ to be a steady state can be formulated using demand functions and relative prices. Thus $(r, \bar{c})$ is a steady state if each consumer maximizes her utility so that

$$f(r, e) = \bar{c} \quad \text{(each generation)}$$

and markets balance

$$f^y(r, e) + f^o(r, e) = e^y + e^o \quad \text{(each date)}$$

Hence, in order to establish the existence of a steady-state, one first finds a relative price such that the equilibrium equation 9.3.B is satisfied and next a consumption bundle by substituting the relative price into the demand function.

Theorem 9.4.B *Each stationary OG economy has a steady state for which* $(r, \bar{c}) = (\rho(e), e)$ *and a steady state for which* $(r, \bar{c}) = (1, f(1, e))$*. The steady states coincide if and only if* $\rho(e) = 1$*.*

Proof: If $(r, \bar{c})$ is a steady state then, according to Walras' law and the equilibrium equation 9.3.B,

$$\begin{aligned} rf^y(r,e) + f^o(r,e) &= re^y + e^o \\ f^y(r,e) + f^o(r,e) &= e^y + e^o \end{aligned}$$

If $f^o(r,e) - e^o$ is eliminated in these equations then the following equation is obtained

$$(r-1)(f^y(r,e) - e^y) = 0$$

This equation has only the solution(s) $r = 1$ and $r = \rho(e)$ corresponding to the steady states $(r, \bar{c}) = (1, f(1, e))$ and $(r, \bar{c}) = (\rho(e), e)$. Thus the economy has a steady state with $r = 1$ and a steady state with $r = \rho(e)$. These are the only steady states and they coincide if $\rho(e) = 1$.

□

A steady state $(r, \bar{c})$ is a real equilibrium if $r = \rho(e)$, in which case $f(r,e) = e$, and it is a nominal equilibrium if $\rho(e) \neq 1$ and $r = 1$, in which case $f(r,e) \neq e$. According to Lemma 9.2.B, on the properties of demand, a nominal steady state is a classical equilibrium if $\rho(e) < 1$ and a Samuelson equilibrium if $\rho(e) > 1$.

No-trade economies

If $e \in \mathbb{R}^2_{++}$ is an endowment such that $\rho(e) = 1$ then the allocation formed by the initial endowments is a strongly optimal allocation, by Corollary 8.5.A, on Cass' criterion, and the real steady state is the unique equilibrium for the economy by Theorem 8.5.B, on the uniqueness of equilibria. Since the consumptions in this equilibrium are just the initial endowments we refer to an economy where $\rho(e) = 1$ as a *no-trade* economy.

9.5 Optimality for Stationary Economies

In Section 8.5 we showed that if the upper contour sets associated with an equilibrium allocation could be approximated by a reduced model made up of hyperbola, then we could derive conditions for an allocation to be a forwards optimal allocation or a strongly optimal allocation.

In this section the purpose is to show that the Maintained Assumptions suffice for the results in Section 8.5 to be applicable to an equilibrium allocation of a stationary economy. The basic idea is to approximate the upper contour sets from the "inside" and the "outside" by hyperbola. But it turns out to be more convenient first to approximate the support functions for the corresponding reduced model. To do so we will need the properties of support functions in general and properties induced by the Maintained Assumptions in particular.

Although we limit the formal results to stationary economies the method is more generally applicable and at the end of this section we will indicate how the reasoning can be generalized to apply to equilibrium allocations of non-stationary economies. Also cf. Exercise 9.I.

Support Functions

A closed, convex set is the intersection of the upper halfspaces containing it. The support function is a concise description of these halfspaces and thus of the set itself.

Definition 9.5.A *Let A be a convex set which is bounded below and satisfies conditions* (R1)-(R3) given on page 252. *The* **support function** *of A is the function* $\tilde{e} : \mathbb{R}^2_+ \longrightarrow \mathbb{R}$ *with values given by*

$$\tilde{e}(p_1, p_2) = \inf_{x \in A} p_1 x_1 + p_2 x_2$$

We warn that there is an abuse of language here since the support function is usually defined as the supremum rather than the infimum.

Properties of support functions

For *positive* prices the values of the support function are given by the values of a linear function at a solution to a minimum problem.

Proposition 9.5.B *Let A be a convex set which is bounded below and satisfies conditions* (R1)-(R3) *and let* $(p_1, p_2) \in \mathbb{R}^2_{++}$. *Then the problem*

$$\text{Min}_{(x_1,x_2)}\ p_1 x_1 + p_2 x_2 \quad s.\ to\ (x_1, x_2) \in A \qquad \textbf{9.5.A}$$

has a solution $\bar{x} = (\bar{x}_1, \bar{x}_2)$ *and* $\tilde{e}(p_1, p_2) = p_1 \bar{x}_1 + p_2 \bar{x}_2$.

Proof: Let $b = (b_1, b_2)$ be a lower bound for A. Let $a \in A$ and let $(p_1, p_2) \in \mathbb{R}^2_{++}$ be given. Consider the set

$$\hat{A} = \{x \in \mathbb{R}^2 \mid p_1 x_1 + p_2 x_2 \leq p_1 a_1 + p_2 a_2, \quad x \in A \text{ and } x \geq b\}$$

$\hat{A}$ is a compact set and the function $p_1 x_1 + p_2 x_2$ attains a minimum at some point $\bar{x} \in \hat{A}$. It is easy to check that $\bar{x}$ is also a solution to the problem 9.5.A.

□

Let A and B be convex subsets of $\mathbb{R}^2$ which satisfy the conditions (R1)-(R3) in the definition of a reduced model and which are bounded below. Rather than comparing A and B with respect to inclusion we may compare the support functions of A and B. This turns out to be useful since it is somewhat easier to approximate the support functions of the sets from a reduced model than to approximate the sets themselves. It is the property asserted in Proposition 9.5.C which makes support functions attractive for approximation purposes.

Proposition 9.5.C *Let A and B be convex sets which are bounded below and satisfy conditions* (R1)-(R3). *Let $\tilde{e}^A$ and $\tilde{e}^B$ be the support functions of A and B. Then $A \subset B$ if and only if*

$$\tilde{e}^A(p_1, p_2) \geq \tilde{e}^B(p_1, p_2) \quad \text{for } (p_1, p_2) \in \mathbb{R}^2_+$$

Proof: To prove the "only if" part assume that $A \subset B$. Then, for $(p_1, p_2) \in \mathbb{R}^2_+$,

$$\tilde{e}^A(p_1, p_2) = \inf_{x \in A} p_1 x_1 + p_2 x_2 \geq \inf_{x \in B} p_1 x_1 + p_2 x_2 = \tilde{e}^B(p_1, p_2)$$

"if". Assume, in order to obtain a contradiction, that for $(p_1, p_2) \in \mathbb{R}^2_{++}$ we have $\tilde{e}^A(p_1, p_2) \geq \tilde{e}^B(p_1, p_2)$ but that there is a point $a \in A$ which does not belong to B. Then by Corollary B, on the separation of a convex set and a point, in Appendix C there is a $\bar{p} \in \mathbb{R}^2, \bar{p} \neq 0$ such that

$$\bar{p}_1 a_1 + \bar{p}_2 a_2 < \inf_{x \in B} \bar{p}_1 x_1 + \bar{p}_2 x_2 < \bar{p}_1 x_1 + \bar{p}_2 x_2 \quad \text{for } x \in B$$

and since $A + \mathbb{R}^2_+ \subset A$ we get, from the right hand inequality that $\bar{p} \in \mathbb{R}^2_+$.

Thus we get the following inequality

$$\tilde{e}^A(\bar{p}_1, \bar{p}_2) \leq \bar{p}_1 a_1 + \bar{p}_2 a_2 < \inf_{x \in B} \bar{p}_1 x_1 + \bar{p}_2 x_2 = \tilde{e}^B(\bar{p}_1, \bar{p}_2)$$

contradicting the assumption $\tilde{e}^A(p_1, p_2) \geq \tilde{e}^B(p_1, p_2)$ for $p \in \mathbb{R}^2_+$.

□

A closed, convex set A which is bounded below and satisfies (R1)-(R3) is completely described by its support function and

$$A = \cap_{p \in \mathbb{R}^2_+ \setminus \{0\}} H^+_{(p,\, \tilde{e}(p))}$$

where $H^+_{(p,\, \tilde{e}(p))}$ is the upper halfspace corresponding to the hyperplane given by $(p, \tilde{e}(p))$.

Support function of an upper contour set

When we consider convex sets occurring as upper contour sets for a utility function satisfying the Maintained Assumptions we also get differentiability properties.

Let U be the image in $\mathbb{R}$ of the mapping $c \longrightarrow u(c)$, where $c \in C$. For $(p_1, p_2, \bar{u}) \in \mathbb{R}^2_{++} \times U$ consider the problem of minimizing the expenditure to attain at least the utility level $\bar{u}$

$$\text{Min}_c \; p_1 c_1 + p_2 c_2 \quad \text{s.to } u(c) \geq \bar{u} \text{ and } c \in C \qquad \textbf{9.5.B}$$

From the strict convexity of preferences follows that the problem has a unique solution. The following proposition gives some of the properties of the solution.

Proposition 9.5.D *Let* $(p, \bar{u}) = (p_1, p_2, \bar{u}) \in \mathbb{R}^2_{++} \times U$.

(a) *The problem* 9.5.B *has a unique solution. Let* $(\hat{\eta}_1, \hat{\eta}_2) : (p, \bar{u}) \longrightarrow \mathbb{R}^2_{++}$ *be the function mapping* $(p, \bar{u})$ *to the solution* $(\hat{\eta}_1(p, \bar{u}), \hat{\eta}_2(p, \bar{u}))$ *Then* $(\hat{\eta}_1, \hat{\eta}_2) \in \mathbb{C}^1(\mathbb{R}^2_{++} \times U,\, \mathbb{R}^2_{++})$.

(b) *The function* $\hat{e} : \mathbb{R}^2_{++} \times U \longrightarrow \mathbb{R}$ *with values*

$$\hat{e}(p,\, \bar{u}) = p_1\, \hat{\eta}_1(p,\, \bar{u}) + p_2\, \hat{\eta}_2(p,\, \bar{u})$$

belongs to $\mathbb{C}^1(\mathbb{R}^2_{++} \times U,\, \mathbb{R})$.

(c) *For* $\bar{u} \in U$, *the function* $\hat{e}(\cdot, \bar{u}) \in \mathbb{C}^2(\mathbb{R}^2_{++},\, \mathbb{R})$.

Proof: To prove **(a)** choose $\bar{c}$ so that $u(\bar{c}) > \bar{u}$. Then

$$A = \{c \in \mathbb{R}^2_{++} \mid p_1 c_1 + p_2 c_2 \le p_1 \bar{c}_1 + p_2 \bar{c}_2 \text{ and } u(c) \ge \bar{u}\}$$

is a non-empty, compact set and thus the linear function $p \longrightarrow p \cdot c$ attains a minimum at some point $\hat{c} \in A$. It is easy to check that $\hat{c}$ is also a solution to 9.5.B. Since the utility function is strictly quasi-concave the solution is unique.

Let $V(p, w) = u(\hat{f}^y(p, w), \hat{f}^o(p, w))$, where $(\hat{f}^y, \hat{f}^o)$ is Walrasian demand as a function of prices and wealth. V is the indirect utility function and by Proposition 1.1.B V is a differentiable function of (p, w). It is easy to prove that

$$\begin{aligned} \hat{\eta}_1(p, V(p, w)) &= \hat{f}^y(p, w) \\ \hat{\eta}_2(p, V(p, w)) &= \hat{f}^o(p, w) \end{aligned}$$

and the reader is asked to show in Exercise 9.B that this implies that the partial derivatives of $(\hat{\eta}_1, \hat{\eta}_2)$ exist and are continuous functions.

(b) Since $\hat{e}(p, \bar{u}) = p_1 \hat{\eta}_1(p, \bar{u}) + p_2 \hat{\eta}_2(p, \bar{u})$ we get from **(a)**

$$\hat{e} \in \mathbb{C}^1(\mathbb{R}^2_{++} \times U, \mathbb{R})$$

(c) We have

$$\begin{aligned} D_1\hat{e}(p, \bar{u}) &= \hat{\eta}_1(p, \bar{u}) + p_1 D_1 \hat{\eta}_1(p, \bar{u}) + p_2 D_1 \hat{\eta}_2(p, \bar{u}) \\ D_2\hat{e}(p, \bar{u}) &= \hat{\eta}_2(p, \bar{u}) + p_1 D_2 \hat{\eta}_1(p, \bar{u}) + p_2 D_2 \hat{\eta}_2(p, \bar{u}) \end{aligned}$$

From the relation $u(\hat{\eta}_1(p, \bar{u}), \hat{\eta}_2(p, \bar{u})) - \bar{u} = 0$ for $p \in \mathbb{R}^2_{++}$ we get with the partial derivatives of u evaluated at the point $\hat{\eta}(p, \bar{u})$,

$$\begin{aligned} D_1 u \cdot D_1 \hat{\eta}_1(p, \bar{u}) + D_2 u \cdot D_1 \hat{\eta}_2(p, \bar{u}) &= 0 \\ D_1 u \cdot D_1 \hat{\eta}_1(p, \bar{u}) + D_2 u \cdot D_1 \hat{\eta}_2(p, \bar{u}) &= 0 \end{aligned}$$

Since $(D_1 u(p, \bar{u}), D_2 u(p, \bar{u}))$ is proportional to (p_1, p_2) we get

$$(D_1\hat{e}(p, \bar{u}), D_2\hat{e}(p, \bar{u})) = (\hat{\eta}_1(p, \bar{u}), \hat{\eta}_2(p, \bar{u}))$$

□

The function $\hat{\eta}$ is the Hicksian demand function and $\hat{e}$ the expenditure function.

Support function of a translated upper contour set

Let $\bar{c} \in C$ be given. We then have the gradient $Du(\bar{c})$ and the upper contour set $\{c \in C \mid u(c) \geq u(\bar{c})\}$. In order to compare the upper contour sets for different values of $\bar{c}$ to sets from the family of hyperbola it is convenient to translate the upper contour set with the vector $\bar{c}$. The set

$$R(\bar{c}) = \{c \in C \mid u(c) \geq u(\bar{c})\} - \{\bar{c}\}$$

is contained in the upper halfspace of the homogenous hyperplane with normal $Du(\bar{c})$.

We can now use the functions $\hat{\eta}$ and $\hat{e}$ to define the support function of the translated upper contour set $R(\bar{c})$. Thus define $\eta : \mathbb{R}^2_{++} \times C \longrightarrow \mathbb{R}^2_{++}$ and $e : \mathbb{R}^2_{++} \times C \longrightarrow -\mathbb{R}_+$ by

$$\begin{cases} \eta_{\bar{c},1}(p) = \hat{\eta}_1(p, u(\bar{c})) - \bar{c}_1 \\ \eta_{\bar{c},2}(p) = \hat{\eta}_2(p, u(\bar{c})) - \bar{c}_2 \end{cases} \quad \text{and} \quad e_{\bar{c}}(p) = p_1\eta_{\bar{c},1}(p) + p_2\eta_{\bar{c},2}(p) \tag{9.5.C}$$

Then $\eta_{\bar{c}}(\cdot)$ is the solution to the minimum problem: $\text{Min}_z\, p_1z_1 + p_2z_2$ s. to $(z_1, z_2) \in R(\bar{c})$ and $e_{\bar{c}}(\cdot)$ is the minimum value.

Price normalization

For $\bar{c} \in C$ let $q(\bar{c}) = (q_1(\bar{c}), q_2(\bar{c})) = Du(\bar{c})/\|Du(\bar{c})\|$ be the normalized gradient. The vector $q(\bar{c})$ is the normal to a homogenous hyperplane whose upper halfspace contains the translated upper contour set $R(\bar{c})$.

Since the function $\eta_{\bar{c}}(\cdot)$ for a given $\bar{c}$ is homogenous of degree 0 and $e_{\bar{c}}(\cdot)$ is homogenous of degree 1 in the prices, we can normalize prices. In order to compare the support functions for translated upper contour sets we restrict attention to prices in the set

$$P_{q(\bar{c})} = \left\{p \in \mathbb{R}^2_{++} \;\middle|\; p_1\frac{1}{q_1(\bar{c})} + p_2\frac{1}{q_2(\bar{c})} = 1\right\}$$

If $p \in P_{q(\bar{c})}$ then there is a unique $\tau \in\,]0,1[$ such that $(\tau q_1(\bar{c}), (1-\tau)q_2(\bar{c})) = (p_1, p_2)$ and the vector $(\tau q_1(\bar{c}), (1-\tau)q_2(\bar{c}))$ is proportional to $q(\bar{c})$ only for $\tau = 1/2$. To save some notation let $\tau \diamond q(\bar{c}) = (\tau q_1(\bar{c}), (1-\tau)q_2(\bar{c}))$.

We can now consider the function $\mathbb{R}^2_{++} \times]0,1[\longrightarrow \mathbb{R}^2$ mapping $(\bar{c},\tau)$ to $\eta_{\bar{c}}(\tau \diamond q(\bar{c}))$. This is a restriction of the mapping η to normalized prices but we will denote also this restriction by η. Analogously the function $\mathbb{R}^2_{++} \times]0,1[\longrightarrow \mathbb{R}$ mapping $(\bar{c},\tau)$ to $e_{\bar{c}}(\tau \diamond q(\bar{c}))$ is a restriction of the function e which will also be denoted by e.

The following proposition gives differentiability properties for η and e which are consequences of the Maintained Assumptions.

Proposition 9.5.E *Let* $e.(:) : \mathbb{R}^2_{++} \times]0,1[\longrightarrow \mathbb{R}$ *be the function mapping* $(\bar{c},\tau)$ *to* $e_{\bar{c}}(\tau \diamond \bar{q})$. *Then*

(a) $e \in \mathbb{C}^1(\mathbb{R}^2_{++} \times]0,1[\,,\mathbb{R})$

(b) *for each* $\bar{c} \in \mathbb{R}^2_{++}$ *the function* $e_{\bar{c}}(:) \in \mathbb{C}^2(]0,1[\,,\mathbb{R})$

(c) *for* $\tau \in]0,1[$, *the second derivative* $D^2_{\tau\tau} e_{\bar{c}}(\tau \diamond q(\bar{c})) < 0$

(d) *for each* $\bar{c} \in \mathbb{R}^2_{++}$ *the function* $e_{\bar{c}}(:)$ *is a strictly concave function attaining its maximum value* 0 *only for* $\tau = 1/2$.

Proof: The properties of e in **(a)** and **(b)** follow from the corresponding properties of $\hat{e}$.

To prove **(c)** let $(q_1(\bar{c}), q_2(\bar{c})) = (\bar{q}_1, \bar{q}_2)$ and since $\bar{c}$ is given we drop the subscript on e and η. We have

$$\begin{aligned} D_\tau e(\tau \diamond \bar{q}) &= \bar{q}_1 D_1 e(\tau \diamond \bar{q}) - \bar{q}_2 D_2 e(\tau \diamond \bar{q}) \\ &= \bar{q}_1 \eta_1(\tau \diamond \bar{q}) - \bar{q}_2 \eta_2(\tau \diamond \bar{q}) \end{aligned}$$

and

$$D^2_{\tau\tau} e(\tau \diamond \bar{q}) = \bar{q}_1 D_\tau \eta_1(\tau \diamond \bar{q}) - \bar{q}_2 D_\tau \eta_2(\tau \diamond \bar{q}) \qquad \textbf{9.5.D}$$

Using the first order conditions for the minimum problem in 9.5.B defining the support function, at $(\bar{c}, p_1, p_2) = (\bar{c}, \tau\bar{q}_1, (1-\tau)\bar{q}_2)$ we get

$$\begin{aligned} \tau\bar{q}_1 D_2 u(\bar{c} + \eta(\tau \diamond \bar{q})) - (1-\tau)\bar{q}_2 D_1 u(\bar{c} + \eta(\tau \diamond \bar{q})) &= 0 \\ u(\bar{c} + \eta(\tau \diamond \bar{q})) - u(\bar{c}) &= 0 \end{aligned}$$

and differentiating these relations with respect to τ we have

$$\begin{aligned} \bar{q}_1 D_2 u + \tau \bar{q}_1 \left[D^2_{21} u \cdot D_\tau \eta_1 + D^2_{22} u \cdot D_\tau \eta_2\right] & \\ + \bar{q}_2 D_1 u - (1-\tau)\bar{q}_2 \left[D^2_{11} u \cdot D_\tau \eta_1 + D^2_{12} u \cdot D_\tau \eta_2\right] &= 0 \qquad \textbf{9.5.E} \\ D_1 u \cdot D_\tau \eta_1 + D_2 u \cdot D_\tau \eta_2 &= 0 \end{aligned}$$

with the derivatives evaluated at $\tau \diamond \bar{q}$ and $\bar{c} + \eta(\tau \diamond \bar{q})$.

The last relation shows that $(D_\tau \eta_1, D_\tau \eta_2)$ is orthogonal to $(D_1 u, D_2 u)$. Since $(D_1 u, D_2 u)$ is proportional to $(\tau \bar{q}_1, (1-\tau)\bar{q}_2)$ there is a $\gamma \in \mathbb{R}$ such that $(D_\tau \eta_1, D_\tau \eta_2) = \gamma(-(1-\tau)\bar{q}_2, \tau \bar{q}_1)$. Substituting into the first relation of 9.5.E and rearranging we have

$$\bar{q}_1 D_1 u + \bar{q}_2 D_2 u + \gamma \left[-(1-\tau)\bar{q}_2,\ \tau \bar{q}_1\right] \begin{bmatrix} D^2_{11} u & D^2_{12} u \\ D^2_{21} u & D^2_{22} u \end{bmatrix} \begin{bmatrix} -(1-\tau)\bar{q}_2 \\ \tau \bar{q}_1 \end{bmatrix} = 0$$

By (C5) the Hessian of u is negative definite on the subspace orthogonal to the gradient of u. Since the partial derivatives of u are positive we have $\gamma > 0$ and using relation 9.5.D we get

$$\begin{aligned} D^2_{\tau\tau} e(\tau \diamond \bar{q}) &= \bar{q}_1 D_\tau \eta_1(\tau \diamond \bar{q}) - \bar{q}_2 D_\tau \eta_2(\tau \diamond \bar{q}) \\ &= \gamma \bar{q}_2 \bar{q}_1 \left[-(1-\tau) - \tau\right] = -\gamma < 0 \end{aligned}$$

(d) Since e is a strictly concave function and $D_\tau e(\tau \diamond \bar{q}) = 0$ only for $\tau = 1/2$ it follows that $e(\tau \diamond \bar{q}) \leq 0$ for $\tau \in [0,1]$ with equality only for $\tau = 1/2$.

Support function of a hyperbola

Let $e_H^{(a,q)} : \mathbb{R}^2_+ \longrightarrow \mathbb{R}$ be the support function of $S_H(a,q)$, where $0 < a < \infty$. For $(p_1, p_2) \in \mathbb{R}^2_{++}$ the value of the function $e_H^{(a,q)}$ is $p_1 \bar{x}_1 + p_2 \bar{x}_2$, where $(\bar{x}_1, \bar{x}_2)$ is the solution to

$$\begin{aligned} \mathrm{Min}_{(x_1,x_2)} \quad p_1 x_1 + p_2 x_2 \quad & \text{s. to } x \in \mathbb{R}^2_+ \\ q_1 x_1 + q_2 x_2 + a(q_1 x_1)(q_2 x_2) &\geq 0 \\ q_1 x_1 + q_2 x_2 &\geq 0 \end{aligned}$$

Since the solution is

$$\begin{cases} \bar{x}_1 = \dfrac{1}{ap_1}\left[\dfrac{(p_1p_2)^{1/2}}{(q_1q_2)^{1/2}} - \dfrac{p_1}{q_1}\right] \\ \bar{x}_2 = \dfrac{1}{ap_2}\left[\dfrac{(p_1p_2)^{1/2}}{(q_1q_2)^{1/2}} - \dfrac{p_2}{q_2}\right] \end{cases}$$

we get the function with values

$$e_H^{(a,q)}(p_1, p_2) = -\frac{1}{a}\left((\frac{p_1}{q_1})^{1/2} - (\frac{p_2}{q_2})^{1/2}\right)^2$$

which can be extended to $\mathbb{R}_+^2$ by continuity.

Since $e_H^{(a,q)}$ is homogenous of degree 0 in (p_1, p_2) we can restrict consideration to prices normalized to belong to the set

$$P_q = \left\{(p_1, p_2) \in \mathbb{R}_+^2 \;\middle|\; p_1\frac{1}{q_1} + p_2\frac{1}{q_2} = 1\right\}$$

Given $(p_1, p_2) \in P_q$ there is a unique $\tau \in [0, 1]$ such that

$$(p_1, p_2) = (\tau q_1, (1 - \tau)q_2)$$

and thus we can restrict attention to the function (the dot indicating the positions of the argument τ)

$$e_H^{(a,q)}(\cdot\, q_1, (1 - \cdot)q_2) : [0, 1] \longrightarrow -\mathbb{R}_+$$

which maps τ to

$$e_H^{(a,q)}(\tau q_1, (1 - \tau)q_2) = -\frac{1}{a}\left[(\tau)^{1/2} - (1 - \tau)^{1/2}\right]^2$$

An inequality for the support function of an hyperbola

The support function of an hyperbola is reasonably simple but for purposes of approximation it is useful to take note of the inequality derived below which allow us to simplify the approximation problem even further. First note that

$$\begin{aligned} 4\left(\tau - \frac{1}{2}\right)^2 &= ((\tau) - (1 - \tau))^2 \\ &= \left((\tau)^{1/2} - (1 - \tau)^{1/2}\right)^2 \left((\tau)^{1/2} + (1 - \tau)^{1/2}\right)^2 \end{aligned}$$

Since $1/2 \leq \left((\tau)^{1/2} + (1-\tau)^{1/2}\right)^2 \leq 1$ we get

$$4\left(\tau - \frac{1}{2}\right)^2 \leq \left((\tau)^{1/2} - (1-\tau)^{1/2}\right)^2 \leq 8\left(\tau - \frac{1}{2}\right)^2$$

and hence

$$\begin{aligned} -\frac{4}{a}\left(\tau - \frac{1}{2}\right)^2 &\geq e_H^{(a,q)}(\tau q_1, (1-\tau)q_2) \\ &= -\frac{1}{a}\left[(\tau)^{1/2} - (1-\tau)^{1/2}\right]^2 \geq -\frac{8}{a}\left(\tau - \frac{1}{2}\right)^2 \end{aligned} \qquad (9.5.F)$$

which suggests that the function $\tau \longrightarrow (\tau - 1/2)^2$ can be used for approximation.

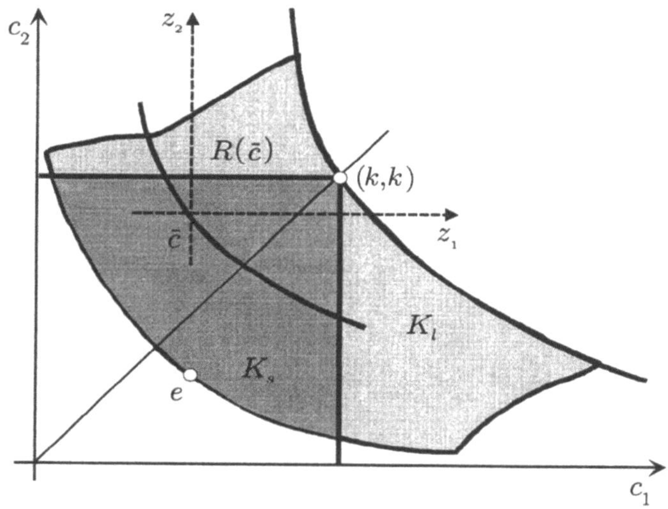

Figure 9.5.A: *For $\bar{c}$ belonging to the "small" compact set K_s we can approximate the (translations) of the upper contour sets $R(\bar{c})$. The role played by the "large" compact set K_l is explained in the proof of* Theorem 9.5.F

Approximation of Upper Contour Sets

The purpose of this section is to show that for an equilibrium allocation for a stationary economy we can approximate the upper contour sets by hyperbola.

Properties of the support functions

The function defined in Proposition 9.5.E is closely related to the expenditure function for the utility level $u(\bar{c})$, which would be the support function of $\hat{R}(\bar{c})$. Since the expenditure function, for the utility level $u(\bar{c})$, is $\eta_{\bar{c}}(p) = e_{\bar{c}}(p) + p\bar{c}$ the two functions differ only by the linear function $\bar{c} \longrightarrow p\bar{c}$.

For $\bar{c} \in K_s$ we want to approximate the functions $e_{\bar{c}}(\cdot\bar{q}_1, (1-\cdot)\bar{q}_2)$ for $\tau \in]0,1[$ by support functions of hyperbola. Cf. Figure 9.5.B.

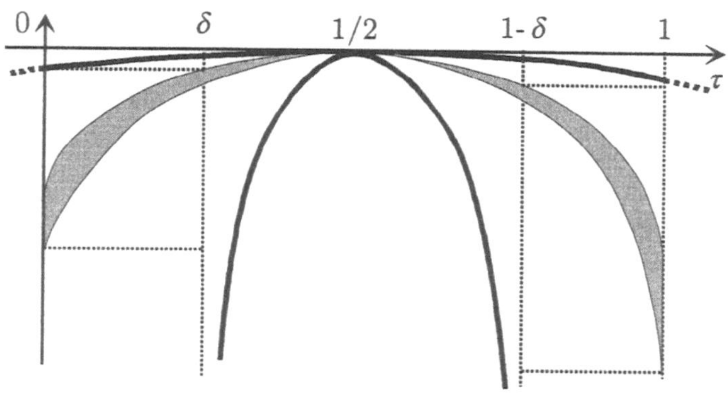

Figure 9.5.B: *The graph of each support function as $c \in K_s$ is contained in the shaded area. These support functions are all approximated from above and below by the functions $-m(\tau-1/2)^2$ and $-M(\tau-1/2)^2$*

For $c \in C$ let $\hat{R}(c) = \{c' \in C | u(c') \geq u(c)\}$. Let $k = 2\max(e^y, e^o)$ and let K_s be the "small" compact set

$$K_s = \{c \in C \mid u(c) \geq u(e) \text{ and } c \leq (k,k)\}.$$

The utility function u satisfies the Maintained Assumptions and thus $R(\bar{c}) = \hat{R}(\bar{c}) - \{\bar{c}\}$ is the translated upper contour set for such a utility function.

Recall that $q(\bar{c})$ is the normalized gradient so that $\|q(\bar{c})\| = 1$

Theorem 9.5.F *There are a and A such that $0 < a \leq A$ and*

$$S_H(A, q(\bar{c})) \subset R(\bar{c}) \subset S_H(a, q(\bar{c})) \quad \textit{for } \bar{c} \in K_s$$

Proof: Let $\bar{q} = q(\bar{c})$ and recall that $\tau \diamond \bar{q} = (\tau \bar{q}_1, (1-\tau)\bar{q}_2)$.

By Proposition 9.5.C, on the relation between support functions and set inclusion, it will suffice to show that there is a $m > 0$ and $M > 0$ such that for $\bar{c} \in K_s$

$$\begin{aligned} e_{\bar{c}}(\tau \diamond \bar{q}) + m\left(\tau - \frac{1}{2}\right)^2 &\leq 0 \quad \text{and} \\ e_{\bar{c}}(\tau \diamond \bar{q}) + M\left(\tau - \frac{1}{2}\right)^2 &\geq 0 \end{aligned} \qquad \textbf{9.5.G}$$

since we can then choose $A = 8/m$ and $a = 4/M$ and by the inequality 9.5.F

$$e_{\bar{c}}(\tau \diamond \bar{q}) \leq -\frac{8}{A}\left(\tau - \frac{1}{2}\right)^2 \leq e_H^{(A,\bar{q})}(\tau \diamond \bar{q})$$

$$e_{\bar{c}}(\tau \diamond \bar{q}) \geq -\frac{4}{a}\left(\tau - \frac{1}{2}\right)^2 \geq e_H^{(a,\bar{q})}(\tau \diamond \bar{q})$$

and hence $S_H(A, q(\bar{c})) \subset R(\bar{c}) \subset S_H(a, q(\bar{c}))$ for $\bar{c} \in K_s$.

The idea is to approximate the functions $e_{\bar{c}}(\tau \diamond \bar{q})$ firstly as $(\bar{c}, \tau) \in K_s \times [\delta, 1-\delta]$ and secondly as $(\bar{c}, \tau) \in K_s \times ([0, \delta] \cup [1-\delta, 1])$ for some $\delta \in]0, 1/2[$. Cf. Figure 9.5.B. Let B be the closed subset defined by

$$B = \{(\bar{c}, c) \in K_s \times \mathbb{R}^2_{++} \mid \bar{c} \in K_s \text{ and } u(c) = u(\bar{c})\}$$

Define the mapping $H : B \longrightarrow K_s \times]0, 1[$ by

$$(\bar{c}, c) \longrightarrow \left(\bar{c}, \frac{\dfrac{D_1 u(c)}{q_1(\bar{c})}}{D_1 u(c)\dfrac{1}{q_1(\bar{c})} + D_2 u(c)\dfrac{1}{q_2(\bar{c})}}\right)$$

where the last component thus equals the uniquely determined τ with the property that $Du(c)$ is proportional to $(\tau q_1(\bar{c}), (1-\tau) q_2(\bar{c}))$. The mapping H is continuous. It has a continuous inverse $H^{-1} : K_s \times]0, 1[\longrightarrow B$ given by

$$(\bar{c}, \tau) \longrightarrow (\bar{c}, \bar{c} + \eta_{\bar{c}}(\tau q_1(\bar{c}), (1-\tau) q_2(\bar{c})))$$

The image of of the compact set $B \cap (K_s \times K_s)$ under the continuous mapping H is a compact set in $K_s \times]0, 1[$ and is thus contained in a set

$K_s \times \left[\hat{\delta}, 1-\hat{\delta}\right]$ with $0 < \hat{\delta} \leq 1/2$. Choose δ so that $0 < \delta < \hat{\delta}$ and let A_δ be the image of the compact set $K_s \times [\delta, 1-\delta]$ under the mapping H^{-1}. In Figure 9.5.A the set K_l is the projection on the second component of this image set.

We now approximate on $K_s \times [\delta, 1-\delta]$ and to this end we consider the restriction of the function $G : K_s \times]0,1[\longrightarrow \mathbb{R}$ defined by

$$G(\bar{c}, \tau) \;=\; e_{\bar{c}}(\tau \diamond \bar{q}) + m\left(\tau - \frac{1}{2}\right)^2$$

to $K_s \times [\delta, 1-\delta]$. We have

$$D^2_{\tau\tau} G(\bar{c}, \tau) \;=\; D^2_{\tau\tau} e_{\bar{c}}(\tau \diamond \bar{q}) + 2m \;<\; 0 \quad \text{if} \; -m \;>\; \frac{1}{2} D^2_{\tau\tau} e_{\bar{c}}(\tau \diamond \bar{q})$$

Hence if

$$0 \;>\; -m \;>\; \frac{1}{2} \max\left\{ D^2_{\tau\tau} e_{\bar{c}}(\tau \diamond \bar{q}) \mid (\bar{c}, \tau) \in K_s \times [\delta, 1-\delta] \right\}$$

then the restrictions of $G(\bar{c}, \cdot)$, that is, the functions $G(\bar{c}, \cdot) : [\delta, 1-\delta] \longrightarrow \mathbb{R}$ for $\bar{c} \in K_s$ are strictly concave functions. Each of these functions attains a maximum equal to 0 only for $\tau = 1/2$.

Next we seek an approximation on $K_s \times ([0,\delta] \cup [1-\delta, 1])$. The convex function $\tau \longrightarrow m(\tau - 1/2)^2$ has a maximum for $\tau \in \{0,1\}$ equal to $(1/4)m$. Thus if we choose m so that

$$0 \;>\; -m > 4 \max\left\{ e_{\bar{c}}(\tau \diamond \bar{q}) \mid (\bar{c}, \tau) \in K_s \times [0,\delta] \cup [1-\delta, 1] \right\}$$

then $G(\bar{c}, \cdot) : [0,\delta] \cup [1-\delta, 1] \longrightarrow \mathbb{R}$ is a nonpositive function for $\bar{c} \in K_s$.

Choose $\bar{m}$ to satisfy both inequalities. Then

$$e_{\bar{c}}(\tau \diamond \bar{q}) \;\leq\; -\bar{m}\left(\tau - \frac{1}{2}\right)^2 \qquad \text{for } (\bar{c}, \tau) \in K_s \times]0,1[$$

An analogous reasoning gives that there is an $M > 0$ satisfying the second inequality in 9.5.G.

□

Note that in the proof part (c) of Proposition 9.5.E, ensuring that the second derivative is negative, rather than merely nonpositive, was essential. For example, the nonpositive concave function $\tau \longrightarrow -(\tau - 1/2)^4$, whose second derivative is 0 for $\tau = 1/2$, is not approximated from above by the function $\tau \longrightarrow -m(\tau - 1/2)^2$ for any $m > 0$.

Equilibrium Prices and Optimal Allocations

The result in the preceding section ensures that the upper contour sets associated with an equilibrium allocation can be approximated from the "inside" and the "outside" by hyperbola. Thus Corollary 8.5.A, on the characterization of optimal allocations, can be applied in order to obtain, for stationary OG economies, a characterization of optimal allocations as conditions on the equilibrium prices.

Corollary 9.5.G *Let* $((p_t)_{t\in\mathbb{Z}}, (\bar{c}_t, \bar{m}_t)_{t\in\mathbb{Z}}, M)$ *be an equilibrium for the stationary OG economy* $\mathcal{E}^s = (C, u, e)$. *Then*

(a) $(\bar{c}_t)_{t\in\mathbb{Z}}$ *is a forwards optimal allocation if and only if*

$$\sum_{t\in\mathbb{N}} \frac{1}{p_t} = \infty$$

(b) $(\bar{c}_t)_{t\in\mathbb{Z}}$ *is a strongly optimal allocation if and only if*

$$\sum_{t\in\mathbb{N}} \frac{1}{p_t} = \sum_{-t\in\mathbb{N}} \frac{1}{p_t} = \infty$$

Proof: For $t \in \mathbb{Z}$, the equilibrium consumption $\bar{c}_t$ belongs to the compact set

$$K = \{c \in \mathbb{R}^2_{++} \mid u(c) \geq u(e) \text{ and } c^y, c^o \leq e^y + e^o\}$$

since trade is voluntary and the total initial endowment is $e^y + e^o$ at each date.

Let $R(\bar{c}_t) = \hat{R}(\bar{c}_t) - \{\bar{c}_t\}, t \in Z$, be the reduced model induced by the equilibrium allocation. For $c \in K_s$ the function $c \longrightarrow D_o u(c)/D_y u(c)$ is bounded above and below from 0. The normalized gradient $q(\bar{c}) = \bar{p}^t/||\bar{p}^t||$ and thus by Theorem 9.5.F, on the "inner" and "outer" approximation by hyperbola, there are $A, a > 0$ such that

$$a \leq \frac{p_{t+1}}{p_t} \leq A \quad \text{and} \quad S_H(A, \frac{p^t}{||p^t||}) \subset R(\bar{c}_t) \subset S_H(a, \frac{p^t}{||p^t||})$$

We can apply Corollary 8.5.A, on the relation between optimal allocations and efficient reduced models, to conclude that the equilibrium allocation $(\bar{c}_t)_{t\in\mathbb{Z}}$ is a forwards optimal allocation if and only if the reduced model is efficient and a strongly optimal allocation if and only if the reduced model is strongly efficient.

The result now follows from Corollary 8.5.A, where the conditions in (a) and (b) were shown to imply the efficiency and strong efficiency, respectively, of a reduced model.

□

Since the optimality criterion in Corollary 9.5.G above follows from Cass' criterion for efficient and strongly efficient reduced models in Corollary 8.5.A we will refer to it as Cass' optimality criterion.

Cass' criterion and relative prices *

Recall that a spot price system $(p_t)_{t\in\mathbb{Z}}$ and a relative price system $(r_t)_{t\in\mathbb{Z}}$ are related through : $r_t = p_t/p_{t+1}$, $p_t = p_0 \prod_{\tau=t}^{0} r_\tau$, for $t \leq 0$, and $p_t = p_0 \prod_{\tau=0}^{t-1} \frac{1}{r_\tau}$ for $t > 0$. Thus the conditions in Corollary 9.5.G can be given using the induced relative equilibrium prices and we have

$$\sum_{t\in\mathbb{N}} \frac{1}{p_t} = \infty \quad \text{if and only} \quad \sum_{t\in\mathbb{N}} \prod_{\tau=1}^{t-1} r_\tau = \infty$$

and

$$\sum_{t\in\mathbb{N}} \frac{1}{p_t} = \sum_{-t\in\mathbb{N}} \frac{1}{p_t} = \infty \quad \text{if and only} \quad \sum_{t\in\mathbb{N}} \prod_{\tau=1}^{t} r_\tau = \sum_{-t\in\mathbb{N}} \prod_{\tau=t}^{0} \frac{1}{r_\tau} = \infty$$

Application to non-stationary economies*

Above an efficiency criterion for stationary OG economies was derived by approximating the upper contour sets by hyperbola.

The reasoning hinged crucially on two features of a stationary economy. Firstly, since endowments are the same for all consumers the consumption bundles from an equilibrium allocation are contained in a compact set. Secondly, the upper contour sets for consumptions in this compact set could be approximated by hyperbola. This second feature made it possible to apply Corollary 8.5.A; the criteria for efficiency and strong efficiency of a reduced model made up by hyperbola.

Also in a non-stationary economy the equilibrium allocations will be contained in a compact set if initial endowments do not grow or decrease to much over time. To carry out an approximation, as in Theorem 9.5.F, we would also need a "uniformity" assumption on utility functions at different dates. To exemplify what might go wrong without such an assumption note that we showed that the second derivative of the support function, as defined in Proposition 9.5.E, was a negative number and used that in our approximation. In a non-stationary economy it might well be that this second derivative is negative but approaches 0 as time goes to infinity, which would destroy the possibilities of an approximation using hyperbola.

The method of approximation can be used also in models with several goods. Given an equilibrium allocation for such an economy, one can obtain embedded one-good models by fixing the amounts of all goods but, say, good 1. Under reasonable assumptions the results for the one-good model studied here would carry over to these embedded one-good models. Cf. Balasko and Shell [1980].

9.6 Summary

By definition a stationary OG economy is given by a single consumer. In an equilibrium demand at the current date stems from the demand of the old consumer, born at the previous date, and the young consumer. Market balance requires demand to be equal to the initial endowment of the consumers, at the current date. This condition was given in the equilibrium equation; a first order, implicit difference equation. The equilibrium price system is a price system satisfying the equilibrium equation at each date. It was shown that the equilibrium equation could be stated using spot prices, relative prices or real interest rates and that the equilibrium equation could also be expressed as a market balance condition for savings.

For stationary OG economies steady state equilibria were defined. At such an equilibrium relative prices are constant over time which induces each of the consumers to choose the same consumption. A stationary economy was shown to have a real steady state, where the initial endowments make up the equilibrium allocation so that there is no trade. Unless the relative price of the real steady state is 1, there is also a nominal steady state, where the relative price is 1, involving the same non-zero net trade for each consumer.

For an equilibrium allocation in a stationary OG economy the upper contour sets, corresponding to the consumptions in equilibrium, were be approximated by hyperbola. The criterion for forwards optimal allocations and strongly optimal allocations from the previous chapter were then be applied to characterize equilibrium allocations with respect to optimality. It was also noted that the results on optimality could be substantially generalized and would apply also to equilibrium allocations for non-stationary economies.

9.7 Exercises

Exercise 9.A Let $\hat{f}$ denote the Walrasian demand function where

$$(p, p') \longrightarrow (\hat{f}^y(p, p', pe^y + p'e^o), \hat{f}^o(p, p', pe^y + p'e^o))$$

and let $f^o(r, e)$ denote demand as old as a function of relative price and endowment.

(a) Show that $f^o(r, e) = \hat{f}^o(r, 1, re^y + e^o) = \hat{f}^o(1, \frac{1}{r}, e^y + \frac{1}{r}e^o)$.

(b) Use the result from (a) to show that the effect on $f^o(r, e)$ of a variation in r can be written

$$\begin{aligned} D_r f^o(r, e) &= D_r \hat{f}^o\left(1, \frac{1}{r}, e^y + \frac{1}{r}e^o\right) \\ &= \left[D_2 \hat{f}^o\left(1, \frac{1}{r}, e^y + \frac{1}{r}e^o\right) + D_3 \hat{f}^o\left(1, \frac{1}{r}, e^y + \frac{1}{r}e^o\right) \cdot e^o\right] \cdot -\frac{1}{r^2} \end{aligned}$$

(c) Interpret the result from (b).

(d) Show that the right-hand member in (b) can be rewritten

$$\left[D_2\hat{f}^o\left(1,\frac{1}{r},e^y+\frac{1}{r}e^o\right)+D_3\hat{f}^o\left(1,\frac{1}{r},e^y+\frac{1}{r}e^o\right)\cdot e^o\right]\cdot-\frac{1}{r^2}$$
$$=\Bigg[\Bigg\{D_2\hat{f}^o\left(1,\frac{1}{r},e^y+\frac{1}{r}e^o\right)$$
$$+D_3\hat{f}^o\left(1,\frac{1}{r},e^y+\frac{1}{r}e^o\right)\cdot\hat{f}^o\left(1,\frac{1}{r},e^y+\frac{1}{r}e^o\right)\Bigg\}$$
$$+D_3\hat{f}^o\left(1,\frac{1}{r},e^y+\frac{1}{r}e^o\right)\cdot\left(e^o-\hat{f}^o(1,\frac{1}{r},e^y+\frac{1}{r}e^o)\right)\Bigg]\cdot-\frac{1}{r^2}$$

(e) Interpret the result from (d) in terms of the income and substitution effect of a variation in price with compensating variation in income and the income effect from the variation in price on the value of initial endowment.

Exercise 9.B Prove the assertion from the proof of Proposition 9.5.D, that is, show that $\hat{\eta}$ is a differentiable function of (p,u) using that

$$\begin{aligned}\hat{\eta}_1(p,V(p,w)) &= \hat{f}^y(p,w)\\ \hat{\eta}_2(p,V(p,w)) &= \hat{f}^o(p,w)\end{aligned}$$

Exercise 9.C Use the Implicit Function Theorem to show that the expenditure function and the Hicksian demand function for a consumer satisfying the Maintained Assumptions are differentiable functions.

Exercise 9.D Consider the steady states of a stationary OG economy $\mathcal{E}^s=(C,u,e)$. Show that

(a) if $\rho(e)<1$ then the real steady state equilibrium allocation $(\bar{c}_t)_{t\in\mathbb{Z}}$ is a finitely optimal allocation but not a forwards optimal allocation.

(b) if $\rho(e)>1$ then the real steady state equilibrium allocation $(\bar{c}_t)_{t\in\mathbb{Z}}$ is a forwards optimal allocation but not a strongly optimal allocation.

(c) if $\rho(e)\neq 1$ then there is a nominal equilibrium allocation which is a strongly optimal allocation.

Exercise 9.E Show that if $(r_t)_{t\in\mathbb{Z}}$ is an equilibrium price system for $\mathcal{E}^s = (C, u, e)$ then $(r_{t+n})_{t\in\mathbb{Z}}$ for $n \in \mathbb{Z}$ is also an equilibrium price system

Exercise 9.F Let $\mathcal{E}^s = (\mathbb{R}^2_{++}, a\ln c^y + b\ln c^o, e)$ where $a, b \in \mathbb{R}_{++}$.

(a) Find the demand function and the savings function.

(b) Find a first-order difference equation which describes equilibria.

(c) Find the steady states.

(d) Characterize the real steady state and the nominal steady state with respect to optimality.

(e) For which values of a and b is the nominal steady state a classical steady state and for which values is the nominal steady state a Samuelson steady state?

Exercise 9.G Let $\mathcal{E}^s = (\mathbb{R}^2_{++}, (c^y)^a + (c^o)^a, e)$ where $0 < a < 1$.

(a) Find the demand function and the savings function.

(b) Find a first-order difference equation which describes equilibria.

(c) Find the steady states.

(d) Characterize the real steady state and the nominal steady state with respect to optimality.

(e) For which values of a is the nominal steady state a classical steady state and for which values is the nominal steady state a Samuelson steady state?

Exercise 9.H Let $\mathcal{E}^s = (\mathbb{R}^2_{++}, (c^y)^a/a + (c^o)^a/a, e)$ where $a < 0$.

(a) Find the demand function and the savings function.

(b) Find a first-order difference equation which describes equilibria.

(c) Find the steady states.

(d) Characterize the real steady state and the nominal steady state with respect to optimality.

(e) For which values of a is the nominal steady state a classical steady state and for which values is the nominal steady state a Samuelson steady state?

Exercise 9.I Let $\mathcal{E} = (C_t, u_t, e_t)_{t\in\mathbb{Z}}$ be a, possibly non-stationary, economy such that there are i types of consumers $(C_i, u_i, e_i)_{i\in\mathbb{I}}$, satisfying the Maintained Assumptions, with the property that, for $t \in \mathbb{Z}$, consumer $(C_t, u_t, e_t) = (C_i, u_i, e_i)$ for some $i \in \mathbb{I}$.

(a) Show that there is a compact set K such that if $(\bar{c}_t)_{t\in\mathbb{Z}}$ is an equilibrium allocation then $\bar{c}_t \in K$ for $t \in \mathbb{Z}$.

(b) Let $(\bar{c}_t)_{t\in\mathbb{Z}}$ be an equilibrium allocation and $\hat{R}_t(\bar{c}_t) - \{\bar{c}_t\}$ the translated upper contour set for consumer t. Can $\hat{R}_t(\bar{c}_t) - \{\bar{c}_t\}$ be approximated by sets from the family of hyperbola from the inside and from the outside?

(c) Can Cass' criterion for optimality be applied to the equilibrium allocation $(\bar{c}_t)_{t\in\mathbb{Z}}$?

Chapter 10

GLOBAL DYNAMICS: TURNPIKE EQUILIBRIA

Introduction

In this chapter the study of stationary economies is continued. We know from the previous chapter that an equilibrium price system can be found by solving the equilibrium equation for each date. In the first part of this chapter we study the pair of relative prices, at the current date and the previous date, clearing the market at the current date. The set of these pairs of prices is the *equilibrium locus* for the economy. The interest in this set is due to the fact that a sequence of relative prices is an equilibrium price system if and only if each consecutive pair of prices belong to the equilibrium locus. We derive properties of the equilibrium locus which will be satisfied by each stationary economy. In particular, we study the problem of solving the equilibrium equation forwards or backwards. In order to obtain an equilibrium price system one needs to solve the equilibrium equation one date forward *and* also make sure that for the price solving one can extend the solution one date further.

A set with the property that for each price in the set there is some price in the same set which solves the equilibrium equation is a, forward or backward, *invariant set*. In Section 10.1 we will prove that OG economies have an invariant set which will in general contain an interval with non-empty interior.

For an Arrow-Debreu (exchange) economy it is generically true that an economy has a *finite* number of equilibria. In this case the fundamentals of the economy; preferences, consumption sets and initial endowments, sharply delimit the equilibria, cf. the discussion in Chapter 5 and 6.

An immediate consequence of the existence of an invariant set is that a stationary OG economy typically has a continuum of equilibria. Hence the fundamentals; consumption set, utility function and endowment of the economy serve to single out the equilibria only to a moderate extent and there is a large amount of *indeterminacy*.

The second part of the chapter is devoted to stationary OG economies where the set of equilibria is easy to characterize. Also here it may be instructive to compare to Arrow-Debreu economies, where preferences and consumption sets are fixed while initial endowments are allowed to vary. If the allocation formed by initial endowments is a Pareto optimal allocation the economy has a unique Walras equilibrium and for reasonable price adjustment rules the equilibrium is stable. Since at the equilibrium there is no trade the intensity of trade is indeed low. The low intensity of trade means that a variation of the prices causes only small variations in the income from the endowment. As the initial endowments are varied so that a Walras equilibrium results in a high intensity of trade the pleasant properties of the equilibria may be lost. This is pursued thoroughly in Balasko [1988].

Is there for OG economies also a family of economies, given by their endowments, where the equilibria are particularly simple? For OG economies there are several notions of optimality corresponding to Pareto optimality in an Arrow-Debreu economy. In the previous chapter we proved that an OG economy where the initial endowments form a strongly optimal allocation has a single equilibrium. Since we have restricted attention to stationary OG economies the assumption that initial endowments form a strongly optimal allocation implies that the economy has no nominal steady state and the real steady state has an equilibrium price system where each price equals 1.

The maximal invariant set for an economy is an invariant set which is not a proper subset of an invariant set. For economies close to no-trade economies we show that the maximal invariant set is the closed interval with the steady state prices as endpoints. The significance of the maximal invariant set is that for a given economy the maximal invariant set necessarily owns each of the equilibrium price systems.

For economies where the initial endowment allocation is close to a strongly optimal allocation, the nominal steady state and the real steady state of the economy are close. From the characterization of the maximal invariant set follows that for such an economy (relative) equilibrium prices will be close to 1 and an equilibrium price system is either a steady state price or has prices converge to *one of the steady state prices,* 1 or $\rho(e)$, as time tends to infinity or minus infinity.

Throughout this chapter we will use J_e to denote the closed interval with endpoints $\min(\rho(e), 1)$ and $\max(\rho(e), 1)$.

10.1 The Equilibrium Equation and the Equilibrium Locus

In this section we study the market balance condition at the current date; the equilibrium equation, and the pairs of relative prices, for the current date and the previous date giving market balance at the current date. These pairs of prices form the equilibrium locus.

The Equilibrium Equation

In Chapter 9 we saw that an equilibrium price system can be found by solving the equilibrium equation at each date. The equilibrium equation for the current date involves the relative price at the current date and the relative price at the previous date, cf. Table 10.1.A where $t + 1$ can be thought of as the current date.

Since we consider only stationary economies the equilibrium equation is independent of time. As a first step towards finding a solution we study the possibilities to solve the equilibrium equation one date forward, given the price at the previous date r, for the price at the current date, r', and backwards, given the price at the current date, r', for the price at the previous date, r.

Table 10.1.A: *Notation for the equilibrium equation and equilibrium locus*

Date	t	$t+1$	$t+2$
Relative prices	r	1 r'	 r
Excess demand	$z^y(r,e)$	$z^o(r,e)$ $z^y(r',e)$	 $z^o(r',e)$

At the current date demand stems from the consumer born at the previous date, who acts (and acted) against the price r, and the consumer born at date $t+1$ who acts against the price r'. To achieve market balance at the current date the relative prices (r, r') must satisfy the equilibrium equation

$$z^o(r,e) + z^y(r',e) = 0$$

so that total excess demand is 0.

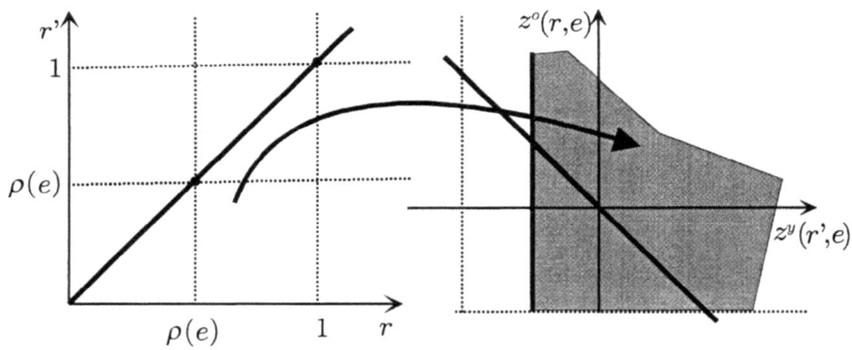

Figure 10.1.A: *The pair of relative prices (r, r') is mapped to the excess demand of the young and old consumer at the current date*

The Equilibrium Locus

Let $H : \mathbb{R}^2_{++} \times \mathbb{R}^2_{++} \longrightarrow \mathbb{R}^2$ be the mapping $(r, r', e) \longrightarrow (z^y(r', e), z^o(r, e))$. Thus H maps the pair of relative prices and endowment (r, r', e) to the excess demands of the young and old consumer at date $t+1$. The pair of relative prices (r, r') clears the market at date $t+1$ if and only if (r, r') is.mapped to the hyperplane $x_1 + x_2 = 0$, as indicated in Figure 10.1.A, where the initial endowment is assumed fixed. The equilibrium locus for the economy e is the set of (r, r') such that (r, r') maps to the hyperplane. Thus the equilibrium locus is the inverse image of the hyperplane under the mapping H.

For the economy defined by e, let $z^y(r', e) + z^o(r, e)$ be the total excess demand at the current date at the relative prices r, at the previous date, and r' at the current date. The equilibrium locus can now be defined.

Definition 10.1.A *The set*

$$G_e = \{(r, r') \in \mathbb{R}^2_{++} \mid z^y(r', e) + z^o(r, e) = 0\}$$

is the **equilibrium locus** *for the economy e.*

The price system $(\bar{r}_t)_{t \in \mathbb{Z}}$ is an equilibrium price system if and only if $(r_t, r_{t+1}) \in G_e$ for $t \in \mathbb{Z}$.

On the one hand it can be argued that it is not important that a price system clearing markets at the current date and forward can be extended backwards to an equilibrium price system since OG economies are believed to have a first date (birth of Adam and Eve). On the other hand it is important that a price system clearing markets at the current date and backwards can be extended forwards to an equilibrium price system for all dates since if this is not possible then the equilibrium will break down at some subsequent date. If a price system can not be extended forwards rational consumers should take this into account and thereby the behavior of consumers as well as the appropriate notion of equilibrium would change.

Excess demand at equal relative prices

It is useful to get an idea about the total excess demand as the price for the young consumer equals the price for the old consumer, that is, $z^y(r, e) + z^o(r, e)$.

We have indicated three possible cases in Figure 10.1.B. The total excess demand is 0 if and only if r is a steady state price system and hence total excess demand can equal 0 for at most for two prices.

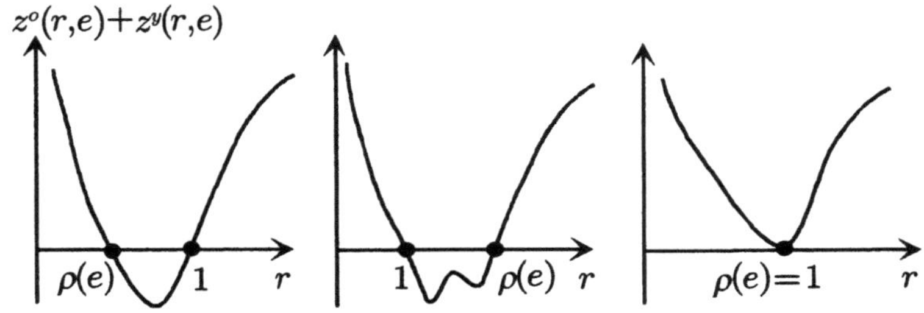

Figure 10.1.B: *In the left panel the nominal equilibrium is Samuelson and in the mid panel it is classical*

By Lemma 9.2.B, on the properties of demand, the excess demand for the young consumer, at the current date, tends to infinity as r tends to 0 and the excess demand for the old consumer tends to infinity as r tends to infinity.

Lemma 10.1.B *Total excess demand at equal prices $z^y(r,e)+z^o(r,e)$ is nonpositive on the interval J_e and positive for $r \notin J_e$.*

Proof: At $(r,r) = (\rho(e), \rho(e))$ excess demand is 0 and since at this point the derivative of the total excess demand is $\alpha(D_o u(e) - D_y u(e)) = \alpha(1-\rho(e))$ with $\alpha < 0$, by Lemma 9.2.C, on the derivatives of demand. Hence, if $\rho(e) \neq 1$ the excess demand takes on nonpositive values on the interval J_e and if $\rho(e) = 1$ total excess demand is nonnegative and takes on the value 0 only for $(r,r) = (\rho(e), \rho(e))$.

Properties of the equilibrium locus

The function H will not be one-to-one in general which means that a single pair of excess demands for the old and young consumer at date $t+1$ can occur for more than a single pair of prices. This in turn makes it difficult to describe the equilibrium locus in detail.

However, we can derive properties which are of interest in the sequel, some of which we have indicated in Figure 10.1.C.

To sum to 0 the excess demand of the old and young consumer have to be both equal to 0 or have different signs. Thus if $(r,r') \in G_e$ then r and r' are both no larger than $\rho(e)$ or both no smaller than $\rho(e)$. From the discussion in the previous section follows that the diagonal of the

(r, r')-space is split into three connected sets where total excess demand is nonpositive on one of the sets and positive on the other two. We summarize these properties in Lemma 10.1.C below.

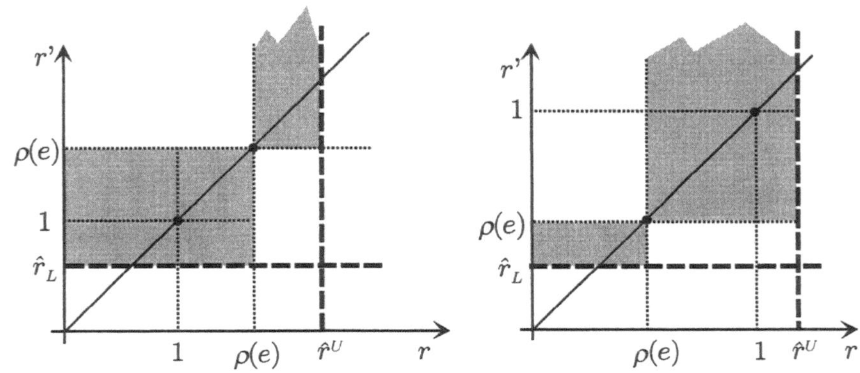

Figure 10.1.C: *The equilibrium locus is contained in the subset marked grey. The excess demand is positive for (r, r') such that $r \leq \hat{r}_L$ or $r' \geq \hat{r}^U$*

Lemma 10.1.C *The equilibrium locus, G_e, satisfies:*

(a) $G_e \subset \,]0, \rho(e)]^2 \cup [\rho(e), \infty[^2$

(b) *there exists $\hat{r}^U, \hat{r}_L \in \mathbb{R}_{++}$ such that $G_e \subset \,]0, \hat{r}^U] \times [\hat{r}_L, \infty[$*

(c) $]0, \max J_e] \subset \operatorname{pr}_1 G_e$ *and* $[\min J_e, \infty[\, \subset \operatorname{pr}_2 G_e$.

Proof: To prove **(a)** note that if $(r, r') \in G_e$ then $z^o(r, e) + z^y(r', e) = 0$ and either both $z^o(r, e)$ and $z^y(r', e)$ are 0, in which case $r = r' = \rho(e)$ or they have different signs which by Lemma 9.2.B, on the basic properties of demand, implies that r and r' are both smaller than $\rho(e)$ or both larger than $\rho(e)$.

(b) By Lemma 9.2.B, on the basic properties of demand, $\lim_{r \to \infty} z^o(r, e) = \infty$ and $\lim_{r' \to 0} z^y(r', e) = \infty$. Hence there are positive numbers $\hat{r}^U$ and $\hat{r}_L$ such that

$$\begin{aligned} z^o(r, e) &> e^y \quad \text{for } r \geq \hat{r}^U \\ z^y(r', e) &> e^o \quad \text{for } r' \leq \hat{r}_L \end{aligned}$$

Since $z^y(r',e) > -e^y$ and $z^o(r,e) > -e^o$ for $r, r' \in \mathbb{R}_{++}$ we get

$$z^o(r,e) + z^y(r',e) > 0 \quad \text{for } r \geq \hat{r}^U$$

$$z^o(r,e) + z^y(r',e) > 0 \quad \text{for } r' \leq \hat{r}_L$$

(c) Let $r \leq \rho(e)$. Then $z^o(r,e) \leq 0$. Since $z^y(\rho(e),e) = 0$ and $\lim_{r' \to 0} z^y(r',e) = \infty$ there is r' such that $z^y(r',e) + z^o(r,e) = 0$. If $r \in [\min J_e, \max J_e]$ then $z^o(r,e) + z^y(r,e) \leq 0$ and since $\lim_{r' \to 0} z^y(r',e) = \infty$ there is r' such that $z^y(r',e) + z^o(r,e) = 0$.

Let $r' \in [\min J_e, \infty[$. If $r' \in [\min J_e, \max J_e]$ then $z^o(r',e) + z^y(r',e) \leq 0$ and since $\lim_{r \to \infty} z^o(r,e) = \infty$ there is r such that $z^y(r',e) + z^o(r,e) = 0$. If $r' \geq \max J_e$ then $z^y(r',e) \leq 0$ and since $\lim_{r \to \infty} z^o(r,e) = \infty$ there is r such that $z^y(r',e) + z^o(r,e) = 0$.

□

Since the equilibrium locus is the contour set of the total excess demand corresponding to the value 0, the gradient of the total excess demand will, unless it is 0, point to values with positive excess demand. From Lemma 9.2.C, on the derivatives of demand, follows that the derivative of the total excess demand, with respect to (r, r'), at the point $(r, r') = (\rho(e), \rho(e))$ is not 0. Hence we can use the Implicit Function Theorem to find, for r in the vicinity of $\rho(e)$, a one-date forward solution r' as function of r.

Lemma 10.1.D *Let $\hat{e}$ be an economy. There is an open interval $I_{\rho(\hat{e})} \subset \mathbb{R}_{++}$ and an open set $E_{\hat{e}} \subset \mathbb{R}^2_{++}$ with $(\rho(\hat{e}), \hat{e}) \in I_{\rho(\hat{e})} \times E_{\hat{e}}$ and a unique function $\gamma \in \mathbb{C}^1(I_{\rho(\hat{e})} \times E_{\hat{e}}, \mathbb{R})$ with $\rho(\hat{e}) = \gamma(\rho(\hat{e}), \hat{e})$ such that $(r, \gamma(r,e)) \in G_e$. The function γ satisfies*

(a) $D_r\gamma(r,e) = -\dfrac{D_r f^o(r,e)}{D_r f^y(r,e)} > 0$ *for* $(r,e) \in I_{\rho(\hat{e})} \times E_{\hat{e}}$

(b) $D_r\gamma(\rho(e),e) = -\dfrac{D_r f^o(\rho(e),e)}{D_r f^y(\rho(e),e)} = \rho(e)$

(c) *if $\rho(\hat{e}) = 1$ then $E_{\hat{e}}$ may be chosen so that $\rho(e), 1 \in I_{\rho(\hat{e})}$ for $e \in E_{\hat{e}}$.*

Proof: Consider the function $F : \mathbb{R}^2_{++} \times \mathbb{R}^2_{++} \longrightarrow \mathbb{R}$ with values

$$F(r, r', e) = (f^y(r', e) - e^y) + (f^o(r, e) - e^0)$$

We have $F(\rho(\hat{e}), \rho(\hat{e}), \hat{e}) = 0$ and the derivative of F with respect to r' is the derivative of excess demand with respect to the relative price, r', for the young consumer. When this derivative is evaluated at $(r, r', e) = (\rho(\hat{e}), \rho(\hat{e}), \hat{e})$ we get

$$D_{r'} F(\rho(\hat{e}), \rho(\hat{e}), \hat{e}) = D_r f^y(\rho(e), e) = -\frac{D_0 u(e)}{\Delta(e, (1, -\rho(e)))}$$

which is not 0. (Here $\Delta(e, (1, -\rho(e)))$ is the value of the quadratic form from Lemma 9.2.C, on the derivatives of demand.) By the Implicit Function Theorem, Theorem C in Appendix A, there is an open interval $I' \subset \mathbb{R}_{++}$, an open set $E' \subset \mathbb{R}^2_{++}$ with $(\rho(\hat{e}), \hat{e}) \in I' \times E'$ and a unique function $\gamma \in \mathbb{C}^1(I' \times E', \mathbb{R}_{++})$ such that $\rho(\hat{e}) = \gamma(\rho(\hat{e}), \hat{e})$ and $F(r, \gamma(r, e), e) = 0$ for $r \in I'$ that is, $(r, \gamma(r, e)) \in G_e$.

To prove **(a)** and **(b)** note that from Lemma 9.2.C, on the derivatives of demand,

$$D_r \gamma(\rho(e), e) = -\frac{D_r f^o(\rho(e), e)}{D_r f^y(\rho(e), e)} = \rho(e) > 0$$

and by continuity of the partial derivatives, there is an open interval $I_{\rho(\hat{e})} \subset I'$ and and an open set $E_{\hat{e}} \subset E'$ such that

$$D_r \gamma(r, e) = -\frac{D_r f^o(r, e)}{D_r f^y(r, e)} > 0 \quad \text{for } (r, e) \in I_{\rho(\hat{e})} \times E_{\hat{e}}$$

(c) Since the function $e \longrightarrow \rho(e)$ is continuous the set

$$E_{\hat{e}} \cap \left\{ e \in \mathbb{R}^2_{++} \mid \rho(e) \in I_{\rho(\hat{e})} \right\}$$

is an open set.

□

Thus given e, there is a small open interval, $I_{\rho(e)}$ owning $\rho(e)$ such that for each price from the previous date, r, there is current price $\gamma(r, e)$ solving the equilibrium equation. Since $D_r \gamma(\rho(e), e) = \rho(e)$ the graph of

$\gamma(\cdot, e)$ is a curve which cuts the diagonal from below if $\rho(e) > 1$ and from above if $\rho(e) < 1$. Cf. Figure 10.3.A

Note that if $(\bar{r}, \bar{r}') \in G_e$ is such that $D_r f^o(r,e)$ or $D_r f^y(r',e)$ is non-zero then the Implicit Function Theorem can be applied get get r' as function of r (or r as a function of r') in the vicinity of $(\bar{r}, \bar{r}')$.

Extending solutions to the equilibrium equation

By Lemma 10.1.C, on properties of the equilibrium locus, we can, given r in $\mathrm{pr}_1\, G_e$, find an r' such $(r, r') \in G_e$ and hence we can solve the equilibrium equation one date forward. Similarly, if $r' \in \mathrm{pr}_2\, G_e$ we can solve the equilibrium equation one date backward, that is, find an r such that $(r, r') \in G_e$. By Lemma 10.1.C the sets $\mathrm{pr}_1 G_e$ and $\mathrm{pr}_2\, G_e$ each contain an interval with non-empty interior. Hence it will often be possible to solve the equilibrium equation one date forward or backward.

Consider the problem of extending the price r at date t to a price system r_τ, for $\tau > t$ clearing markets for date t and onward. If $r \in \mathrm{pr}_1\, G_e$ we can extend the price system one date forward with, say, $r' = r_{t+1}$. However, it may be occur that $r' = r_{t+1} \notin \mathrm{pr}_1\, G_e$, in which case the process can not be continued. Hence in order to extend forward to a system of equilibrium prices for the economy we need to ascertain the existence of a subset, A, of $\mathrm{pr}_1\, G_e$ with the property that $r \in A$ implies that there is $r' \in A$ such that $(r, r') \in \mathrm{pr}_1\, G_e$. Such a set is a *forward invariant set.* The significance of such a set is that given an initial relative price in the set we can find a forward solution to the equilibrium equation where each of the future prices belong the same invariant set.

Similar considerations apply to the problem of extending the price r' backwards, that is, we need to find a set $B \subset \mathrm{pr}_2\, G_e$ such that $r' \in B$ implies that there is $r \in B$ such that $(r, r') \in G_e$. The set B is then a *backward invariant set.*

Definition 10.1.E *Let e be an economy with equilibrium locus G_e.*

(a) *A set S_e^f is a* **forward invariant set** *for e if for each $r \in S_e^f$ there is $r' \in S_e^f$ with $(r, r') \in G_e$.*

(b) *A set* S_e^b *is a* **backward invariant set** *for* e *if for each* $r' \in S_e^b$ *there is* $r \in S_e^b$ *with* $(r, r') \in G_e$.

(c) *A set* S_e *is an* **invariant set** *for* e *if* S_e *a forward invariant set and a backward invariant set for* e.

(d) *An invariant set,* S_e*, is a* **maximal invariant set** *for* e *if* $S_e \subset B$ *and* $S_e \neq B$ *implies that* B *is not an invariant set for* e.

The significance of a maximal invariantset for the economy e is that for each equilibrium price system each price belongs to the maximal invariant set. We have the following lemma.

Lemma 10.1.F *Each economy* e *has a maximal invariant set,* M_e.

Proof: Let S_e and S_e' be two invariant sets for e. Then $S_e \cup S_e'$ is an invariant set. Thus take M_e to be the union of all invariant sets for e.

□

In view of Lemma 10.1.F we can refer to *the* maximal set for e. We define a maximal forward invariant subset for e to be a forward invariant set which is not a proper subset of a forward invariant set for e and analogously for a maximal backward invariant set. We can in fact describe the maximal forward and maximal backward invariant sets for e. Note that the intersection of the maximal forward invariant set and the maximal backward invariant set is the maximal invariant set.

By Walras law, $rz^y(r,e) = -z^o(r,e)$ so that the equilibrium equation may be given as

$$z^y(r',e) - rz^y(r,e) = 0$$

This is useful since it is easy to compare the functions $r \longrightarrow rz^y(r,e)$ and $z^y(\cdot,e)$. Cf. Figure 10.1.D which also suggests the main ideas for the proof of Proposition 10.1.G. Note that in the proof of the proposition we have a sequence of prices, which may well fail to be monotone, but the corresponding excess demand for the young or for the old consumer is monotone. As will be seen the same idea is applied in Proposition 10.3.B below.

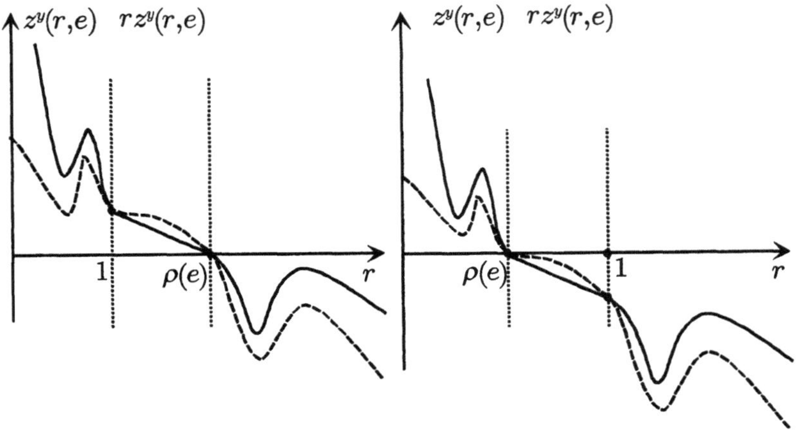

Figure 10.1.D: *The equilibrium equation may be written $z^y(r', e) - rz^y(r, e) = 0$. The solid curve is the graph of $r \to z^y(r, e)$ and the dashed curve the graph of $r \to rz^y(r, e)$*

Proposition 10.1.G *Let e be an economy.*

(a) *The set*

$$M_e^f = \left\{ \bar{r} \in \mathbb{R}_{++} \;\middle|\; \bar{r}z^y(\bar{r}, e) \geq \min_{r \in]0, \max J_e]} z^y(r, e) \right\}$$

is the maximal forward invariant set. The interval $]0, \max J_e]$ is a subset of M_e^f.

(b) *The set*

$$M_e^b = \left\{ \bar{r} \in \mathbb{R}_{++} \;\middle|\; z^y(\bar{r}, e) \leq \max_{r \in [\min J_e, \infty[} rz^y(r, e) \right\}$$

is the maximal backward invariant set. The interval $[\min J_e, \infty[$ is a subset of M_e^b.

Proof: To prove **(a)** let $\bar{r}$ satisfy

$$\bar{r}z^y(\bar{r}, e) \geq \min_{r \in]0, \max J_e]} z^y(r, e)$$

Then, since $z^y(\cdot, e)$ is continuous and $\lim_{r \to 0} z^y(r, e) = \infty$, there is an $r \in]0, \max J_e]$ such that $z^y(r, e) - \bar{r}z^y(\bar{r}, e) = 0$. If $\bar{r} \in]0, \max J_e]$ then $\bar{r}$ satisfies the inequality above and hence $]0, \max J_e] \subset M_e^f$.

On the other hand, let $\bar{r}$ be such that $\bar{r}z^y(\bar{r},e) < \min_{r\in]0,\max J_e]} z^y(r,e)$ and assume, in order to obtain a contradiction, that there is a forward solution, $(r_n)_{n\in\mathbb{N}}$, with $r_1 = \bar{r}$ and $(r_n, r_{n+1}) \in G_e$ so that $z^y(r_{n+1},e) - r_n z^y(r_n,e) = 0$ for $n \in \mathbb{N}$.

Since $r_1 = \bar{r}$ it follows that $r_1 > \max J_e \geq 1$ and hence

$$z^y(r_2,e) \;=\; r_1 z^y(r_1,e) \;=\; z^y(r_1,e) + (r_1-1)z^y(r_1,e) \qquad \textbf{10.1.A}$$

Since $z^y(r_1,e) < 0$ we get $z^y(r_2,e) < z^y(r_1,e)$. Furthermore, r_2 belongs to the set

$$\{r \in \mathbb{R}_{++} \mid rz^y(r,e) \leq \bar{r}z^y(\bar{r},e)\}$$

which is a subset of an interval $[a,\infty[$ with $a > \max J_e$. Let

$$\delta \;=\; \max_{r\in[a,\infty[} (r-1)z^y(r,e)$$

Then $\delta < 0$ and from relation 10.1.A it is seen that $z^y(r_2,e) \leq z^y(r_1,e)+\delta$. By induction we get $z^y(r_{n+1},e) \leq z^y(r_n,e) + n\delta$ for $n \in \mathbb{N}$ which can not be true since the function $z^y(\cdot,e)$ is bounded below by $-e^y$. Hence there can be no forward solution not belonging to M_e^f and thus M_e^f is the maximal forward invariant set.

To prove **(b)**, let $\bar{r}$ satisfy

$$z^y(\bar{r},e) \;\leq\; \max_{r\in[\min J_e,\infty[} rz^y(r,e)$$

Then, since $r \longrightarrow rz^y(r,e)$ is continuous and $\lim_{r\to\infty} rz^y(r,e) = -\infty$, there is an $r \in [\min J_e,\infty[$ such that $z^y(r,e) - \bar{r}z^y(\bar{r},e) = 0$. If $\bar{r} \in [\min J_e,\infty[$ then $\bar{r}$ satisfies the inequality and hence $[\min J_e,\infty[\subset M_e^b$.

On the other hand, let $\bar{r}$ be such that $z^y(\bar{r},e) > \max_{r\in[\min J_e,\infty[} rz^y(r,e)$ and assume, in order to obtain a contradiction, that there is a backward solution, $(r_{-n})_{n\in\mathbb{N}}$, with $r_{-1} = \bar{r}$ and $(r_{-n-1}, r_{-n}) \in G_e$ so that $z^y(r_{-n},e) - r_{-n-1}z^y(r_{-n-1},e) = 0$ for $n \in \mathbb{N}$.

Since $r_{-1} = \bar{r}$ it follows that $r_{-1} < \min J_e \leq 1$ and hence

$$z^y(r_{-1},e) \;=\; r_{-2}z^y(r_{-2},e) \;=\; z^y(r_{-2},e)+(r_{-2}-1)z^y(r_{-2},e) \;\leq\; z^y(r_{-2},e) \qquad \textbf{10.1.B}$$

Since $z^y(r_{-1}, e) > 0$ we get $z^y(r_{-2}, e) > z^y(r_{-1}, e)$. Furthermore, r_{-2} belongs to the set

$$\{r \in \mathbb{R}_{++} \mid rz^y(r, e) \geq \bar{r}z^y(\bar{r}, e)\}$$

which is a subset of an interval $]0, a]$ with $a < \min J_e$. Let

$$\delta = \max_{r \in]0,a]} (r - 1)z^y(r, e) < 0$$

Then $\delta < 0$ and by relation 10.1.B we have $z^y(r_{-1}, e) \leq z^y(r_{-2}, e) + \delta$. By induction we get $z^y(r_{-1}, e) \leq z^y(r_{-n}, e) + (n-1)\delta$ for $n \in \mathbb{N}$ so that for n large enough we have $z^y(r_{-n}, e) \geq e^o$ which implies that there can be no backward solution form r_{-n}. Hence there can be no backward solution not belonging to M_e^b and thus M_e^b is the maximal backward invariant set

□

Since the maximal invariant set, M_e, for the economy e is the intersection $M_e^f \cap M_e^b$ we get the following corollary.

Corollary 10.1.H *The maximal invariant set, M_e, for the economy e contains the interval J_e. If $\rho(e) \leq 1$ then $M_e \subset [\rho(e), \infty[$ and if $\rho(e) \geq 1$ then $M_e \subset]0, \rho(e)]$.*

10.2 Indeterminacy of Equilibrium

Proposition 10.1.G, on solutions to the equilibrium equation, immediately gives that there is large amount of indeterminacy of an equilibrium (relative) price system in case $\rho(e) \neq 1$. Since we are using relative prices the different equilibrium price systems will be associated with different equilibrium allocations. This indeterminacy can be seen as a lack of market clearing in the indefinite future and past. Thus, if an OG economy either has a starting date or is influenced by a shock then fundamentals, that is, consumption sets, endowments and utility functions, do not necessarily determine prices. Therefore expectations and economic policy become important in the determination of equilibrium price systems. Recall that M_e is the maximal invariant set.

Theorem 10.2.A *Let $\mathcal{E}^s = (C, u, e)$ be a stationary OG economy with $\rho(e) \neq 1$ and let $\bar{r} \in M_e$. Then there is an equilibrium price system, $(\bar{r}_t)_{t\in\mathbb{Z}}$, with $\bar{r}_0 = \bar{r}$. Hence there is a continuum of equilibria.*

Proof: For each $\bar{r} \in M_e$ there exists an equilibrium price system $(\bar{r}_t)_{t\in\mathbb{Z}}$, with $\bar{r}_0 = \bar{r}$ and the equilibrium allocations are all different since the relative price for consumer 0 differs between equilibria. Since $\rho(e) \neq 1$ the interval J_e has non-empty interior and is, by Proposition 10.1.G, contained in the maximal invariant set. Hence there is a continuum of equilibria.

□

An example indicating the indeterminacy result in Theorem 10.2.A was given by Geanakoplos and Polemarchakis [1991]. In the following example it is easy to find the maximal invariant set.

Example 10.2.A Let $\mathcal{E}^s = (\mathbb{R}^2_{++}, \ln c^y + \ln c^o, (3,1))$ then

$$G_e = \left\{ (r, r') \in \mathbb{R}^2_{++} \;\middle|\; \frac{1}{r'} - 4 + 3r = 0 \right\} \quad \text{and} \quad J_e = \left[\frac{1}{3}, 1\right]$$

Solving for r and r' we get

$$r' = \frac{1}{4 - 3r}$$
$$r = \frac{4 - r'}{3}$$

The maximal forward invariant set is $]0, 1]$ and the maximal backward invariant set is $[1/3, \infty[$. Thus, the maximal invariant set is $M_e = [1/3, 1]$ and for $\bar{r} \in M_e$ there exists a unique equilibrium price system, $(\bar{r}_t)_{t\in\mathbb{Z}}$ with $r_0 = \bar{r}$.

The theory of dynamical systems

In Appendix D some results are given from the theory of dynamical systems which are relevant for stationary, one-dimensional, first-order difference equations. These results can be applied in order to study equilibria of an OG economy provided that its equilibria can be given as are solutions of stationary, one-dimensional, first-order difference equations.

10.3 Turnpike Equilibria

In growth theory an economy has the turnpike property if each equilibrium converges to a steady state. Often economies are studied where time extends into the indefinite future but not backwards which means that initial conditions have to be specified. In this case an economy has the turnpike property if each equilibrium, independent of initial conditions, converges to a steady state. Hence prices as well as consumptions approach the values of a steady state as time tends to infinity. In this section we study equilibria of economies where the endowments form a strongly optimal allocation, in which case $\rho(e) = 1$, and the equilibria of economies where the endowment is close to such an endowment. From Theorem 10.3.A we know that if $\rho(e) \neq 1$ then the economy has a continuum of equilibria. But we will see that each of these equilibria are close to the equilibrium of an economy with $\rho(e) = 1$ and prices and consumptions converge to a steady state price and a steady state consumption as time goes to infinity forwards or backwards.

Uniqueness of Equilibrium and No-trade Endowment

Since we study stationary OG economies where time extends forward and backward the initial endowment takes the place of the initial conditions. By Theorem 8.5.B, on the uniqueness of equilibrium for economies with strongly optimal endowments, an economy where the initial endowment is a no-trade endowment has a unique equilibrium. The initial endowment allocation of stationary economy is a strongly optimal allocation if and only if the real steady state price $\rho(e) = 1$. Thus Theorem 8.5.B applied to stationary economies gives us the theorem below.

Theorem 10.3.A *Suppose that* $\rho(\hat{e}) = 1$. *Then the steady state*

$$(\bar{r}, \bar{c}) \;=\; (1, f(1, \hat{e})) \;=\; (\rho(\hat{e}), \hat{e})$$

is the unique equilibrium.

The theorem implies that the maximal invariant set for an economy $\hat{e}$ with $\rho(\hat{e}) = 1$ is the one-point set $M_e = \{1\}$.

No-trade endowment, the equilibrium locus and the maximal invariant set

In the following we want to consider the equilibria for economies which are close to no-trade economies. The results to be derived will be immediate consequences of the following proposition on the maximal invariant set for such economies.

Proposition 10.3.B *Let $\hat{e}$ be an economy with $\rho(\hat{e}) = 1$. Then there is an open set $E_{\hat{e}}$ owning $\hat{e}$ such that for $e \in E_{\hat{e}}$ the maximal invariant set for e is the interval J_e.*

Proof: Define the non-increasing, continuous functions $h, \hat{h} : \mathbb{R}_{++} \times \mathbb{R}^2_{++} \longrightarrow \mathbb{R}$ by

$$\begin{aligned} h(r,e) &= \min\nolimits_{\alpha \leq r} z^y(\alpha, e) - e^o \\ \hat{h}(r,e) &= \max\nolimits_{\beta \geq r} \beta z^y(\beta, e) + e^y \end{aligned}$$

and put

$$\begin{aligned} c &= \max\{r \in \mathbb{R}_{++} \mid h(r,\hat{e}) \geq 2\} \\ d &= \min\left\{r \in \mathbb{R}_{++} \mid \hat{h}(r,\hat{e}) \leq -2\right\} \end{aligned}$$

Consider the continuous function $\mathbb{R}^2_{++} \longrightarrow \mathbb{R}$ given by

$$e \longrightarrow \max_{r \in [c,d]} |h(r,e) - h(r,\hat{e})| + \max_{r \in [c,d]} \left|\hat{h}(r,e) - \hat{h}(r,\hat{e})\right|$$

The set

$$\left\{e \in \mathbb{R}^2_{++} \;\middle|\; \max_{r \in [c,d]} |h(r,e) - h(r,\hat{e})| + \max_{r \in [c,d]} \left|\hat{h}(r,e) - \hat{h}(r,\hat{e})\right| < 1\right\}$$

is an open set owning $\hat{e}$. From this set we can choose a compact subset, $E''_{\hat{e}}$, such that its interior owns $\hat{e}$. For $e \in E''_{\hat{e}}$ we have $z^y(r,e) - e^o \geq 1$ for $r \in]0, c[$ and $rz^y(r,e) + e^y \leq 1$ for $r \in]d, \infty[$ which implies that the maximal invariant set for e, the set $M_e \subset [c,d]$.

Since $\rho(\hat{e}) = 1$ and $D_r z^y(1,\hat{e}) < 0$, there is, by continuity, an open interval $I_1 =]a,b[\subset [c,d]$ owning 1 and an open set, $E'_{\hat{e}}$, owning $\hat{e}$ such that for $(e,r) \in E'_{\hat{e}} \times I_1$ we have $\rho(e) \in I_1$ and

$$D_r z^y(r,e) < 0, D_r r z^y(r,e) < 0 \quad \text{for } r \in I_1$$

There is a number $\delta > 0$ such that

$$\begin{aligned} \max\{z^y(r,e) \mid r \in [b,d]\} &\leq -\delta \quad \text{and} \\ \min\{rz^y(r,e) \mid r \in [c,a]\} &\geq \delta \end{aligned}$$

Since the function $z^y(1,\cdot)$ is continuous the set $\{e \in \mathbb{R}^2_{++} \mid |z^y(1,e)| < \delta/2\}$ is an open set which owns $\hat{e}$. Let $E_{\hat{e}}$ be the intersection of this set with the interior of $E''_{\hat{e}}$ and $E'_{\hat{e}}$.

For $e \in E_{\hat{e}}$ we have

$$\begin{aligned} \min\nolimits_{r \in]0, \max J_e]} z^y(r,e) &= z^y(\max J_e, e) \\ \max\nolimits_{r \in [\min J_e, \infty[} rz^y(r,e) &= \min J_e \cdot z^y(\min J_e, e) \end{aligned}$$

and thus by Proposition 10.1.G

$$\begin{aligned} M^f_e = \{\bar{r} \in \mathbb{R}^2_{++} \mid \bar{r}z^y(\bar{r},e) \geq z^y(\max J_e, e)\} &=]0, \max J_e] \\ M^b_e = \{\bar{r} \in \mathbb{R}^2_{++} \mid \bar{r}z^y(\bar{r},e) \leq \min J_e \cdot z^y(\min J_e, e)\} &= [\min J_e, \infty[\end{aligned}$$

which implies $M^f_e \cap M^b_e = M_e = J_e$.

□

From the proof of Proposition 10.3.B above it is seen that the functions $z^y(\cdot,e)$ and $r \longrightarrow rz^y(r,e)$ are decreasing on J_e. It follows that the one-date forward and backward solutions on J_e are unique. On the other hand Lemma 10.1.D, on the local properties of the equilibrium locus, shows that the forward and backward solutions vary differentiably with $r \in J_e$. Thus we get the following corollary.

Corollary 10.3.C *Let $\hat{e}$ be an economy with $\rho(\hat{e}) = 1$. Then there is an open set $E_{\hat{e}}$ owning $\hat{e}$, such that for $e \in E_{\hat{e}}$ the maximal invariant set for e is the interval J_e and there is function*

$$\gamma : \{(r,e) \in \mathbb{R}_{++} \times E_{\hat{e}} \mid r \in J_e\} \longrightarrow \mathbb{R}$$

such that $r, r' \in J_e$ and $(r,r') \in G_e$ if and only if $r' = \gamma(r,e)$. For each $e \in E_{\hat{e}}$ the function $\gamma(\cdot,e)$ is an increasing function.

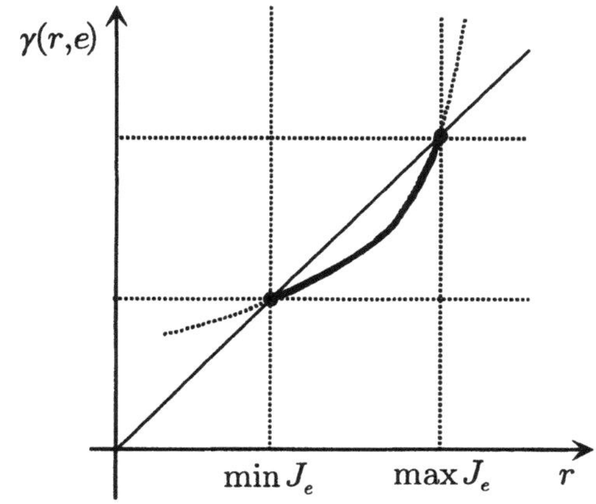

Figure 10.3.A: *For each $e \in E_{\hat{e}}$ the function $\gamma(\cdot, e)$ is an increasing function on J_e*

Equilibria for Economies Close to No-Trade Economies

If $\rho(\hat{e}) = 1$ so that the endowment $\hat{e}$ forms a strongly optimal allocation and the endowment e is sufficiently close to $\hat{e}$ then the stationary economy defined by e has a continuum of equilibrium price systems but each of these price systems have all their prices close to 1.

Corollary 10.3.D *Let $\hat{e}$ be an economy with $\rho(\hat{e}) = 1$ and let I_1 be an open interval owning 1. There is an open set $E_{\hat{e}}$ owning $\hat{e}$ such that if $e \in E_{\hat{e}}$ and $(\bar{r}_t)_{t\in\mathbb{Z}}$ is an equilibrium price system for e then $\bar{r}_t \in I_1$ for $t \in \mathbb{Z}$.*

Proof: By Proposition 10.1.G, on the maximal invariant set, there is an open set $E_{\hat{e}}$ owning $\hat{e}$, such that J_e is the maximal invariant set for $e \in E_{\hat{e}}$ and, by continuity of the functions z^y and z^o, the set $E_{\hat{e}}$ may be chosen so that $J_e \subset I_1$. For $e \in E_{\hat{e}}$ each equilibrium price system $(\bar{r}_t)_{t\in\mathbb{Z}}$ for the economy e has $\bar{r}_t \in J_e$ and hence $\bar{r}_t \in I_1$ for $t \in \mathbb{Z}$.

Characterization of Turnpike Equilibria

If an initial endowment is sufficiently close to a no-trade endowment then each of the equilibrium price systems for the economy has each price close to 1 according to Corollary 10.3.D, on equilibrium price systems. In particular, an economy e where e is sufficiently close to a no-trade economy has its steady state(s) equilibrium prices close to 1.

It is difficult to explicitly describe the equilibrium locus in general. However if we restrict attention to economies with initial endowments close to some initial endowment where $\rho(\hat{e}) = 1$ then we can use Corollary 10.3.C to get the following result concerning equilibria for such an economy.

Theorem 10.3.E *Let $\hat{e}$ be an economy with $\rho(\hat{e}) = 1$. Then there is an open set, $E_{\hat{e}}$, owning $\hat{e}$ such that if $e \in E_{\hat{e}}$ and $((\bar{r}_t)_{t\in\mathbb{Z}}, (\bar{c}_t)_{t\in\mathbb{Z}})$ is an equilibrium for e then one and only one of the alternatives* (a) *and* (b) *holds:*

(a) *the equilibrium is a steady state so that*

$$(\bar{r}_t, \bar{c}_t) = (\rho(e), e) \text{ for } t \in \mathbb{Z} \quad \text{or} \quad (\bar{r}_t, \bar{c}_t) = (1, f(1, e) \text{ for } t \in \mathbb{Z}$$

(b) *the equilibrium converges to one of the steady states in forward time and the other steady state in backward time so that*

$$\lim_{t\to\infty}(\bar{r}_t, \bar{c}_t) = (\min J_e, f(\min J_e, e)) \quad \text{and}$$
$$\lim_{t\to-\infty}(\bar{r}_t, \bar{c}_t) = (\max J_e, f(\max J_e, e))$$

Proof: By Proposition 10.3.B there is an open set, $E_{\hat{e}}$, owning $\hat{e}$ such that $e \in E_{\hat{e}}$ implies that J_e is the maximal invariant set for e and such that the function $\gamma(\cdot, e)$ restricted to J_e is an increasing function.

To prove **(b)** let $e \in E_{\hat{e}}$ and let $((\bar{r}_t)_{t\in\mathbb{Z}}, (\bar{c}_t)_{t\in\mathbb{Z}})$ be an equilibrium for e which is not a steady state. Then $\bar{r}_t \in J_e$ for $t \in \mathbb{Z}$ but there is a t' such that $r_{t'} \notin \{\rho(e), 1\}$. For $r \in \operatorname{int} J_e$ we have $\gamma(r, e) < r$ and hence

$$\max J_e \geq \ldots r_{t-1} > r_t > r_{t+1} > r_{t+2} \ldots \geq \min J_e$$

which shows that $\underline{r} = \lim_{t\to\infty}\bar{r}_t$ and $\bar{r} = \lim_{t\to-\infty}\bar{r}_t$ exist. By continuity of $\gamma(\cdot, e)$ we get $\gamma(\underline{r}, e) = \underline{r}$ and $\gamma(\bar{r}, e) = \bar{r}$ which implies that that $\underline{r}$ and $\bar{r}$ are steady state prices so that $\underline{r}, \bar{r} \in \{\rho(e), 1\}$. Hence $\underline{r} = \min J_e$ and $\bar{r} = \max J_e$.

Since the demand function is continuous the assertions about the convergence of the consumptions follow from the convergence of the prices.

Summary

The problem of finding an equilibrium price system for a stationary OG economy is equivalent to solving the equilibrium equation at each date. We first studied the solutions to the equilibrium equation one date forward or one date backward. The pairs of relative prices, at the current date and the previous date, satisfying the equilibrium equation was defined to be the equilibrium locus. We noted that, in general, the relative price at the previous date did not determine the relative price at the current date uniquely so that there might be more than one forward solution and analogously for a backward solution, given the price at the current date.

It was noted that to find an equilibrium price system one needed to solve the equilibrium equation forward and ensure that the price solving was again such that the equation again could be solved forward. This was captured by the notion of a forward invariant set. In the same way a backward invariant set described prices which could be extended backwards indefinitely. We were able to characterize the maximal invariant set and to show that this set contained a non-empty open interval unless the economy was a no-trade economy. An immediate consequence was that a stationary OG economy has a continuum of equilibria, unless it is a no-trade economy.

For economies close to no trade economies the maximal invariant set was seen to be the closed interval with the steady state prices as endpoints. The equilibria for such economies could then be described. Each equilibrium was either a steady state or had prices converging to the smallest steady state price in forward time and to the largest steady state price in backward time.

Exercises

Exercise 10.A Consider the functions $h, \hat{h} : \mathbb{R}_{++} \times \mathbb{R}^2_{++} \longrightarrow \mathbb{R}$ from the proof of Proposition 10.3.B with values given by

$$\begin{aligned} h(r,e) &= \min_{\alpha \le r} z^y(\alpha, e) - e^o \\ \hat{h}(r,e) &= \max_{\beta \ge r} \beta z^y(\beta, e) + e^y \end{aligned}$$

Show that the functions $h(\cdot, e), \hat{h}(\cdot, e) : \mathbb{R}_{++} \longrightarrow \mathbb{R}$ are non-increasing and continuous.

Exercise 10.B Let $\mathcal{E}^s = (C, u, e)$ be a stationary economy.

(a) State a necessary and sufficient condition for the economy to have a continuum of equilibria.

(b) Assume that $\rho(e) < 1$. Show that consumer 0 orders equilibria so that if $\bar{r}_0$ and $\hat{r}_0$ are the date 0 prices for two equilibrium price systems then: $\bar{r}_0 > \hat{r}_0$ implies $u(\bar{c}_0) > u(\hat{c}_0)$ where $\bar{c}_0$ and $\hat{c}_0$ are the equilibrium consumptions.

(c) Give a sufficient condition on the equilibrium locus which implies that all consumers will agree on the ordering of equilibrium allocations based on their consumptions.

Exercise 10.C Let G_e be the equilibrium locus for a stationary economy $\mathcal{E}^s = (C, u, e)$. Define

$$G_e \circ G_e = \{(r, r'') \mid \text{there exists } r' \text{ so that } (r, r') \in G_e \text{ and } (r', r'') \in G_e\}$$

(a) Let $((\bar{r}_t)_{t\in\mathbb{Z}}, (\bar{c}_t)_{t\in\mathbb{Z}})$ be an equilibrium. Show that $(\bar{r}_t, \bar{r}_{t+k}) \in G_e \circ G_e \circ \ldots \circ G_e$ (k times) for $k \in \mathbb{N}$ and $t \in \mathbb{Z}$.

(b) For the bounds $\hat{r}^U$ and $\hat{r}_L$ defined in Lemma 10.1.C show that

$$r \in \left]0, \hat{r}^U\right] \text{ implies } r \in \mathrm{pr}_1\, G_e \circ G_e \circ \ldots \circ G_e \ (k \text{ times, each } k \in \mathbb{N})$$

$$r' \in \left[\hat{r}_L, \infty\right[\text{ implies } r' \in \mathrm{pr}_2\, G_e \circ G_e \circ \ldots \circ G_e \ (k \text{ times, each } k \in \mathbb{N})$$

Exercise 10.D Let $\mathcal{E}^s = \left(\mathbb{R}^2_{++}, ac^y - (1/2)a(c^y)^2 + c^o, (0, e^o)\right)$ with $e^o > 1$ and $0 < a < 4$.

(a) Does the economy satisfy the Maintained Assumptions?

(b) Find the demand function for $0 < r < a$.

(c) Plot the functions $z^y(r,e)$ and $rz^y(r,e)$.

(d) Show that $]0,a]$ is an invariant set.

(e) Find $G_e \cap (]0,a] \times]0,a])$.

Exercise 10.E Let G_e be the equilibrium locus for a stationary economy $\mathcal{E}^s = (C, u, e)$. Assume that $(r, r') \in G_e$, $(\tilde{r}, \tilde{r}') \in G_e$ and $r < \tilde{r}$ implies $r' < \tilde{r}'$.

(a) Assume that $\rho(e) < 1$. Show that if $(r_k)_{k \in \mathbb{N}}$ is a sequence such that $(r, r_k) \in G_e \circ G_e \circ \ldots \circ G_e$ (k times) for each $k \in \mathbb{N}$ then

$$\lim_{k \to \infty} r_k = \min(\rho(e), 1)$$

(b) Assume that $\rho(e) < 1$. Show that if $(r_k)_{k \in \mathbb{N}}$ is a sequence such that $(r_k, r) \in G_e \circ G_e \circ \ldots \circ G_e$ (k times) for each $k \in \mathbb{N}$ then

$$\lim_{k \to \infty} r_k = \max(\rho(e), 1)$$

Exercise 10.F Let $\mathcal{E}^s = (\mathbb{R}^2_{++}, a \ln c^y + b \ln c^o, e,)$ where $a, b \in \mathbb{R}_{++}$.

(a) Find a first-order difference equation for an equilibrium price system in relative prices.

(b) Find the steady states.

(c) Find the equilibrium locus.

Exercise 10.G Let $\mathcal{E}^s = (\mathbb{R}^2_{++}, e, (c^y)^a + (c^o)^a)$ where $0 < a < 1$.

(a) Find a first-order difference equation an equilibrium price system in relative prices.

(b) Find the steady states.

(c) Find the equilibrium locus.

(d) Draw a diagram illustrating the relation between r_t and r_{t+1} and discuss properties of equilibria.

Exercise 10.H Show that, if $\rho(e) \leq 1$, then there are no classical equilibria.

Exercise 10.I Show that, if $\rho(e) \geq 1$, then there are no Samuelson equilibria.

Chapter 11

GLOBAL DYNAMICS: CYCLICAL EQUILIBRIA

Introduction

In this chapter we continue the study of stationary OG economies and consider the simplest type of equilibria where equilibrium prices and consumptions fluctuate over time in a regular way, rather than being constant or converging. These fluctuations may manifest themselves as deterministic or stochastic cycles.

In the previous chapter we studied economies where the endowment was close to a no-trade endowment so that the intensity of trade was low at an equilibrium. For such economies the equilibrium equation could be solved uniquely forwards and backwards. Equivalently this could be seen from the equilibrium locus which showed that next date relative price was increasing in the relative price at the current date. The assumption of perfect foresight for the consumers was then easy to accept since the consumers could infer the next date price from the price at the current date .

In case the equilibrium equation has more than one forward solution, it is not evident which price the young consumer at the current date should expect at the next date. The assumption of perfect foresight does not single out any particular price. Hence the economy may, for a given date 0 price, have more than one equilibrium price system, each resulting in a perfect foresight equilibrium. In the simplest kind of cyclic equilibrium, referred to as a 2-cycle, prices fluctuate between two different values and each consumer expects and acts accordingly so that expectations are fulfilled at every date.

In case the equilibrium equation has more than one forward solution the possibility arises that the realized equilibrium will be influenced by the expectations of the consumers. In particular the economy may have no intrinsic uncertainty, that is, uncertainty in the *fundamentals* such as consumption sets, utility functions or endowments. However, despite the lack of intrinsic uncertainty consumers may believe that the next date price is uncertain and act on these believes. It turns out that the feedback from expectations to the realized equilibrium may be such that the consumers' expectations of random prices are in fact realized. Hence consumers may perceive that the next date price is influenced by some extrinsic event; the outcome of the toss of a coin or the number of observed "sunspots". In the realized equilibrium prices will fluctuate and confirm the beliefs of the consumers. Such an equilibrium is referred to as a sunspot equilibrium, to remind of the fact that the realized equilibrium is influenced by some seemingly irrelevant extrinsic uncertainty. The simplest type of such an equilibrium is a 2-state sunspot equilibrium. In the equilibrium the price at the current date can be succeeded by one of two different prices at the next date. The young consumer at the current date has rational expectations and plans accordingly.

The results from Chapter 10 shows that cycles do not occur for economies where the no-trade equilibrium is close to the nominal equilibrium. For such economies price variations induce the intended substitution effects between consumption as young and as old and the intensity of trade is low. For economies with a high intensity of trade, price variations give rise to strong income effects through variations in the value of the initial endowment, which may counteract the intended substitution effects. There then arises the possibility of deterministic or stochastic fluctuations. These may be viewed as business cycles which result from the use of the market institution.

11.1 Cycles

Cycles in stationary OG economies are remarkable since there are cycles in prices and consumption bundles although fundamentals such as consumption sets, initial endowments and utility functions are stationary. Cycles may be seen as indications of market instability as they are caused by the markets themselves.

Existence of Cycles

2-cycles defined

Deterministic fluctuations where prices may take two different values are referred to as 2-cycles since prices are identical at all even dates and at all odd dates and consumption bundles are identical for all even generations and for all odd generations. Such a 2-cycle equilibrium, or 2-cycle for short, is a particular instance of an equilibrium as defined in Section 7.2

Definition 11.1.A *An* **2-cycle equilibrium** *for the stationary economy* $\mathcal{E}^s = (C, e, u)$ *is an equilibrium* $((p_t)_{t\in\mathbb{Z}}, (\bar{c}_t)_{t\in\mathbb{Z}})$ *such that there exist a pair of prices* (p, p') *and a pair of consumption bundles* $(\bar{c}, \bar{c}'))$ *such that*

$$(p_t, \bar{c}_t) = \begin{cases} (p, \bar{c}) & \text{for } t \text{ even} \\ (p', \bar{c}') & \text{for } t \text{ odd} \end{cases}$$

A 2-cycle equilibrium is **non-trivial** *if* $p \neq p'$.

At a 2-cycle consumers expect the next date price to be p' if it is p at the current date and the next date price to be p if it is p' today and these expectations turn out to be correct.

2-cycles and the equilibrium equation in relative prices

Let $((p, p'), (\bar{c}, \bar{c}'))$ be a 2-cycle equilibrium. Then, for $t \in \mathbb{Z}$ and t even, consumer t acts against the prices (p, p') at dates t and $t+1$ and her demand is $\bar{c} = (\bar{c}^y, \bar{c}^o) = f(p/p', e)$ while for t odd, consumer t acts against the prices (p', p) and her demand is $\bar{c}' = (\bar{c}^{y\prime}, \bar{c}^{o\prime}) = f(p'/p, e)$.

Let $r = p/p'$ and $r' = p'/p$ be the relative price for even and odd consumers, respectively.

Since $r' = 1/r$ a 2-cycle is given by the single number r and a pair of consumptions $(\bar{c}, \bar{c}')$ such that each consumer maximizes her utility,

$$\begin{cases} f(r,e) = \bar{c} & \text{(even generations)} \\ f\left(\frac{1}{r}, e\right) = \bar{c}' & \text{(odd generations)} \end{cases} \qquad \textbf{11.1.A}$$

and markets balance,

$$\begin{cases} f^y(r,e) + f^o\left(\frac{1}{r}, e\right) = e^y + e^o & \text{(even dates)} \\ f^y\left(\frac{1}{r}, e\right) + f^o(r,e) = e^y + e^o & \text{(odd dates)} \end{cases} \qquad \textbf{11.1.B}$$

Hence, in order to establish the existence of a 2-cycle, first a relative price such that the equations in relation 11.1.B are satisfied is found and next consumption bundles are derived by substituting the relative price in the equations in 11.1.A. The lemma below shows that although there are two equations and a single unknown the problem of finding a solution to the equations in 11.1.B can be reduced to one equation in a single unknown.

Lemma 11.1.B *If*

$$f^y(\bar{r}, e) + f^o\left(\frac{1}{\bar{r}}, e\right) = e^y + e^o$$

then both equations in 11.1.B *are satisfied for* $r = \bar{r}$.

Proof: Clearly,

$$r f^y(r,e) + f^o(r,e) = r e^y + e^o$$

by Walras' law. Using this relation for $r = \bar{r}$ and $r = 1/\bar{r}$ and replacing, in the market balance condition for even dates,

$$f^y(r,e) - e^y \quad \text{with} \quad \frac{1}{r}(e^o - f^o(r,e))$$

and

$$f^o\left(\frac{1}{r}, e\right) - e^o \quad \text{with} \quad \frac{1}{r}\left(e^y - f^y\left(\frac{1}{r}, e\right)\right)$$

the market balance condition for odd dates is obtained for $r = \bar{r}$.

A sufficient condition for existence of a 2-cycle equilibrium

Below we will see that $D_r f^y(1,e) > D_r f^o(1,e)$ is a sufficient condition for the existence of a 2-cycle equilibrium. After stating and proving the theorem we will discuss the interpretation of this condition and in Exercise 11.I the reader is asked to relate this condition on demand as a function of relative price to the Walrasian demand function.

Theorem 11.1.C *Let $\mathcal{E}^s = (C, e, u)$ be a stationary OG economy where*

$$D_r f^y(1,e) > D_r f^o(1,e)$$

Then $\mathcal{E}^s$ has a non-trivial 2-cycle equilibrium.

Proof: Consider the function $F \in \mathbb{C}^1(\mathbb{R}_{++}, \mathbb{R})$ defined by

$$F(r) = f^y(r,e) + f^o\left(\frac{1}{r}, e\right) - (e^y + e^o).$$

By Lemma 11.1.B, on the solution of the equilibrium equation, the relative price r is a non-trivial 2-cycle equilibrium price if $r \neq 1$ and $F(r) = 0$.

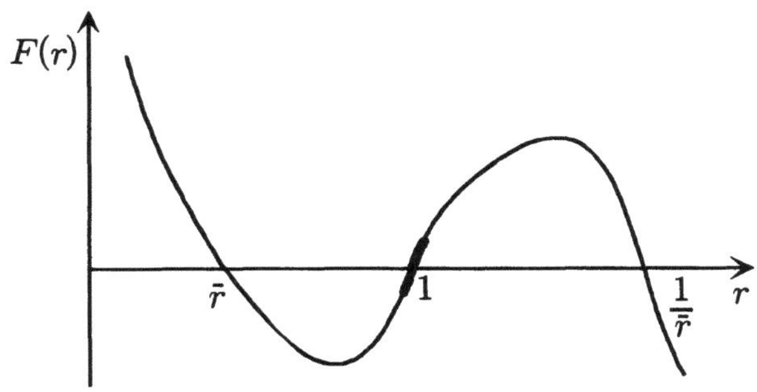

Figure 11.1.A: *The assumptions imply that F has a positive derivative for $r = 1$. Since $\lim_{r \to 0} F(r) = \infty$ there is a $\bar{r} < 1$ such that $F(\bar{r}) = 0$*

By Theorem 9.4.B, on the steady states, $F(1) = 0$. The derivative $D_r F(1) = D_r f^y(1,e) - D_r f^o(1,e) > 0$ by assumption and Lemma 9.2.B, on the basic properties of demand, gives that $\lim_{r \to 0} F(r) > 0$. Thus there exists $r < 1$ such that $F(r) < 0$. Since F is continuous there is

$r < 1$ such that $F(r) = 0$. Then $(r, 1/r, f(r,e), f(1/r,e))$ defines a 2-cycle equilibrium.

□

An example of a 2-cycle was first given by Gale [1973]. In the papers by Benhabib and Day [1982], Ghiglino and Tvede [1995] and Grandmont [1985] cyclical equilibria are studied.

Note that an equilibrium allocation associated with a 2-cycle equilibrium is a strongly optimal allocation according to Corollary 9.5.G since the price for the good oscillates between two values or equivalently the relative price oscillates between one value and its inverse.

Example 11.1.A Let $\mathcal{E}^s = \left(\mathbb{R}^2_+, (6,2), -(6-c^y)^2 - \alpha(3-c^o)^2\right)$. Then $u^{-1}(a)$ is not a closed subset of $\mathbb{R}^2$ for $a \in \mathbb{R}$ and $Du(c)$ fails to be a positive vector for some $c \in \mathbb{R}^2_{++}$. Hence the Maintained Assumptions are not satisfied.

If $\alpha > 1/9$, then, for each $r > 0$, there exists a unique solution to the problem

$$\text{Max}_{(c^y,c^o)} \quad -(6-c^y)^2 - \alpha(3-c^o)^2$$

$$\text{subject to} \quad \begin{cases} rc^y + c^o \le 6r+2 \\ \text{and } (c^y,c^o) \in X \end{cases}$$

The solution is

$$f_\alpha(r) = \left(6 - \frac{\alpha r}{\alpha r^2+1}, 2 + \frac{\alpha r^2}{\alpha r^2+1}\right)$$

and the derivatives are

$$D_r f^y_\alpha(r) = \frac{\alpha(\alpha r^2-1)}{(\alpha r^2+1)^2} \quad \text{and} \quad D_r f^o_\alpha(r) = \frac{2\alpha r}{(\alpha r^2+1)^2}$$

Furthermore, $\lim_{r\to 0} f^y_\alpha(r) = 6$ and $\lim_{r\to\infty} f^o_\alpha(r) = 3$ which implies $\lim_{r\to 0} f^y_\alpha(r) + f^o_\alpha(1/r) - (e^y + e^o) = 1 > 0$. Since

$$D_r f^y_\alpha(1) > D_r f^o_\alpha(1) \quad \text{if and only if} \quad \alpha > 3,$$

one can prove, by the same reasoning as in the proof of Theorem 11.1.C, that if $\alpha > 3$ then there is a 2-cycle.

Properties of Excess Demand and Dynamics

A reformulation of the sufficient condition

Since $rf^y(r,e) + f^o(r,e) = re^y + e^o$ for $r \in \mathbb{R}_{++}$ we get by Walras' law the relation

$$f^y(r,e) + rD_r f^y(r,e) + D_r f^o(r,e) \;=\; e^y \quad \text{for } r \in \mathbb{R}_{++}$$

for the derivatives $D_r f^y(r,e)$ and $D_r f^o(r,e)$. Then for $r = 1$

$$D_r f^y(1,e) + D_r f^o(1,e) \;=\; -(f^y(1,e) - e^y) \;=\; f^o(1,e) - e^o$$

so that we have the following three equivalent conditions

$$\begin{aligned}
&\text{(i)} \quad D_r f^y(1,e) \;>\; -\frac{f^y(1,e) - e^y}{2} \\
&\text{(ii)} \quad D_r f^o(1,e) \;<\; \frac{f^o(1,e) - e^o}{2} \\
&\text{(iii)} \quad D_r f^y(1,e) \;>\; D_r f^o(1,e)
\end{aligned}$$

Implications for excess demand and real savings

Assume that (iii) holds so that

$$D_r f^y(1,e) \;>\; D_r f^o(1,e)$$

If $\rho(e) < 1$ then $f^y(1,e) - e^y < 0$, by Lemma 9.2.B, on the basic properties of demand, and hence $D_r f^y(1,e) > 0$, by (i). In this case excess demand as young is increasing in its own-price, r, at $r = 1$. Equivalently, real savings $\hat{g}(r,e) = -(f^y(1,e) - e^y)$ are decreasing in r at $r = 1$.

On the other hand, if $\rho(e) > 1$ then $f^o(1,e) - e^o < 0$ by the same lemma. Part (ii) then implies $D_r f^o(1,e) < 0$. The own-price for consumption as old is $1/r$ and we get

$$D_{\frac{1}{r}}(f^o(r,e) - e^o) \;=\; D_r f^o(r,e) \cdot \frac{\partial r}{\partial \frac{1}{r}} \;=\; D_r f^o(r,e) \cdot -r^2 \;>\; 0$$

so that demand for consumption as old is an increasing function of the own-price, $1/r$, at $1/r = 1$. Alternatively this can be formulated as a condition on the dividends of real savings, $f^o(r,e) - e^o = r\hat{g}(r,e)$, which is thus a decreasing function of r at $r = 1$.

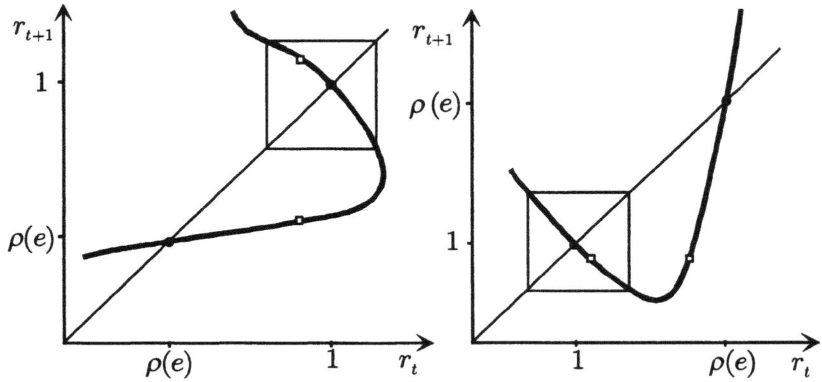

Figure 11.1.B: *The equilibrium locus for two economies with 2-cycle equilibria. In the left hand panel the nominal steady state is Samuelson. For r_t close to 1 there are two forward solutions to the equilibrium equation. In the right hand panel the nominal steady state is classical and for r_t close to 1 there are two backward solutions to the equilibrium equation.*

Forward solutions to the equilibrium equation

If $\rho(e) < 1$ then $f^y(1,e) - e^y < 0$ and by (i), $D_r f^y(1,e) > 0$ and again by (i)

$$-2 < \frac{f^y(1,e) - e^y}{D_r f^y(1,e)} < 0$$

Applying the Implicit Function Theorem to the equilibrium equation we can consider r_{t+1} as a function of r_t in the vicinity of $(r_t, r_{t+1}) = (1,1)$ and calculate the derivative

$$\begin{aligned}\left.\frac{\partial r_{t+1}}{\partial r_t}\right|_{(r_t,r_{t+1})=(1,1)} &= -\frac{D_r f^o(1,e)}{D_r f^y(1,e)} \\ &= \frac{(f^y(1,e) - e^y) + D_r f^y(1,e)}{D_r f^y(1,e)} \\ &= \frac{(f^y(1,e) - e^y)}{D_r f^y(1,e)} + 1 \in \,]-1,1[\end{aligned}$$

Therefore the relation between r_t and r_{t+1} may be as illustrated in the left hand panel of Figure 11.1.B. In Figure 11.1.B for r_t in a neighborhood of 1 there exist r_{t+1} in a neighborhood of 1 as well as r'_{t+1} smaller than r_t such that both (r_t, r_{t+1}) and (r_t, r'_{t+1}) belong to the equilibrium locus so that

$$f^y(r_{t+1}, e) + f^o(r_t, e) \;=\; f^y(r'_{t+1}, e) + f^o(r_t, e) \;=\; e^y + e^o$$

Thus the forward equilibrium price system, that is prices from date $t+1$ and forward, is indeterminate in the sense that r_{t+1} is not determined by r_t in a neighborhood of $r = 1$. Expectations or economic policy may influence whether r_{t+1} or r'_{t+1} is realized as a successor of r_t.

Backward solutions to the equilibrium equation

Still under the assumption $D_r f^y(1, e) > D_r f^o(1, e)$ we get that if $\rho(e) > 1$ then $f^o(1, e) - e^o < 0$, by Lemma 9.2.B, on the basic properties of demand, and $D_r f^o(1, e) < 0$ by relation (ii) which also implies

$$2 > \frac{f^o(1, e) - e^o}{D_r f^o(1, e)} > 0$$

We can apply the Implicit Function Theorem to solve "backwards" for r_t as a function of r_{t+1} in neighborhood of $(r_t, r_{t+1}) = (1, 1)$ and calculate the derivative of the equilibrium locus at $(1, 1)$

$$\begin{aligned}\left.\frac{\partial r_t}{\partial r_{t+1}}\right|_{(r_t, r_{t+1})=(1,1)} &= -\frac{D_r f^y(1, e)}{D_r f^o(1, e)} \\ &= -\frac{(f^o(1, e) - e^o) - D_r f^o(1, e)}{D_r f^o(1, e)} \\ &= -\frac{(f^o(1, e) - e^o)}{D_r f^o(1, e)} + 1 \in \,]-1, 1[\end{aligned}$$

Thus the relation between r_t and r_{t+1} may be as illustrated in the right hand panel of Figure 11.1.B. There the forward equilibrium price system, that is prices from date $t+1$ and forward, is determinate in the sense that r_{t+1} is determined by r_t and so on. However the backward equilibrium price system is indeterminate in the sense that r_t is not determined by r_{t+1} in a neighborhood of 1

11.2 Sunspot Equilibrium

In all previous sections consumers were assumed to have perfect foresight, that is, every young consumer knew the price at the first date of her life, she expected some single price at the second date of her life and this expectation was correct. If consumers have perfect foresight, an equilibrium is a sequence of prices and a sequence of consumption bundles such that each consumer maximizes her utility given prices and markets balance.

In the present section consumers are supposed to have rational expectations, rather than perfect foresight, so that every young consumer, at the first date of her life, knows the price at the first date of her life and the probability distribution on prices at the second date of her life. When consumers have rational expectations an equilibrium is a sequence of probability distributions on prices and a sequence of probability distributions on consumption bundles, such that each consumer maximizes her expected utility given prices at the first date of her life and the probability distribution on prices at the last date of her life, and markets balance.

A rational expectations equilibrium is referred to as a sunspot equilibrium in a stationary OG economy, where there is no "real" uncertainty, in order to indicate that some extrinsic variable such as the number of sunspots influences the equilibrium. Sunspot equilibria that are not perfect foresight equilibria are remarkable since there is uncertainty about prices although there is no uncertainty about fundamentals such as consumption sets, initial endowments or utility functions.

Uncertainty about prices at a rational expectations equilibrium arises since consumers expect that random variables seemingly unrelated to the economy influence prices. Thus uncertainty at a sunspot equilibrium stems from consumers' expectations since no random variables influence prices unless consumers believe that random variables influence prices. At a sunspot equilibrium consumers' beliefs concerning the relation between random variables and prices may represent "strange" theories held by the consumers and these theories are confirmed in equilibrium. Hence a sunspot equilibrium may be seen as an example of stochastic fluctuations that are driven by "animal spirits" and "expectations volatility".

Uncertainty and the Maintained Assumptions

In this section we extend preferences in an OG economy to take uncertainty into account. The consumer is assumed to have preferences over lotteries between consumptions. These preferences are in turn assumed to have an expected utility representation. In order to achieve this we have to strengthen the Maintained Assumptions somewhat.

Utility function and state utility function

In what follows we will consider a consumer acting under uncertainty about the price at the next date. Attention is restricted to the case where the price at the next date can take at most two values. The consumer has a state utility function $\bar{u} : \mathbb{R}^2_{++} \longrightarrow \mathbb{R}$. In the interpretation the value $\bar{u}(c_0, c_1)$ is the utility of *sure* consumption in the amount c_0, at the current date, and the amount c_1, at the next date. It is assumed that the consumer's preferences extend to lotteries, with two prizes, of current date-next date consumption. The utility ascribed to the lottery, with prize (x_0, x_1) with probability $\pi \in [0, 1]$ and prize (z_0, z_1) with probability $1 - \pi$, is given by a function $\hat{u}_\pi : \mathbb{R}^4_{++} \longrightarrow \mathbb{R}$ defined by

$$\hat{u}_\pi((x_0, x_1), (z_0, z_1)) \; = \; \pi \bar{u}(x_0, x_1) + (1 - \pi)\bar{u}(z_0, z_1)$$

which is the expected utility of the lottery. There is no uncertainty about the amount consumed at the current date, which corresponds to $x_0 = z_0$, and we can then consider the function $u_\pi : \mathbb{R}^3_{++} \longrightarrow \mathbb{R}$ with values

$$u_\pi(c_0, c_1, c_2) \; = \; \hat{u}_\pi((c_0, c_1), (c_0, c_2)) \; = \; \pi \bar{u}(c_0, c_1) + (1 - \pi)\bar{u}(c_0, c_2)$$

We need to strengthen the Maintained Assumptions somewhat in order for the consumer acting under uncertainty to satisfy (C1)-(C5). Thus we will assume in the following that the consumer and the stationary economy satisfy also assumption (C6) below.

(C6) *The state utility function,* $\bar{u} : \mathbb{R}^2_{++} \longrightarrow \mathbb{R}$ *satisfies*

(a) *the Maintained Assumptions* (C3)-(C5)

(b) $h^T D^2 \bar{u}(c_1, c_2) h \; < \; 0$ *for* $h \in \mathbb{R}^2 \setminus \{0\}$

(c) $\inf_{c \in \mathbb{R}^2_{++}} \bar{u}(c) \; = \; -\infty$

We can then define the economy as before $\mathcal{E}^s = (C, \bar{u}, e)$. On the one hand, we can interpret this as an economy with no uncertainty. On the other hand, interpreting C as the set of prizes, $\bar{u}$ as the state utility function and the endowment $e = (e^y, e^o)$ as inducing the endowment $(e^y, e^s, e^d) = (e^y, e^o, e^o)$, we get an OG economy with uncertainty. The endowment (e^y, e^s, e^d) is interpreted as endowment e^y as young, e^s for the state "same" the next date and e^d for the state "different" at the next date. The reason for naming the states "same" and "different" will become apparent in the sequel.

Exercise 11.C suggests that (C6) will in fact be satisfied if we take a consumer acting under uncertainty and satisfying (C1)-(C5) as the point of departure. The properties assumed for the state utility function in (C6) will then follow from the assumption that the consumer, acting under uncertainty, satisfies (C1)-(C5).

Properties of utility functions

When the function $\bar{u} : \mathbb{R}^2_{++} \longrightarrow \mathbb{R}$ satisfies the Maintained Assumptions including (C6) its boundary behavior is described by the following lemma.

Lemma 11.2.A *Let* $\bar{u} : \mathbb{R}^2_{++} \longrightarrow \mathbb{R}$ *satisfy the Maintained Assumptions. Let* $(c_1^n, c_2^n)_{n \in \mathbb{N}}$ *be a sequence converging to a boundary point* $(\bar{c}_1, \bar{c}_2)$ *of* $\mathbb{R}^2_{++}$. *Then* $\lim_{n \to \infty} \bar{u}(c_1^n, c_2^n) = -\infty$.

Proof: Let $k \in \mathbb{R}$. The set $\{c \in \mathbb{R}^2_{++} \mid \bar{u}(c) \geq k\}$ is contained in $\mathbb{R}^2_{++}$ and non-empty if, say, $k \leq \bar{u}(1,1)$.Since $(c_1^n, c_2^n)_{n \in \mathbb{N}}$ converges to a boundary point $(\bar{c}_1, \bar{c}_2)$ of $\mathbb{R}^2_{++}$ there is $\bar{n} \in \mathbb{N}$ such that $(c_1^n, c_2^n) \notin \{c \in \mathbb{R}^2_{++} \mid \bar{u}(c) \geq k\}$, and hence $\bar{u}(c_1^n, c_2^n) < k$, for $n \geq \bar{n}$. Since this is true for each $k \in \mathbb{R}$ we have $\lim_{n \to \infty} \bar{u}(c_1^n, c_2^n) = -\infty$.

□

The purpose of assumption (C6) was to make sure that the utility function induced by the state utility function satisfies (C3)-(C5). The lemma below shows that u_π will in fact be concave rather than merely quasi-concave.

Lemma 11.2.B *For* $\pi \in]0,1[$ *the function* $u_\pi : \mathbb{R}^3_{++} \longrightarrow \mathbb{R}$ *with values*

$$u_\pi(c^y, c^s, c^d) = \pi\bar{u}(c^y, c^s) + (1-\pi)\bar{u}(c^y, c^d)$$

is a strictly concave function which satisfies the Maintained Assumptions (C3)-(C5).

Proof: Let $(c_0, c_1) = c_{0,1}$ and $(c_0, c_2) = c_{0,2}$. Calculating the matrix of second derivatives, $D^2u_\pi(c)$, we get

$$\begin{bmatrix} \pi D^2_{11}\bar{u}(c_{0,1}) + (1-\pi)D^2_{11}\bar{u}(c_{0,2}) & \pi D^2_{12}\bar{u}(c_{0,1}) & (1-\pi)D^2_{12}\bar{u}(c_{0,2}) \\ \pi D^2_{21}\bar{u}(c_{0,1}) & \pi D^2_{22}\bar{u}(c_{0,1}) & 0 \\ (1-\pi)D^2_{21}\bar{u}(c_{0,2}) & 0 & (1-\pi)D^2_{22}\bar{u}(c_{0,2}) \end{bmatrix}$$

which equals

$$\pi \begin{bmatrix} D^2_{11}\bar{u}(c_{0,1}) & D^2_{12}\bar{u}(c_{0,1}) & 0 \\ D^2_{21}\bar{u}(c_{0,1}) & D^2_{22}\bar{u}(c_{0,1}) & 0 \\ 0 & 0 & 0 \end{bmatrix}$$

$$+ (1-\pi) \begin{bmatrix} D^2_{11}\bar{u}(c_{0,2}) & 0 & D^2_{12}\bar{u}(c_{0,2}) \\ 0 & 0 & 0 \\ D^2_{21}\bar{u}(c_{0,2}) & 0 & D^2_{22}\bar{u}(c_{0,2}) \end{bmatrix}$$

Let $h = (h_1, h_2, h_3) \in \mathbb{R}^3$, $h \neq 0$. It will be proved that $h^T D^2 u_\pi(c) h < 0$. At least one of the vectors (h_1, h_2) and (h_1, h_3) is different from 0. Since

$$\begin{aligned} &h^T D^2 u_\pi(c) h \\ &= \pi \,[h_1, h_2]\, D^2\bar{u}(c_{0,1}) \begin{bmatrix} h_1 \\ h_2 \end{bmatrix} + (1-\pi)(h_1, h_3) D^2\bar{u}(c_{0,2}) \begin{bmatrix} h_1 \\ h_3 \end{bmatrix} \end{aligned}$$

and $D^2\bar{u}(c_{0,1})$ and $D^2\bar{u}(c_{0,2})$ by (C6) are negative definite matrices it follows that $D^2u_\pi(c)$ is a negative definite matrix. This in turn implies that u_π is a strictly concave function.

To prove that u_π satisfies (C3) note that $\bar{u} \in \mathbb{C}^2(\mathbb{R}^2_{++}, \mathbb{R})$ implies $u_\pi \in \mathbb{C}^2(\mathbb{R}^3_{++}, \mathbb{R})$. To prove that each contour set of u_π is a closed subset of $\mathbb{R}^3$ consider the contour set

$$K = \left\{ (c^y, c^s, c^d) \in \mathbb{R}^3_{++} \mid \pi\bar{u}(c^y, c^s) + (1-\pi)\bar{u}(c^y, c^d) = k \right\}$$

where $k \in u_\pi(\mathbb{R}^3_{++})$. Let $(c^y_n, c^s_n, c^d_n)_{n\in\mathbb{N}}$ be a sequence in K converging to $(\bar{c}^y, \bar{c}^s, \bar{c}^d)$. We will prove that the limit point belongs to K and hence to $\mathbb{R}^3_{++}$. By continuity the limit belongs to K if $(\bar{c}^y, \bar{c}^s)$ and $(\bar{c}^y, \bar{c}^d)$ both belong to $\mathbb{R}^2_{++}$. Assume, in order to obtain a contradiction, that, say, $(\bar{c}^y, \bar{c}^s) \notin \mathbb{R}^2_{++}$. Then $\lim_{n\to\infty} \bar{u}(c^y_n, c^s_n) = -\infty$ and since $\bar{u}(c^y_n, c^d_n)$, $n \in \mathbb{N}$, is bounded above the points (c^y_n, c^s_n, c^d_n) can not belong to K for each $n \in \mathbb{N}$. Hence $(\bar{c}^y, \bar{c}^s)$ and $(\bar{c}^y, \bar{c}^d)$ both belong to $\mathbb{R}^2_{++}$ which gives $(\bar{c}^y, \bar{c}^s, \bar{c}^d) \in \mathbb{R}^3_{++}$ as was to be proved.

Demand under Uncertainty

The consumer problem

We now consider the Consumer Problem for a consumer acting under uncertainty. We have introduced the notation in Figure 11.2.A. Recall that the two states at date $t+1$ are denoted "same" and "different". The consumer has an initial endowment $e = (e^y, e^s, e^d)$ and chooses a consumption plan, $c = (c^y, c^s, c^d)$. The spot-prices are given by $p = (p^y, p^s, p^d)$. At present we allow for $e^s \neq e^d$ although it will suffice for our purpose to consider the case $e^s = e^d = e^o$, so that there is no uncertainty about the initial endowment.

In Chapter 9 we introduced relative prices by normalizing so that the price for the consumer as young was r and as old 1. Hence the consumer acted against prices $(r, 1)$ conveniently described by the single number r.

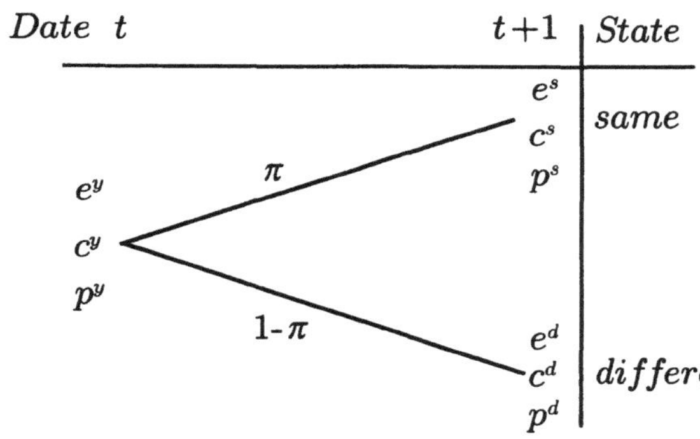

Figure 11.2.A: *The Consumer Problem under uncertainty; notation*

With uncertainty we can not, in general, get by with a single relative price. To do so we have to restrict attention to the case where one of the expected next date prices is the same as the price today. Formally we consider price systems (p^y, p^s, p^d) where $p^y = p^s$ so that $(p^y, p^s, p^d) = (p, p, p')$ for some p and p'. Putting $r = p/p'$ we have $(p^y, p^s, p^d) = (1/p')(r, r, 1)$ and by the homogeneity of demand we can restrict attention to price systems of the form $(r, r, 1)$ Cf. the left and mid panel of Figure 11.2.B on page 352.

The market structure is spot-markets with money which implies that the Consumer Problem can be formulated as a problem of how much money (or debt) to carry over to date $t + 1$.

In order for the consumption to belong to $\mathbb{R}^3_{++}$ we must require that the amount of money chosen belongs to the interval

$$I_r = \{m \in \mathbb{R} \mid m \in \,]\min(-re^s, -e^d),\ re^y[\ \}$$

The Consumer Problem given the prices $(p^y, p^s, p^d) = (r, r, 1)$ is

$$\text{Max}_{(c,m)\in\mathbb{R}^3_{++}\times I_r} \quad \pi\bar{u}(c^y, c^s) + (1-\pi)\bar{u}(c^y, c^d)$$

$$\text{subject to} \quad \begin{cases} (1) & rc^y \leq re^y - m \\ (2) & rc^s \leq re^s + m \\ (3) & c^d \leq e^d + m \end{cases} \qquad \textbf{11.2.A}$$

Let $(c, m) \in \mathbb{R}^3_{++} \times I_r$ satisfy the restrictions (1)-(3). Then c satisfies

$$\begin{cases} (1') & c^y + c^s \leq e^y + e^s \\ (2') & rc^y + c^d \leq re^y + c^d \end{cases} \qquad \textbf{11.2.B}$$

On the other hand if $c \in \mathbb{R}^3_{++}$ and satisfies (1') and (2') then $(c, m) = (c, r(e^y - c^y))$ satisfies (1)-(3).

States with probability 0 or 1

The Consumer Problem is well-defined also for $\pi = 0$ or $\pi = 1$. In spite of the fact that it would seem reasonable to disregard, for example, restriction (3) in 11.2.A if $\pi = 1$ we will insist that all three restrictions be satisfied also in case $\pi = 0$ or $\pi = 1$. Further on we will see that for a suitable set of prices this restriction will not be binding.

For $\pi \in]0,1[$ the solution to the Consumer Problem with the restrictions from 11.2.B is unique and it follows that also the solution to 11.2.A is unique. Let $(\bar{c}, \bar{m})$ be a solution to the Consumer Problem 11.2.A as $\pi = 1$. Then clearly $(\bar{c}^y, \bar{c}^s, \hat{c}^d, \bar{m})$ with $0 < \hat{c}^d \leq \bar{c}^d$ is also a solution. Since we are not interested in finding all the solutions to the Consumer Problem 11.2.A we restrict attention to the unique solution satisfying the restrictions with equality.

Solutions to the consumer problem

Restricting attention to the solutions satisfying the restriction in 11.2.A with equality we can formulate the Consumer Problem as a problem with a single decision variable m. We have to pay particular attention to the cases where $\pi = 0$ or $\pi = 1$ since then the Maintained Assumptions are not satisfied for the function u_π.

Proposition 11.2.C *Let $(r, e, \pi) \in \mathbb{R}_{++} \times \mathbb{R}^3_{++} \times [0,1]$. The problem*

$$\text{Max}_{\, m \in I_r} \; \pi \bar{u}\left(e^y - \frac{m}{r}, e^s + \frac{m}{r}\right) + (1-\pi)\bar{u}\left(e^y - \frac{m}{r}, e^d + \frac{m}{1}\right) \qquad \textbf{11.2.C}$$

has a unique solution. Let

$$\mu : \mathbb{R}_{++} \times \mathbb{R}^3_{++} \times [0,1] \longrightarrow \mathbb{R}$$

be the function mapping (r, e, π) to the solution. Then $\mu \in \mathbb{C}^1(\mathbb{R}_{++} \times \mathbb{R}^3_{++} \times [0,1], \mathbb{R})$.

Proof: Let $\hat{F} : I_r \longrightarrow \mathbb{R}$ be the function with values given by

$$m \longrightarrow \pi \bar{u}\left(e^y - \frac{m}{r}, e^s + \frac{m}{r}\right) + (1-\pi)\bar{u}\left(e^y - \frac{m}{r}, e^d + \frac{m}{1}\right)$$

The set

$$A = \left\{m \in \mathbb{R} \mid \hat{F}(m) \geq \hat{F}(0)\right\} \cap I_r$$

is a compact subset of I_r. Hence $\hat{F}$ attains a maximum at a point $\bar{m} \in A$. It is easy to see that $\hat{F}(\bar{m}) \geq \hat{F}(m)$ for $m \in I_r$ and since $\hat{F}$ is a strictly concave function $\bar{m}$ is unique.

To save notation let

$$\bar{c}^{y,s} = \left(e^y - \frac{\bar{m}}{r}, e^s + \frac{\bar{m}}{r}\right) \quad \text{and} \quad \bar{c}^{y,d} = \left(e^y - \frac{\bar{m}}{r}, e^d + \frac{\bar{m}}{1}\right)$$

The point $\bar{m}$ satisfies the first order condition

$$\begin{aligned} D_m\hat{F}(\bar{m}, r, e, \pi) &= \pi D_m\bar{u}(\bar{c}^{y,s}) + (1-\pi)D_m\bar{u}(\bar{c}^{y,d}) \\ &= \pi D_1\bar{u}(\bar{c}^{y,s}) \cdot -\frac{1}{r} + \pi D_2\bar{u}(\bar{c}^{y,s})\frac{1}{r} \\ &\quad + (1-\pi)D_1\bar{u}(\bar{c}^{y,d}) \cdot -\frac{1}{r} + (1-\pi)D_2\bar{u}(\bar{c}^{y,d}) = 0 \end{aligned} \qquad \textbf{11.2.D}$$

A calculation shows that the second derivative of $\hat{F}$ with respect to m evaluated at the point $(\bar{m}, r, e, \pi)$ is

$$\begin{aligned} D^2_{mm}\hat{F}(\bar{m}, r, e, \pi) &= \pi \left[-\frac{1}{r}, \frac{1}{r}\right] D^2\bar{u}(\bar{c}^{y,s}) \begin{bmatrix} -\frac{1}{r} \\ \frac{1}{r} \end{bmatrix} \\ &\quad + (1-\pi)\left[-\frac{1}{r}, 1\right] D^2\bar{u}(\bar{c}^{y,d}) \begin{bmatrix} -\frac{1}{r} \\ 1 \end{bmatrix} \end{aligned}$$

By Assumption (C6), $D^2\bar{u}(\bar{c}^{y,s})$ and $D^2\bar{u}(\bar{c}^{y,d})$ are negative definite matrices. Hence $D^2_{mm}\hat{F}(\bar{m}, r, e, \pi) < 0$. (Note that this holds also for $\pi = 0$ and $\pi = 1$.)

The first order condition, relation 11.2.D, makes sense not only for $\pi \in [0, 1]$ but for $\pi \in \mathbb{R}$. Let F denote the function, with values as $\hat{F}$ but with the domain of $\hat{F}$ extended to $\mathbb{R}$ for the variable π. Let $(\hat{m}, \hat{r}, \hat{e}, \hat{\pi})$ be an arbitrary point of $\mathbb{R} \times \mathbb{R}_{++} \times \mathbb{R}^3_{++} \times [0, 1]$ where $\hat{F}(\hat{m}, \hat{r}, \hat{e}, \hat{\pi}) = 0$. Since $D^2_{mm}\hat{F}(\hat{m}, \hat{r}, \hat{e}, \hat{\pi})$ we can apply the Implicit Function Theorem, Theorem C in Appendix A, to conclude that there is a neighborhood, $\mathcal{N}_{(\hat{r},\hat{e},\hat{\pi})}$, of $(\hat{r}, \hat{e}, \hat{\pi})$ and a unique function $\mu : \mathcal{N}_{(\hat{r},\hat{e},\hat{\pi})} \longrightarrow \mathbb{R}$ with $\mu(\hat{r}, \hat{e}, \hat{\pi}) = \hat{m}$ such that

$$F(\mu(r, e, \pi), r, e, \pi) = 0 \quad \text{for } (r, e, \pi) \in \mathcal{N}_{(\hat{r},\hat{e},\hat{\pi})}$$

By the same theorem the function $\mu \in \mathbb{C}^1(\mathbb{R}_{++} \times \mathbb{R}^3_{++} \times [0, 1], \mathbb{R})$.

The demand function under uncertainty

Since the problem given in 11.2.C has a unique solution we can define the demand function under uncertainty which gives demand as a function of relative price for prices such that the price at the current date equals the price at the next date if the state "same" occurs.

Definition 11.2.D *The* **demand function under uncertainty** *as a function of relative price is the function* $h = (h^y, h^s, h^d) : \mathbb{R}_{++} \times \mathbb{R}^2_{++} \times [0,1] \longrightarrow \mathbb{R}^3_{++}$ *where*

$$\begin{cases} h^y(r,e,\pi) &= e^y - \dfrac{\mu(r,e,\pi)}{r} \\ h^s(r,e,\pi) &= e^s + \dfrac{\mu(r,e,\pi)}{r} \\ h^d(r,e,\pi) &= e^d + \mu(r,e,\pi) \end{cases}$$

Note that we can not reconstruct the demand function under uncertainty as a function of nominal prices from the demand function under uncertainty as a function of relative price.

For emphasis we will in the sequel refer to the demand function, f, derived in Chapter 9 as the demand function under certainty. From the definition and Proposition 11.2.C we have the following corollary.

Corollary 11.2.E *The demand function under uncertainty, h, satisfies*

(a) $h = (h^y, h^s, h^d) \in \mathbb{C}^1(\mathbb{R}_{++} \times \mathbb{R}^3_{++} \times [0,1], \mathbb{R}^3_{++})$

(b) *for* $(r,e,\pi) \in \mathbb{R}_{++} \times \mathbb{R}^3_{++} \times [0,1]$

$$\begin{cases} h^y(r,e,\pi) + h^s(r,e,\pi) &= e^y + e^s \\ rh^y(r,e,\pi) + h^d(r,e,\pi) &= re^y + e^d \end{cases}$$

The last two relations follow from the relation between the Consumer Problem 11.2.A and the Consumer Problem 11.2.B.

Relating Demand under Certainty and under Uncertainty

We have derived the demand under uncertainty for a consumer who knows the price at the current date and plans against two possible prices at the next date. The consumer plans against a price system $(p^y, p^s, p^d) = (r, r, 1)$.

From now on we will restrict attention to the case where the endowment as old, e^o, is certain so that $e^d = e^s = e^o$ where the "o" refers to the endowment as old.

We noted above that the Consumer Problem 11.2.A was somewhat contrived in case $\pi = 0$ or $\pi = 1$ since we required that the consumer should abide also by the restriction concerning the state of Nature, which had 0 probability. The proposition below shows that for a subset of relative prices a consumer who is not forced to abide by both restrictions will in fact do so anyway. This allows us to relate demand under uncertainty to demand under certainty for those prices.

The $\rho(e)$ in the proposition below refers to the real steady state price of the stationary OG economy $\mathcal{E}^s = (\bar{u}, C, e)$ where $\bar{u}$ is the state utility function and $e = (e^y, e^o)$. The demand function under certainty, f, is the demand function for this economy. If $\pi = 0$ and price today is r then the consumer expects the next date price to be 1 for sure. The result below corresponds to the case where $\pi = 0$, implying "sure" transition to the state "different". There is an analogus result for $\pi = 1$ but it will not be needed.

Proposition 11.2.F *Let $\rho(e) < 1$ and let $r \geq \rho(e)$. The problem*

$$\mathrm{Max}_{(c^y,c^d)\in\mathbb{R}^2_{++}} \ \bar{u}(c^y, c^d) \quad s.\ to \quad rc^y + c^d \leq re^y + e^o \qquad \mathbf{11.2.E}$$

has a unique solution $(\bar{c}^y, \bar{c}^d)$ and

$$(\bar{c}^y, \bar{c}^d) = (h^y(r, e, 0), h^d(r, e, 0)) = (f^y(r, e), f^o(r, e))$$

Proof: The proposition is a consequence of Lemma 9.2.B, on the properties of demand under certainty, which implies that for $r \geq \rho(e)$ the consumer chooses a nonnegative amount of money

The problem 11.2.E has the unique solution $(\bar{c}^y, \bar{c}^d) = (f^y(r, e), f^o(r, e))$ and since $\rho(e) < 1$ and $r \geq \rho(e)$ the corresponding $\bar{m} = r(e^y - f^y(r, e))$ is a nonnegative number. It follows that $(\bar{c}^y, \bar{c}^d)$ is a solution to 11.2.A as restriction (2) is disregarded. Put $\bar{c}^s = e^o + \bar{m}$, then $(\bar{c}^y, \bar{c}^s, \bar{c}^d)$ satisfies (1)-(3) of 11.2.A and is the unique solution satisfying the inequalities of that problem with equality. However this solution is, by the definition of demand under uncertainty, $h(r, e, 0)$ and hence

$$(h^y(r,e,0), h^d(r,e,0)) = (\bar{c}^y, \bar{c}^d) = (f^y(r,e), f^o(r,e))$$

□

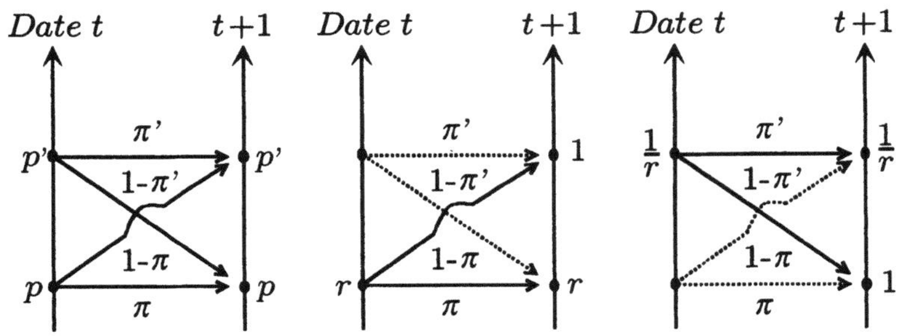

Figure 11.2.B: *The Consumer Problem at a 2-state sunspot equilibrium at nominal prices (p, p'). The second panel shows the relative prices in the state where the current nominal price is p and the third panel shows the relative prices as the current nominal price is p'*

Existence of a Sunspot Equilibrium

In this section we define an OG economy with uncertainty. We show that the problem of finding a 2-state sunspot equilibrium can be reduced to finding a solution to a single equation with one relative price as unknown.

OG economy with uncertainty

The stationary OG economy defined by a single consumer, $\mathcal{E}^s = (\bar{u}, C, (e^y,))$ with $C = \mathbb{R}^2_{++}$, induces a consumer acting under uncertainty with $C = \mathbb{R}^3_{++}$, $e = (e^y, e^s, e^d)$ with $e^s = e^d = e^o$ and utility function $u_\pi : \mathbb{R}^3_{++} \longrightarrow \mathbb{R}$ for evaluating lotteries having $\bar{u}$ as state utility function.

To determine consumer behaviour we also need assumptions about how the consumer forms expectations. We assume that

- there exist two points such that the support of the probability distribution is either one of the points or both points

- if prices at date t and t' are identical then the probability distributions on prices at date $t+1$ and $t'+1$ are identical.

Prices and relative prices at a sunspot equilibrium

In a 2-state sunspot equilibrium the nominal price at each date is either p or p'. If the current price is p the consumer born at date t expects the price at the next date to be either p or p'. When the current price is p, the consumer assigns probability π to the event that the price will be the same, that is p, at date $t+1$ and probability $1-\pi$ to the event that the price will be p'. Cf. the left panel in Figure 11.2.B. With $r = p/p'$ the consumer acts against the relative prices $(p^y, p^s, p^d) = (r, r, 1)$ and her demand is given by $h(r, e, \pi)$. Cf. the mid panel in Figure 11.2.B.

On the other hand, if the current price is p' then the consumer assigns the probabilities π' and $1-\pi'$ to the events that the price is going to be p' or p respectively at the next date. We can then normalize prices so that the current price is $1/r$ implying expected prices of $1/r$ or 1 at the next date. With $1/r = p'/p$ the consumer acts against the relative prices $(p^y, p^s, p^d) = (1/r, 1/r, 1)$ and her demand is given by $h(1/r, e, \pi)$. Cf. the right panel in Figure 11.2.B.

Consider the market balance conditions at date t. At date t the price may be p or p' but the demand of the old consumer at date t depends on whether the price was p or p' when she was young. Hence there are four events at date t and thus four market balance conditions corresponding to the values taken by the price at the current date and the price at the previous date.

Definition of a 2-state sunspot equilibrium

We can now define a 2-state sunspot equilibrium. We give the definition in nominal prices but with demand under uncertainty expressed as a function of relative price.

Definition 11.2.G *A* **2-state sunspot equilibrium** *is a pair of prices, a pair of consumption bundles and a pair of probabilities* $((p, p'), (\bar{c}, \bar{c}'), (\bar{\pi}, \bar{\pi}'))$ *such that, for* $t \in \mathbb{Z}$,

(a) *consumer t maximizes her utility*

$$h\left(\frac{p}{p'}, e, \bar{\pi}\right) = \bar{c} \quad (p_t = p)$$

$$h\left(\frac{p'}{p}, e, \bar{\pi}'\right) = \bar{c}' \quad (p_t = p')$$

(b) *the market balances*

$$\begin{aligned} \bar{c}^y + \bar{c}^s &= e^y + e^o & (p_{t-1}, p_t) &= (p, p) \\ \bar{c}'^y + \bar{c}^d &= e^y + e^o & (p_{t-1}, p_t) &= (p, p') \\ \bar{c}^y + \bar{c}'^d &= e^y + e^o & (p_{t-1}, p_t) &= (p', p) \\ \bar{c}'^y + \bar{c}'^s &= e^y + e^o & (p_{t-1}, p_t) &= (p', p') \end{aligned}$$

A 2-state sunspot equilibrium is **non-trivial** *if $p' \neq p$ and $\bar{\pi}, \bar{\pi}' \in]0,1[$.*

A 2-state sunspot equilibrium is a particular instance of an equilibrium under uncertainty in an OG model; here assumed stationary. It would take us too far afield to relate it to a general notion of equilibrium under uncertainty for OG models. Suffices it to remark that the equilibrium concept introduced in Chapter 7 is too restrictive to encompass sunspot equilibria.

The equilibrium equations

Let $((p, p'), (\bar{c}, \bar{c}'), (\bar{\pi}, \bar{\pi}'))$ be a 2-state sunspot equilibrium. If the current price is p the young consumer acts against the prices $(p^y, p^s, p^d) = (p, p, p')$ or the relative prices $(r, r, 1)$. If the current price is p' then the relevant prices are $(p^y, p^s, p^d) = (p', p', p)$ and $(1/r, 1/r, 1)$

Using the demand function under uncertainty and the relative price r the 2-state sunspot equilibrium is given by $(r, (\bar{c}, \bar{c}'), (\bar{\pi}, \bar{\pi}'))$, where the price $1/r$ is suppressed. The conditions defining a 2-state sunspot equilibrium are that each consumer maximizes her utility, that is, for consumer $t \in \mathbb{Z}$,

$$\begin{aligned} h(r, e, \bar{\pi}) &= \bar{c} \quad (p_t = p) \\ h\left(\frac{1}{r}, e, \bar{\pi}'\right) &= \bar{c}' \quad (p_t = p') \end{aligned} \qquad \textbf{11.2.F}$$

and the date t market balances for $t \in \mathbb{Z}$,

$$\begin{aligned}
h^y(r,e,\bar{\pi}) &+ h^s(r,e,\bar{\pi}) &= e^y + e^o \qquad (p_{t-1}, p_t) &= (p,p) \\
h^y\left(\frac{1}{r}, e, \bar{\pi}'\right) &+ h^d(r,e,\bar{\pi}) &= e^y + e^o \qquad (p_{t-1}, p_t) &= (p,p') \\
h^y(r,e,\bar{\pi}') &+ h^d(\frac{1}{r}, e, \bar{\pi}) &= e^y + e^o \qquad (p_{t-1}, p_t) &= (p',p) \\
h^y\left(\frac{1}{r}, e, \bar{\pi}'\right) &+ h^s\left(\frac{1}{r}, e, \bar{\pi}'\right) &= e^y + e^o \qquad (p_{t-1}, p_t) &= (p',p')
\end{aligned} \tag{11.2.G}$$

Hence, in order to establish the existence of a 2-state sunspot equilibrium, one first finds a relative price and a pair of probabilities such that equations 11.2.G are satisfied, and next, a pair of consumption bundles by substituting the relative price in equations 11.2.F. The lemma below shows that although there are four equation and only one variable it is enough to find a solution to one of the market balance conditions.

Lemma 11.2.H *If for* $(p_{t-1}, p_t) = (p', p)$ *the market balance condition*

$$h^y(r,e,\bar{\pi}) + h^d\left(\frac{1}{r}, e, \bar{\pi}'\right) = e^y + e^o \tag{11.2.H}$$

is satisfied for $r = \bar{r}$ *then the equations* 11.2.G *are all satisfied for* $\bar{r}$.

Proof: From 11.2.B follows that demand under uncertainty as a function of relative price satisfies, for $r > 0$,

$$\begin{aligned}
h^y(r,e,\bar{\pi}) &+ h^s(r,e,\bar{\pi}) = e^y + e^o \\
rh^y(r,e,\bar{\pi}) &+ h^d(r,e,\bar{\pi}) = re^y + e^o
\end{aligned}$$

Substituting the values $r = \bar{r}$ and $r = 1/\bar{r}$ in the first relation we obtain the market balance conditions for $(p_{t-1}, p_t) = (p,p)$ and for $(p_{t-1}, p_t) = (p',p')$.

Using the second relation for $r = \bar{r}$ and $r = 1/\bar{r}$ we get

$$\begin{aligned}
h^y(\bar{r},e,\bar{\pi}) &- e^y = \frac{1}{\bar{r}}(e^o - h^d(\bar{r},e,\bar{\pi})) \\
h^d\left(\frac{1}{\bar{r}}, e, \bar{\pi}'\right) &- e^o = \frac{1}{\bar{r}}\left(e^y - h^y\left(\frac{1}{\bar{r}}, e, \bar{\pi}'\right)\right)
\end{aligned}$$

Substitution in the assumed market balance condition for $(p_{t-1}, p_t) = (p', p)$, given in 11.2.H gives the market balance condition for $(p_{t-1}, p_t) = (p', p)$.

A sufficient condition for the existence of a sunspot equilibrium

A 2-sunspot equilibrium may fail to be non-trivial for several reasons. Firstly, the price may be the same in both states. In this case, by the strict concavity of the state utility function, the solution to the Consumer Problem is $(\bar{c}^y, \bar{c}^s, \bar{c}^d)$ with $\bar{c}^s = \bar{c}^d$ and $(\bar{c}^y, \bar{c}^s)$ is the nominal steady state consumption for the economy (with no uncertainty) $\mathcal{E}^s = (\bar{u}, C, e)$. If $\pi = \pi' = 1$ then the equilibrium may disintegrate into (two) nominal steady states for $\mathcal{E}^s = (\bar{u}, C, e)$ differing only in the price level.

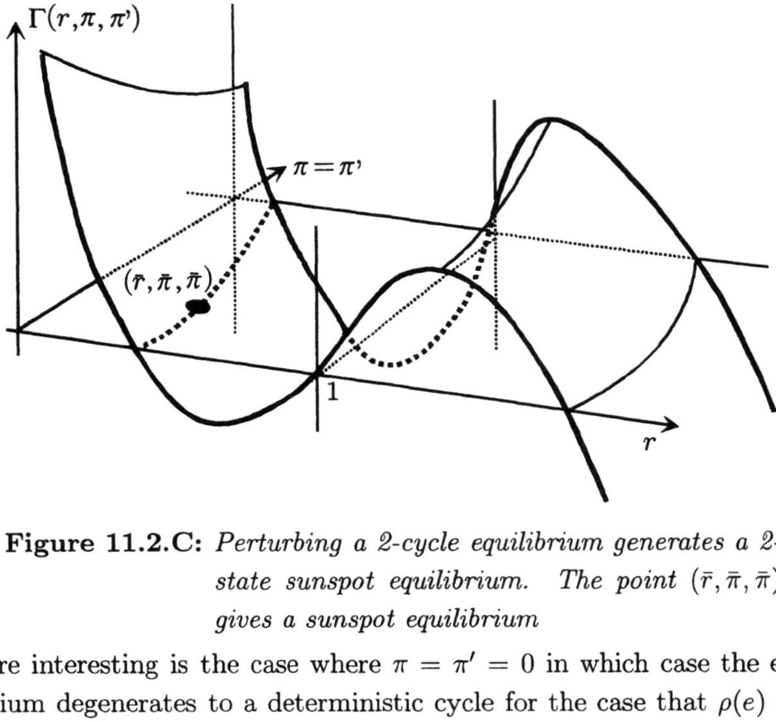

Figure 11.2.C: *Perturbing a 2-cycle equilibrium generates a 2-state sunspot equilibrium. The point* $(\bar{r}, \bar{\pi}, \bar{\pi})$ *gives a sunspot equilibrium*

More interesting is the case where $\pi = \pi' = 0$ in which case the equilibrium degenerates to a deterministic cycle for the case that $\rho(e) < 1$, as studied in Section 11.1. The idea of the proof for the existence of a 2-state sunspot equilibrium in the theorem below is to begin with the degenerate case of a deterministic cycle. Here the consumers believe that only the "different" price is possible at the next date. By perturbing the probabilities somewhat so that consumers believe that the price may, at least with some small probability, be the same as at the current date a sunspot equilibrium is obtained. Cf. Figure 11.2.C where the case of $\pi = \pi'$ is illustrated.

Theorem 11.2.I *Assume that* $\rho(e) < 1$ *and*

$$D_r f^y(1, e) \;>\; D_r f^o(1, e)$$

Then the OG economy has a non-trivial 2-state sunspot equilibrium.

Proof: Consider the function $\Gamma \in \mathbb{C}^1([\rho(e), 1] \times [0, 1]^2, \mathbb{R})$ defined by

$$\Gamma(r, \pi, \pi') \;=\; h^y(r, e, \pi) + h^d\left(\frac{1}{r}, e, \pi'\right) - (e^y + e^o)$$

According to Lemma 11.2.H, on the market balance conditions, $r \neq 1$ is a relative price and $\bar{\pi}, \bar{\pi}' \in\,]0, 1[$ is a pair of probabilities associated with a non-trivial 2-state sunspot equilibrium if and only if $\Gamma(r, \bar{\pi}, \bar{\pi}') = 0$

For $\pi = \pi' = 0$ we have by Proposition 11.2.F, relating demand under certainty to demand under uncertainty,

$$\begin{aligned}\Gamma(r, 0, 0) \;&=\; h^y(r, e, 0) + h^d\left(\frac{1}{r}, e, 0\right) - (e^y + e^o)\\ &=\; f^y(r, e) + f^o\left(\frac{1}{r}, e\right) - (e^y + e^o)\end{aligned}$$

We have $D_r\Gamma(1, 0, 0) = D_r f^y(1, e) - D_r f^o(1, e)$, which by assumption is a positive number. It follows that there is $\hat{r}$ such that $\rho(e) < \hat{r} < 1$ and $\Gamma(\hat{r}, 0, 0) < 0$. Since $\rho(e) < 1$ we have $1/\rho(e) > \rho(e)$ and from Lemma 9.2.B, on the basic properties of demand, we get

$$\begin{aligned} f^y(\rho(e), e) \;&-\; e^y \;=\; 0\\ f^o\left(\frac{1}{\rho(e)}, e\right) \;&-\; e^o \;>\; 0\end{aligned}$$

which implies $\Gamma(\rho(e), 0, 0) > 0$.

Since Γ is a continuous function there exists $\varepsilon > 0$ such that

$$\Gamma(\rho(e), \pi, \pi') \;>\; 0 \;>\; \Gamma(\hat{r}, \pi, \pi') \quad \text{for} \quad (\pi, \pi') \in [0, \varepsilon]^2$$

and, again by the continuity of Γ, there exists a triple $(r, \bar{\pi}, \bar{\pi}') \in\,]\rho(e), 1[\, \times\,]0, \varepsilon[^2$ such that $\Gamma(r, \bar{\pi}, \bar{\pi}') = 0$. The triple $(r, \bar{\pi}, \bar{\pi}')$ and the associated consumptions is a non-trivial sunspot equilibrium.

Sunspot equilibria and incomplete markets

Shell [1987] gives an informal introduction to sunspot equilibria, for economies with an asset structure. For such economies with two dates and financial markets, a necessary condition for the existence of a sunspot equilibrium is that financial markets are incomplete.

In an OG economy, with no uncertainty, where spot markets with money is the market structure, financial markets are complete at an equilibrium since there is one state at every date and one asset. However, financial markets are incomplete at a sunspot equilibrium since there are at least two states at every date and a single asset. Markets are also incomplete because of the demographic structure. For consumer t there are four relevant events, namely

$$(p_t, p_{t+1}) \in \{(p,p), (p,p'), (p',p), (p',p')\},$$

but she is only able to distinguish two events, namely

$$(p_t, p_{t+1}) \in \{(p,p), (p,p')\} \quad (p \text{ when young})$$
$$(p_t, p_{t+1}) \in \{(p',p), (p',p')\} \quad (p' \text{ when young})$$

Hence, consumer t is born too late to consider all relevant events and can not make her consumption contingent on the price prevailing at the date when she is born.

11.3 Endowments, Cycles and Sunspot Equilibria

The purpose of this section is to relate the sufficient condition on demand under certainty for the existence of a cycle or a sunspot equilibrium

$$D_r f^y(1,e) > D_r f^o(1,e)$$

to a condition on the initial endowments for the family of economies $\mathcal{E}^s = (\bar{u}, C, e)$, with $(\bar{u}, C)$ fixed and e varying in $\mathbb{R}^2_{++}$.

For $a > 0$ consider economies with the total endowment equal to a, that is, the set

$$\mathbb{E}(a) = \{e \in \mathbb{R}^2 \mid e^y + e^o = a\} \cap \mathbb{R}^2_{++}$$

which is the intersection of the affine subspace where $e^y + e^o = a$ and $\mathbb{R}^2_{++}$.

Each $e \in \mathbb{E}(a)$ induces an economy such that at the relative price $r = 1$ income is $re^y + e^o = e^y + e^o = a$. Hence the function $f(1, \cdot) : \mathbb{R}^2_{++} \longrightarrow \mathbb{R}^2_{++}$ is constant equal to, say, $\bar{c}_a = (\bar{c}^y_a, \bar{c}^o_a)$ on $\mathbb{E}(a)$.

The family of sets $(\mathbb{E}(a))_{a \in \mathbb{R}_{++}}$ is a partition of the set of initial endowments, $\mathbb{R}^2_{++}$. Recall that according to Lemma 9.2.C, on the differentiability properties of demand, the derivative of demand with respect to the relative price is

$$\begin{aligned} D_r f^y(1,e) &= -\frac{(e^y - \bar{c}^y_a)\left[D^2_{yo}u(\bar{c}_a) - D^2_{oo}u(\bar{c}_a)\right] - D_o u(\bar{c}_a)}{\Delta(\bar{c}_a, (1,-1))} \\ D_r f^o(1,e) &= \frac{(e^y - \bar{c}^y_a)\left[D^2_{yy}u(\bar{c}_a) - D^2_{oy}u(\bar{c}_a)\right] - D_y u(\bar{c}_a)}{\Delta(\bar{c}_a, (1,-1))} \end{aligned} \qquad \textbf{11.3.A}$$

where $\Delta(\bar{c}_a, (1,-1))$ is a determinant which by Lemma 9.2.C is negative. The relation 11.3.A shows that the mapping

$$(e^y, e^o) \longrightarrow e^y \longrightarrow D_r f(1,e)$$

is an affine function on $\mathbb{E}(a)$.

Since $\Delta(\bar{c}_a, (1,-1)) < 0$ and $D^2_{oy}u(\bar{c}_a) = D^2_{yo}u(\bar{c}_a)$ relation 11.3.A also implies that for $e \in \mathbb{E}(a)$ the condition

$$D_r f^y(1,e) > D_r f^o(1,e)$$

is equivalent to

$$D_y u(\bar{c}_a) + D_o u(\bar{c}_a) < (e^y - \bar{c}^y_a)(D^2_{yy}u(\bar{c}_a) - D^2_{oo}u(\bar{c}_a)) \qquad \textbf{11.3.B}$$

If $e \in \mathbb{E}(a)$ then $e^y \in \,]0, a[$ and obviously relation 11.3.B is most likely to be satisfied for e^y in a neighborhood of either zero or a depending on the sign of $D^2_{yy}u(\bar{c}_a) - D^2_{oo}u(\bar{c}_a)$.

Proposition 11.3.A *Assume that condition* (a) *or* (b) *holds where*

(a) $D_y u(\bar{c}_a) + D_o u(\bar{c}_a) < -\bar{c}^y_a(D^2_{yy}u(\bar{c}_a) - D^2_{oo}u(\bar{c}_a))$

(b) $D_y u(\bar{c}_a) + D_o u(\bar{c}_a) < (a - \bar{c}^y_a)(D^2_{yy}u(\bar{c}_a) - D^2_{oo}u(\bar{c}_a))$

Then there exists an initial endowment $\hat{e} \in E(a)$ such that the OG economy $\mathcal{E}^s = (\bar{u}, C, \hat{e})$ has a non-trivial 2-cycle and the induced OG economy with uncertainty has a continuum of non-trivial 2-state sunspot equilibria.

The partition $(\mathbb{E}(a))_{a \in \mathbb{R}_{++}}$ of the set of feasible endowments, $\mathbb{R}^2_{++}$, is convenient due to the fact that for $r = 1$ demand is constant and its derivative is an affine function as $e \in \mathbb{E}(a)$. Partitions of endowments with these properties are applied to simple OG economies in Balasko and Ghiglino [1996] and to OG economies with many goods at each date and many consumers in each generation in Ghiglino and Tvede [1995].

An Example

We have given sufficient conditions for 2-cycles and sunspot equilibria to arise. We now give an example of an economy which has a 3-cycle equilibrium. By Theorem B in Appendix D this implies that the economy has an equilibrium price system with a k-cycle for each $k \in \mathbb{N}$.

Example 11.3.A Let $\mathcal{E}^s = (ac^y - (1/2)a(c^y)^2 + c^o, (0, e^o))$ with $e^o > 1$. Then u violates the Maintained Assumptions (C3) and (C4) since $u^{-1}(c)$ is not a closed subset in $\mathbb{R}^2$ for each $c \in \mathbb{R}$ and $Du(c)$ fails to be positive for some $c \in C$ but in the sequel we only consider consumption bundles which belong to $\mathbb{R}^2_{++}$.and where $Du(c) \in \mathbb{R}^2_{++}$. Solving the Consumer Problem we get for $0 < r < a$.

$$\begin{aligned} c^y &= \frac{a-r}{a} \\ c^o &= -r\frac{a-r}{a} + e^0 \end{aligned}$$

The equilibrium equation is

$$z^o(r,e) + z^y(r',e) = 0 \quad \text{or} \quad -r\frac{a-r}{a} + \frac{a-r'}{a} = 0$$

which gives the equilibrium locus

$$G_e = \{(r,r') \in \mathbb{R}^2_{++} \mid r' - [a - r(a-r)] = 0\}$$

For $a = 3.9$ the function $r \longrightarrow r' = [a - r(a-r)]$ has a minimum for $a/2 = 1.95$ with the value of the function equal to 0.0975. The interval

[0.0975, 3.9] is an invariant set. The points

$$r_0 = 3,1941, \quad r_1 = 1,6454, \quad r_2 = 0,1902$$

form a 3-cycle so that $r_0 \longrightarrow r_1 \longrightarrow r_2 \longrightarrow r_0 \longrightarrow \ldots$.
By Sarkowski's theorem, Theorem A in Appendix D, we can for each $k \in \mathbb{N}$ find a point $r \in [0.0975,\ 3.9]$ such that r is a k-cycle.

Summary

We have shown that when the price at the current date does not determine the price at the next date uniquely then a cyclic equilibrium might emerge together with the steady states for the economy. A sufficient condition for this to occur was given in terms of the demand function.

The cyclic equilibrium could be a deterministic 2-cycle equilibrium or a stochastic 2-state sunspot equilibrium. The 2-state sunspot equilibrium was obtained by a perturbation of a deterministic 2-cycle. The sufficient condition on the demand function for cyclic equilibria was seen to be related to the strong income effects arising from the variation in the value of the initial endowment induced by a variation in prices.

Exercises

Exercise 11.A Let G_e be the equilibrium locus for a stationary economy $\mathcal{E}^s = (C, u, e)$. Let $G_e^2 = G_e \circ G_e$ be defined by

$$G_e^2 = \{(r, r'') \in \mathbb{R}^2_{++} \mid \text{there exists } r' \text{ with } (r, r') \in G_e \text{ and } (r', r'') \in G_e\}$$

(a) Explain why r defines a non-trivial 2-cycle if and only if $(r, r) \in G_e^2$ and $(r, r) \notin G_e$.

(b) Construct G_e^2 for the equilibrium locus in the right hand panel of Figure 11.1.B.

(c) For $k \in \mathbb{N}$ how should G_e^k reasonably be defined? Assume that $(r, r) \in G_e^k$. Under what conditions does r generate a k-cycle but not a k'-cycle for $k' < k$?

Exercise 11.B Show that the condition $D_r f^y(1,e) > D_r f^o(1,e)$ is equivalent to $2D_r f^y(1,e) > -(f^y(1,e) - e^y)$. Sketch a graph of $z^y(r,e)$ and $-z^o(r,e) = rz^y(r,e)$ such that the condition is satisfied. What is the maximal invariant set?

Exercise 11.C Let $\bar{u} : \mathbb{R}^2_{++} \longrightarrow \mathbb{R}$ and assume that for $\pi \in]0,1[$ the function $\hat{u}_\pi : \mathbb{R}^4_{++} \longrightarrow \mathbb{R}$ with values

$$\hat{u}_\pi((x_0, x_1), (z_0, z_1)) = \pi\bar{u}(x_0, x_1) + (1-\pi)\bar{u}(z_0, z_1)$$

satisfies the Maintained Assumptions.

(a) Give an interpretation of the function $\hat{u}_\pi$.

(b) Show that $\bar{u}$ is a strictly concave function and that for $x \in \mathbb{R}^2_{++}$ the matrix $D^2\bar{u}(x)$ is a negative definite matrix. (*Hint*: Consider the function $\hat{u}_\pi$ as $(x_0, x_1) = (z_0, z_1)$, $\pi = 1/2$ and use (C5).)

(c) Show that $\inf \left\{\bar{u}(x) \in \mathbb{R}^2_{++} \mid x \in \mathbb{R}^2_{++}\right\} = -\infty$.

(d) Show that $\bar{u} : \mathbb{R}^2_{++} \longrightarrow \mathbb{R}$ satisfies the Maintained Assumptions.

Exercise 11.D Let $\mathcal{E}^s = (C, u, e)$ be a stationary economy where $C = \mathbb{R}^2_{++}$, $e = (1,0)$ and $u : \mathbb{R}^2_{++} \longrightarrow \mathbb{R}$ where $u(c^y, c^o) = c^y - \delta\frac{1}{c^o}$ and $\delta > 0$.

(a) Do the consumers satisfy the Maintained Assumptions?

(b) Find the demand function for $r > \delta$.

(c) Show that the equilibrium locus is $r' = 1/r$.

(d) Conclude that the economy has a continuum of 2-cycles.

Exercise 11.E Let $\mathcal{E}^s = (\mathbb{R}^2_{++}, e, c^y - (c^o)^{-\alpha}/\alpha)$ where $e^y + e^o > 1$ and $\alpha > 0$.

(a) Does the utility function satisfy the Maintained Assumptions?

(b) Find the demand function and the savings function.

(c) Find a first-order difference equation satisfied by equilibrium price systems.

(d) What are the steady states?

(e) Show that if $\omega^y + \omega^o > 1$ and

$$\frac{\alpha - 1}{\alpha + 1} > e^o$$

then the sufficient condition for the existence of a 2-cycle from Theorem 11.1.C is satisfied.

Exercise 11.F Let $((p, p'), (\bar{c}, \bar{c}'), (\bar{\pi}, \bar{\pi}'))$ be a 2-state sunspot equilibrium and let $\rho(e) < 1$ for the induced OG economy $\mathcal{E}^s = (C, \bar{u},, e)$, with no uncertainty. Show that if $\pi = \pi' = 1$ then the 2-state sunspot equilibrium disintegrates into (two) nominal equilibria for $\mathcal{E}$.

Exercise 11.G Find a sufficient condition on the saving function for the existence of a 2-cycle.

Exercise 11.H Discuss optimality in relation to sunspot equilibria.

The condition in the following exercise is given in Balasko and Ghiglino [1996].

Exercise 11.I Let $\hat{f}$ be the Walrasian demand function taking (r, e) to $(\hat{f}^y(r, 1, re^y + e^o), \hat{f}^o(r, 1, re^y + e^o))$ and let f be the demand function as a function of relative price and endowment, taking (r, e) to $(f^y(r, e)$, $f^o(r, e))$, for a consumer satisfying the Maintained Assumptions.

(a) Show that

$$\begin{aligned} D_r f^y(r, e) &= D_1 \hat{f}^y(r, 1, re^y + e^o) + D_3 \hat{f}^y(r, 1, re^y + e^o) \cdot \hat{f}^y \\ &\quad + D_3 \hat{f}^y(r, 1, re^y + e^o)(e^y - \hat{f}^y) \end{aligned}$$

and

$$\begin{aligned} &D_r f^o(r, e) \\ &= -\frac{1}{r^2}\Big[D_2 \hat{f}^o(1, \frac{1}{r}, e^y + \frac{1}{r}e^o) + D_3 \hat{f}^o(1, \frac{1}{r}, re^y + \frac{1}{r}e^o)\hat{f}^o \\ &\quad + D_3 \hat{f}^o(1, \frac{1}{r}, e^y + \frac{1}{r}e^o)(e^o - \hat{f}^o)\Big] \end{aligned}$$

with $(\hat{f}^y, \hat{f}^o)$ evaluated at $(r, 1, re^y + e^o)$. Cf. Exercise 9.A

(b) Show that $D_3\hat{f}^y(1,1,w) + D_3\hat{f}^o(1,1,w) = 1$.

(c) Assume that $D_3\hat{f}^y(1,1,w)$ and $D_3\hat{f}^o(1,1,w)$ are both positive. Show that if $D_3\hat{f}^y(1,1,w) \neq D_3\hat{f}^o(1,1,w)$ then the condition $D_r f^y(1,e) > D_r f^o(1,e)$ will be satisfied by some e with $e^y + e^o = w$. Here e may fail to belong to $\mathbb{R}^2_{++}$.

(d) Assume that $\hat{f}$ and f are derived from the utility function $u(c) = \ln c^y + \ln c^0$. Show that then $D_3\hat{f}^y(1,1,w) = D_3\hat{f}^o(1,1,w)$ and that $D_r f^y(1,e) > D_r f^o(1,e)$ can not hold for any choice of e where $e^y + e^o > 0$.

Exercise 11.J Let J be a closed interval and let $g : J \longrightarrow J$ be continuous. If there is a point r such that

$$g^3(r) \leq r < g(r) < g^2(r)$$

then there is a 3-cycle for the difference equation $r_{t+1} = g(r_t)$. Cf. Theorem B in Appendix D.

(a) Consider the mapping $r \longrightarrow r(a-r)$ with $a = 3.9$ and show that the point $r = 0,6$ satisfies the condition above. (*Hint:* The values should be $g^3(0,6)$= $0,374077$, $g(0,6) = 1,98$ and $g^2(0,6) = 3.806$)

(b) Let r_0, r_1 and r_2 be a 3-cycle for the mapping $r \longrightarrow r(a-r)$ with $a > 0$. Show that $a - r_0, a - r_1$ and $a-$ r_2 is a 3-cycle for the mapping $r \longrightarrow a - r(a-r)$. Cf. Example 11.3.A.

APPENDIX

Appendix A Some Results from Calculus

Proofs of the results in this section can be found in Apostol [1974]. Mas-Colell et al. [1995] contains a discussion and often proofs of these and many other results as does Sydsaeter and Hammond [1995].

In economics one often has to prove the existence of a "best" action from a choice set and the following result is useful.

Theorem A Maximum Theorem. *Let A be a non-empty, compact subset of $\mathbb{R}^n$ and $h : A \longrightarrow \mathbb{R}$ a continuous function. Then there is $\bar{x} \in A$ such that $h(\bar{x}) \geq h(x)$ for $x \in A$.*

In applications there is often given a continuous function $g : B \longrightarrow \mathbb{R}$ where the set B is not compact. The theorem is applied to a subset $A \subset B$ which is non-empty and compact and h is taken to be the restriction of g to A. The restriction of a continuous function is again a continuous function.

Let S be an open set of $\mathbb{R}^n$. A function $h : S \longrightarrow \mathbb{R}$ is *differentiable at $x \in S$* if h has continuous first order partial derivatives at x. The function h is *differentiable* on S if h is differentiable at every point in S. A function $g = (g_1, \ldots, g_m) : S \longrightarrow \mathbb{R}^m$ is differentiable on S if each g_i, $i = 1, \ldots, m$, is differentiable on S.

The use of the following theorem by Lagrange is pervasive in economics.

Theorem B Lagrange's Theorem. *Let S be an open subset of $\mathbb{R}^n$ and let $h : S \longrightarrow \mathbb{R}$ be a function which is differentiable on S.*

Let $m < n$ and let $g = (g_1, \ldots, g_m) : S \longrightarrow \mathbb{R}^m$ be a differentiable function on S. Assume that $x \in S$ and

(a) $g(\bar{x}) = 0$

(b) *the rank of the $m \times n$ matrix $[Dg(\bar{x})]$ is m*

(c) $h(\bar{x}) \geq h(x)$ *for x such that $g(x) = 0$*

Then there exists a vector $\lambda = (\lambda_1, \lambda_2, \ldots, \lambda_m) \in \mathbb{R}^m$ such that

$$Dh(\bar{x}) = \lambda_1 Dg_1(\bar{x}) + \lambda_2 Dg_2(\bar{x}) + \ldots + \lambda_m Dg_m(\bar{x})$$

Thus if h attains a maximum at $\bar{x} \in S$ subject to the restrictions given by the function g then the gradient of h, evaluated at $\bar{x}$, is a linear combination of the gradients of $g_1, \ldots, g_m$, evaluated at the same point The rank condition on the matrix $[Dg(\bar{x})]$ expresses a kind of "local independence at $\bar{x}$".

Let A be an $n \times n$ matrix and B an $n \times k$ matrix. If the matrix A is invertible then there is for each $q \in \mathbb{R}^k$ a unique solution to the equation $Ax - Bq = 0$. For A an invertible matrix the solution is $x = A^{-1}Bq$, which gives x as a differentiable function of q. The matrix $[A, -B]$ defines a linear function from $\mathbb{R}^n \times \mathbb{R}^k$ to $\mathbb{R}^n$ taking (x, q) to $Ax - Bq$. Loosely speaking, the Implicit Function Theorem states that the result for linear functions carries over "locally" to non-linear functions.

Theorem C The Implicit Function Theorem. *Let S be an open subset of $\mathbb{R}^n \times \mathbb{R}^k$ and let $(\bar{x}, \bar{q}) \in S$. Let $F : \mathbb{R}^n \times \mathbb{R}^k \longrightarrow \mathbb{R}^n$ be differentiable on S and $F(\bar{x}, \bar{q}) = 0$. Assume that the determinant of the $n \times n$ matrix*

$$\begin{bmatrix} D_1F_1(\bar{x}, \bar{q}) & D_2F_1(\bar{x}, \bar{q}) & \cdots & D_nF_1(\bar{x}, \bar{q}) \\ D_1F_2(\bar{x}, \bar{q}) & D_2F_2(\bar{x}, \bar{q}) & \cdots & D_nF_2(\bar{x}, \bar{q}) \\ \vdots & \vdots & \ddots & . \\ D_1F_n(\bar{x}, \bar{q}) & D_2F_n(\bar{x}, \bar{q}) & \cdots & D_nF_n(\bar{x}, \bar{q}) \end{bmatrix}$$

is not 0.

Then there exists an open set $T \subset \mathbb{R}^k$ which owns $\bar{q}$ and a unique function $h : T \longrightarrow \mathbb{R}^n$ such that

(a) $h(\bar{q}) = \bar{x}$

(b) h *is differentiable on* T

(c) $F(h(q), q) = 0$ *for* $q \in T$

In applications it is often known that the equation $F(x, q) = 0$ has a unique solution in x for a given q. For $q \in T$ this is also the content of (c) together with the assertion that the function h is unique. However, it is often (b) which tells us that the solution varies differentiably with q which is of main importance since it is often hard to prove differentiability without referring to the Implicit Function Theorem.

Appendix B Quadratic Forms and Bordered Matrices

The following theorem relates a quadratic form which is negative definite on some linear subspace to the determinant of a matrix formed by the coefficients of the quadratic form bordered with the vectors defining the linear subspace. A proof of this and related results can be found in Debreu [1952].

Theorem A *Let* A *be an* $n \times n$ *matrix and* B *an* $n \times m$ *matrix with* rank $B = m$, *where* $1 \leq m \leq n$. *Assume that*

$$B^t x = 0 \text{ and } x \neq 0 \text{ implies } x^t A x < 0$$

so that the quadratic form induced by A *is negative definite on* $\langle B^t \rangle^{\perp}$ *which is the subspace orthogonal to the row vectors of* B^t.

Let E *be the* $(n+m) \times (n+m)$ *matrix formed by bordering the matrix* A *with the matrix* B.

$$E = \left[\begin{array}{c|c} A & B \\ \hline B^t & 0 \end{array}\right] = \left[\begin{array}{cccc|ccc} a_{11} & a_{12} & \cdots & a_{1n} & b_{11} & \cdots & b_{1m} \\ a_{21} & a_{22} & \cdots & a_{2n} & b_{21} & & b_{2m} \\ \vdots & & \ddots & & \vdots & & \vdots \\ a_{n1} & a_{n2} & & a_{nn} & b_{n1} & & b_{nm} \\ \hline b_{11} & b_{21} & \cdots & b_{n1} & 0 & \cdots & 0 \\ \vdots & & & \vdots & \vdots & & \vdots \\ b_{1m} & b_{2m} & \cdots & b_{nm} & 0 & \cdots & 0 \end{array}\right]$$

Then $\det E \neq 0$.

Proof: We will show that the equations

$$E\begin{bmatrix} x \\ y \end{bmatrix} = \left[\begin{array}{c|c} A & B \\ \hline B^t & 0 \end{array}\right]\begin{bmatrix} x \\ y \end{bmatrix} = \begin{bmatrix} Ax & + & By \\ B^t x & & \end{bmatrix} = \begin{bmatrix} 0 \\ 0 \end{bmatrix}$$

have the unique solution $(x, y) = (0, 0)$. This implies that the column vectors of E are linearly independent and hence $\det E \neq 0$.

Let $(\bar{x}, \bar{y})$ be any solution. Premultiplying the first n equations by $\bar{x}^t$ we get

$$\begin{bmatrix} \bar{x}^t A\bar{x} & + & \bar{x}^t B\bar{y} \\ B^t \bar{x} & & \end{bmatrix} = \begin{bmatrix} 0 \\ 0 \end{bmatrix}$$

The last equation shows that the $m \times 1$ vector $B^t x$ is the 0 vector in $\mathbb{R}^m$ and thus $\bar{x} \in \langle B^t \rangle^\perp$. It follows that $\bar{x}^t B = 0$, where 0 is the $1 \times m$ vector in $\mathbb{R}^m$. Hence $\bar{x}^t B\bar{y} = 0$ and the equations imply $\bar{x}^t A\bar{x} = 0$. Since A is negative definite on $\langle B^t \rangle^\perp$ and $\bar{x} \in \langle B^t \rangle^\perp$ we get $\bar{x} = 0$. We then get $B\bar{y} = 0$; again by the initially given equations . Since $\operatorname{rank} B = m$ the column vectors of B are linearly independent and hence $\bar{y} = 0$. Thus $(\bar{x}, \bar{y}) = 0$ and it follows that $\det E \neq 0$.

□

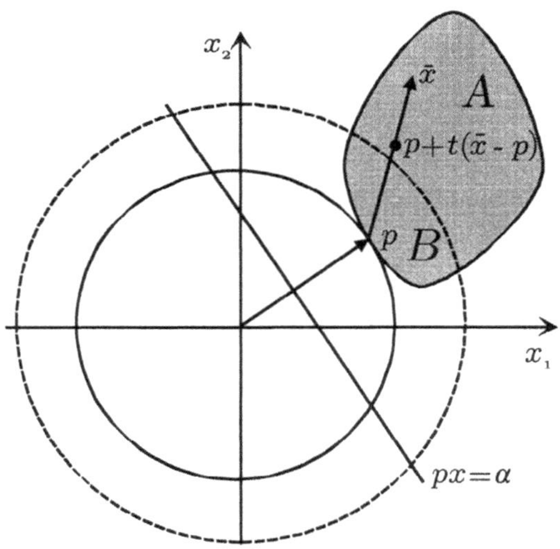

Figure A: *The point 0 belongs to the lower halfspace and the set A to the upper halfspace of the hyperplane given by the vector p and the number α*

Appendix C Separation Theorems and Farkas' Lemma

The purpose of this section is to give a few basic results which are important in convex analysis. In economics separation theorems can often be used to find prices in order to characterize solutions to optimization problems.

Separating a point from a convex set

The following theorem is the starting point for a number of results on convex sets with applications to economic problems.

Theorem A *Let A be a closed, convex set in $\mathbb{R}^n$ such that $0 \notin A$. Then there is a hyperplane, with normal $p \in \mathbb{R}^n$, $p \neq 0$, and $\alpha \in \mathbb{R}_{++}$ such that*

$$p \cdot x \;>\; \alpha \;>\; p \cdot 0 \;=\; 0 \qquad \textit{for } x \in A$$

(Thus the set A is contained in *the upper halfspace* determined by the vector p and the number α. The point 0 belongs to *the lower halfspace* as illustrated in Figure A.)

Proof: If A is the empty set the conclusion is trivially true. Thus we assume that A is non-empty. Let $a \in A$ and put

$$B \;=\; \{x \in A \mid \|x\| \;\leq\; \|a\|\}$$

The point a belongs to the set B so B is a non-empty set. B is also a closed, convex and bounded set; hence a compact set. Consider the problem

$$\text{Min}\, \|x\| \quad \text{s. to } x \in B$$

Since $x \longrightarrow \|x\|$ is a continuous function from $\mathbb{R}^n$ to $\mathbb{R}$, its restriction to the compact set B is also continuous. Since a continuous function defined on a non-empty, compact set attains its minimum there is a vector $x^* \in B$ such that

$$\|x^*\| \;\leq\; \|x\| \quad \text{for } x \in B$$

In particular, $\|x^*\| \leq \|a\|$ and $x^* \neq 0$ since $0 \notin B$. Choose $x \in A \backslash B$ then $\|x\| > \|a\|$ and hence $\|x^*\| \leq \|x\|$. Thus

$$\|x^*\| \ \leq \ \|x\| \quad \text{for } x \in A$$

Put $p = x^*$. Let $\bar{x} \in A$. Then, for $t \in [0,1]$, the point

$$x \ = \ p + t(\bar{x} - p) \ = \ (1-t)p + t\bar{x} \quad \text{for } t \in [0,1]$$

belongs to A since A is a convex set. We have, for $t \in [0,1]$,

$$x \cdot x \ = \ [p + t(\bar{x} - p)] \cdot [p + t(\bar{x} - p)] \ \geq \ p \cdot p$$

if and only if

$$p \cdot p \ + \ 2p \cdot t(\bar{x} - p) \ + \ t(\bar{x} - p) \cdot t(\bar{x} - p) \ \geq \ p \cdot p$$

if and only if

$$2t(p \cdot \bar{x} - p \cdot p) \ + \ t^2(\bar{x} - p) \cdot (\bar{x} - p) \ \geq \ 0$$

Hence, for $t \in]0,1]$,

$$2(p \cdot \bar{x} - p \cdot p) \ + \ t(\bar{x} - p) \cdot (\bar{x} - p) \ \geq \ 0$$

and this can be true only if

$$p \cdot \bar{x} \ \geq \ p \cdot p$$

Put $\alpha = \frac{1}{2} p \cdot p$. Since $\bar{x}$ was an arbitrary point of A we have

$$p \cdot x \ > \ \alpha \ > \ p \cdot 0 \ = \ 0 \quad \text{for } x \in A$$

as asserted.

□

It is now easy to extend the result to an arbitrary point not belonging to a closed, convex set.

Corollary B *Let A be a closed, convex set in $\mathbb{R}^n$ and d a point not belonging to A. Then there is a hyperplane with normal $p \in \mathbb{R}^n$, $p \neq 0$, and $\beta \in \mathbb{R}$ such that*

$$p \cdot x \ > \ \beta \ > \ p \cdot d \quad \textit{for } x \in A$$

Proof: The condition $d \notin A$ is equivalent to the condition $0 \notin A - \{d\}$. The set $A - \{d\}$ is a translation of A and it is easy to prove that $A - \{d\}$ is a convex, closed set.

By Theorem A there is a $p \neq 0$ and $\alpha > 0$ such that

$$p \cdot z \;>\; \alpha \;>\; p \cdot 0 \quad \text{for } z \in A - \{d\}$$

But this implies, by the definition of $A - \{d\}$,

$$p \cdot (x - d) \;>\; \alpha \;>\; p \cdot 0 \quad \text{for } x \in A$$

or, equivalently

$$p \cdot x \;>\; \alpha + p \cdot d \;>\; p \cdot d \quad \text{for } x \in A$$

Put $\beta = \alpha + p \cdot d$ to get the desired result.

□

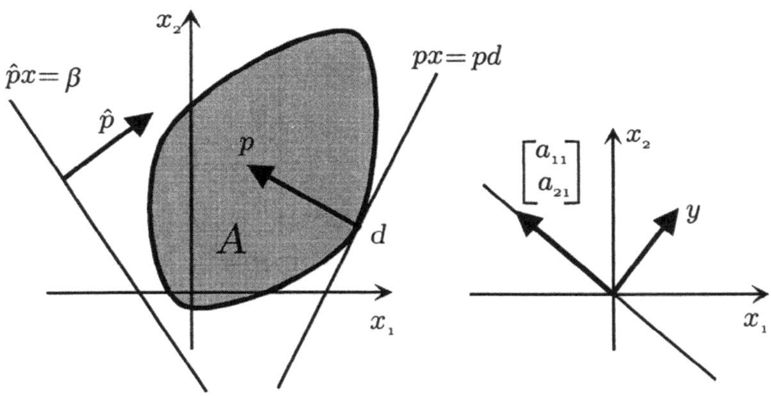

Figure A: *On the left a bounding and a supporting hyperplane for the set A. On the right Farkas' lemma is illustrated for the simplest possible case*

Supporting hyperplane theorem

Let A be a compact set in $\mathbb{R}^n$ and let $p \in \mathbb{R}^n, p \neq 0$ and $\alpha \in \mathbb{R}$ be such that

$$p \cdot x \;\geq\; \alpha \quad \text{for } x \in A$$

Then p and α define a *bounding hyperplane* for the set A. Cf. Figure A. Geometrically the set A is "on or above" the hyperplane. If $p \cdot x > \alpha$

for $x \in A$, then we there is a hyperplane "closer to" A. If, on the other hand, $p \cdot x = \alpha$ for some $x \in A$, then this is not the case.

If p and α are such that $p \cdot x \geq \alpha$ for $x \in A$ and $p \cdot \bar{x} = \alpha$ for some $\bar{x} \in A$ then the hyperplane, defined by p and α is a supporting hyperplane. If the set A is closed, but not necessarily compact, a hyperplane defined by $(p, \alpha) \in \mathbb{R}^n \times \mathbb{R}$, $p \neq 0$, is, by definition, a *supporting hyperplane,* if $\inf_{x \in A} px = \alpha$.

Theorem C *Let A be a closed, convex set in $\mathbb{R}^n$ and let d be a point on the boundary of A. Then there is a $p \in \mathbb{R}^n, p \neq 0$ such that*

$$p \cdot x \geq p \cdot d \quad \textit{for } x \in A$$

(Here we do not need a constant since the point d determines the constant $p \cdot d$.)

Proof: Since d is a boundary point of A there is a sequence $(d^n)_{n \in \mathbb{N}}$ converging to d with $d^n \notin A$, for $n \in \mathbb{N}$. For each $n \in \mathbb{N}$, we apply Corollary B above to get $p^n \neq 0$ such that

$$p^n \cdot x \geq p^n \cdot d^n \quad \text{for } x \in A$$

Without loss of generality we may assume that $\|p^n\| = 1$ for $n \in \mathbb{N}$. Then the sequence $(p^n)_{n \in \mathbb{N}}$ is a sequence on the unit sphere $\{z \in \mathbb{R}^n \mid \|z\| = 1\}$ which is a compact set. This implies that $(p^n)_{n \in \mathbb{N}}$ has a subsequence which converges to, say, p where p also belongs to the unit sphere. Without loss of generality we may assume that the sequence itself converges to p.

Choose a fixed but arbitrary vector $x \in A$. The function

$$(\hat{p}, \hat{d}) \longrightarrow \hat{p} \cdot x - \hat{p} \cdot \hat{d} \quad \text{where } \hat{p} \in \mathbb{R}^n, \|\hat{p}\| = 1 \text{ and } \hat{d} \in \mathbb{R}^n$$

is continuous in $(\hat{p}, \hat{d})$. Since $(p^n, d^n)_{n \in \mathbb{N}}$ converges to (p, d) we get by continuity

(1) $p^n \cdot x - p^n \cdot d^n$ converges to $p \cdot x - p \cdot d$

(2) $p^n \cdot x - p^n \cdot d^n \geq 0$ for $n \in \mathbb{N}$ implies $p \cdot x - p \cdot d \geq 0$

Since x was an arbitrary point of A we have from (2)

$$p \cdot x \geq p \cdot d \quad \text{for } x \in A$$

Separating hyperplane theorem

Above we stated a result on (the strict) separation of a closed, convex set and a point. The theorem below shows that we can substitute a compact set for the point.

Theorem D *Let A and B be non-empty, closed, convex sets in $\mathbb{R}^n$ such that B is compact and $A \cap B = \varnothing$. Then there is a vector $p \in \mathbb{R}^n, p \neq 0$, and a number β such that*

$$p \cdot x > \beta > p \cdot y \quad \textit{for } x \in A \textit{ and } y \in B$$

Proof: It is easy to check, using that B is a compact set, that $A - B$ is a closed set . The condition $A \cap B = \varnothing$ is equivalent to $0 \notin A - B$. By Theorem A there is a vector $p \neq 0$ and number α such that

$$p \cdot z > \alpha > 0 \quad \text{for } z \in A - B$$

which implies

$$p \cdot (x - y) > \alpha > 0 \quad \text{for } x \in A \text{ and } y \in B$$

But this is equivalent to

$$p \cdot x > \alpha + p \cdot y > p \cdot y \quad \text{for } x \in A \text{ and } y \in B$$

Since B is a compact set there is a vector $\bar{y} \in B$ such that $p \cdot \bar{y} \geq p \cdot y$ for $y \in B$. We get

$$p \cdot x > \alpha + p \cdot \bar{y} \geq \alpha + p \cdot y > p \cdot y \quad \text{for } x \in A \text{ and } y \in B$$

and the proof is finished by choosing $\beta = \alpha + p \cdot \bar{y}$.

□

If we weaken the concept of separation we do not have to require one of the sets to be a compact set.

Theorem E *Let A and B be non-empty, disjoint, convex sets in $\mathbb{R}^n$. Then there is a vector $p \in \mathbb{R}^n, p \neq 0$ and a number β such that*

$$p \cdot x \geq \beta \geq p \cdot y \quad \textit{for } x \in A \textit{ and } y \in B$$

Proof: The set $A - B$ is a convex set which does not contain 0. Hence 0 is an exterior point or a boundary point of $A - B$. But then 0 is also a boundary point or an exterior point of the closure of $A - B$ and by Corollary B or Theorem C there is a $p \in \mathbb{R}^n, p \neq 0$ such that

$$p \cdot z \geq 0 \quad \text{for } z \in A - B$$

and hence

$$p \cdot x \geq p \cdot y \quad \text{for } x \in A \text{ and } y \in B$$

Let $\beta = \sup_{y \in B} p \cdot y$. Since A is a non-empty set β the function $y \longrightarrow p \cdot y$ with $y \in B$ is bounded above by $p \cdot x$ for $x \in A$ which implies that β is a real number. Clearly $p \cdot x \geq \beta$ for $x \in A$ and $\beta \geq p \cdot y$ for $y \in B$.

Farkas' lemma

We use Theorem D to prove the following result, referred to as Farkas' lemma or Stiemke's lemma.

Lemma F Farkas' Lemma. *Let A be a $k \times \ell$ matrix. One and only one of the following two alternatives is true:*

(I) *there exists a vector $x \in \mathbb{R}^\ell$ such that $Ax > 0$*

(II) *there exists a vector $y \in \mathbb{R}^k_{++}$ such that $yA = 0$*

(A geometric interpretation is the following: Let $\langle A \rangle$ be the subspace spanned by the column vectors of A. Either $\langle A \rangle$ intersects $\mathbb{R}^k_+$ at a point $x \neq 0$, in which case (I) is true, or $\langle A \rangle \cap \mathbb{R}^k_+ = \{0\}$. In the latter case there exists a vector $y \in \mathbb{R}^k_{++}$ which belongs to the orthogonal subspace of $\langle A \rangle$. Note that $yA = 0$ if and only if y is orthogonal to each column vector of A. This is illustrated in Figure A for the simplest possible case where $k = 2$ and $l = 1$.)

Proof: Assume that **II** is true with $y \in \mathbb{R}^k_{++}$, so that $yA = 0$. Then $yAx = 0$ for every $x \in \mathbb{R}^\ell$ and since y is a positive vector $Ax > 0$ can not be true for any $x \in \mathbb{R}^\ell$. Hence **I** is false. The converse then gives: If **I** is true then **II** is false.

Assume that **I** is false. We show that **II** is then true.

Since **I** is false we have $\langle A\rangle \cap \mathbb{R}^k_+ = \{0\}$. Thus the unit simplex,

$$\Delta = \left\{x \in \mathbb{R}^k_+ \mid \sum_{i=1}^k x_i = 1\right\}$$

in $\mathbb{R}^k$ does not intersect the closed set $\langle A\rangle$. The set $\Delta - \langle A\rangle$ does not contain 0 and using that Δ is a compact set, it is easy to show that $\Delta - \langle A\rangle$ is a closed set. By Theorem D there is $y \in \mathbb{R}^k$ and $\alpha \in \mathbb{R}$ such that

$$y \cdot (z - x) > \alpha > 0 \quad \text{for } z \in \Delta \text{ and } x \in \langle A\rangle$$

and since $0 \in \langle A\rangle$, this implies $y \cdot z > \alpha > 0$ for $z \in \Delta$.

Let $h \in \{1, 2, ..., k\}$ and let e_h be the vector in $\mathbb{R}^k$ which is 0 in each component except the h'th, where it is 1. The vector $e_h \in \Delta$ and we get

$$y \cdot e_h = y_h > 0$$

which shows that y_h is positive. Hence $y \in \mathbb{R}^k_{++}$.

Choose $\bar{z} \in \Delta$. Then $y \cdot \bar{z} > 0$ and from above we have

$$y \cdot \bar{z} > y \cdot x + \alpha > y \cdot x \quad \text{for } x \in \langle A\rangle$$

Since $\langle A\rangle$ is a linear subspace, the inequality "$y \cdot \bar{z} > y \cdot x$ for $x \in \langle A\rangle$" can be true only if $y \cdot x = 0$ for $x \in \langle A\rangle$. Hence **II** is true.

Appendix D Dynamical Systems

The results in this appendix can be found in Devaney [1986].

Consider a one-dimensional difference equation

$$\tau_{n+1} = \phi(\tau_n) \quad \text{for } n \in \mathbb{Z}$$

Throughout this appendix $I \subset \mathbb{R}$ is a closed interval and $\phi : I \longrightarrow I$ is a continuous function. Since ϕ is a continuous function it can always be extended to a continuous function from $\mathbb{R}$ to $\mathbb{R}$. A sequence $(\tau_n)_{n\in\mathbb{Z}}$ is a *solution* to the difference equation if $\tau_{n+1} = \phi(\tau_n)$ for $n \in \mathbb{Z}$. A point, $\tilde{\tau} \in I$, which is mapped to itself by ϕ so that $\tilde{\tau} = \phi(\tilde{\tau})$ is a *fixed point* for ϕ.

For $m \in \mathbb{N}$ let $\phi^m(\cdot)$ denote the m-fold composition of ϕ so that

$$\tau \longrightarrow \phi(\tau) \longrightarrow \phi \circ \phi(\tau) \longrightarrow \ldots \longrightarrow \underbrace{\phi \circ \phi \circ \ldots \circ \phi}_{m \text{ times}}(\tau) = \phi^m(\tau)$$

A point, $\tilde{\tau}$, with $\tilde{\tau} = \phi^m(\tilde{\tau})$ is a *periodic point of period* m for ϕ. Given a periodic point, $\tilde{\tau}$, the smallest m such that $\tilde{\tau}$ is a periodic point of period m is the *prime period* of $\tilde{\tau}$. We will refer to a periodic point, $\tilde{\tau}$, of prime period m as an *m-cycle* or a cycle of *order* m. Thus $\tilde{\tau}$ is an m-cycle if $\tilde{\tau} = \phi^m(\tilde{\tau})$ and $\tilde{\tau} \neq \phi^m(\tilde{\tau})$ for $n \in \{1, \ldots, m-1\}$.

Linear, one-dimensional difference equations are easy to solve and one can describe all the solutions and their properties. Cf. Sydsaeter and Hammond [1995] or Azariadis [1993]. This simplicity is lost for non-linear one-dimensional difference equations. Indeed, non-linear difference equations may have cycles of all periods as a theorem by Sarkovskii asserts. Consider the following ordering or the natural numbers: the odd numbers $3, 5, 7, \ldots$ precede the numbers which are of the form 2^n times an odd number in the way indicated in the table below and last in the ordering are the even numbers and 1.

$$\begin{array}{ccccccccccc}
3 & \rhd & 5 & \rhd & 7 & \ldots \rhd & 2l-1 & \rhd & 2l+1 & \rhd \ldots \\
2\cdot 3 & \rhd & 2\cdot 5 & \rhd & 2\cdot 7 & \ldots \rhd & 2\cdot(2l-1) & \rhd & 2\cdot(2l+1) & \rhd \ldots \\
\vdots & & & & & & & & & \\
2^n\cdot 3 & \rhd & 2^n\cdot 5 & \rhd & 2^n\cdot 7 & \ldots \rhd & 2^n\cdot(2l-1) & \rhd & 2^n\cdot(2l+1) & \rhd \ldots \\
2^{n+1}\cdot 3 & \rhd & 2^{n+1}\cdot 5 & \rhd & 2^{n+1}\cdot 7 & \ldots \rhd & 2^{n+1}\cdot(2l-1) & \rhd & 2^{n+1}\cdot(2l+1) & \rhd \ldots \\
\vdots & & & & & & & & & \\
\ldots & \rhd & 16 & \rhd & 8 & \rhd & 4 & \rhd & 2 & \rhd\ 1
\end{array}$$

Theorem A (Sarkovskii) *If $\phi : I \longrightarrow I$ has an m-cycle and $m \rhd n$ then ϕ also has an n-cycle.*

This surprising theorem was discovered and established in the early sixties by Sarkovskii. A consequence of this result is that a one-dimensional difference equation which has a 3-cycle also has an m-cycle for each $m \in \mathbb{N}$.

If there exists a 3-cycle the solutions to the difference equation may be very complicated indeed. This is highlighted by the following theorem which gives a sufficient condition for the existence of a 3-cycle and some consequences of the occurrence of a 3-cycle for existence of complicated solutions. Let P denote the set of points with the property that, for some $m \in \mathbb{N}$, the point $\tilde{\tau}$ is an m-cycle. A set $S \subset I$ such that $\phi(S) \subset S$ is an *invariant* set.

Theorem B (Li and Yorke [1975]) *Assume that there is $\bar{\tau} \in I$ such that*

$$\phi^3(\bar{\tau}) \leq \bar{\tau} < \phi(\bar{\tau}) < \phi^2(\bar{\tau})$$

then ϕ has a 3-cycle.

If $\phi : I \longrightarrow I$ has a 3-cycle. Then there exists an uncountable, invariant set S such that

(a) *for $\tau \in S$ and $\tau' \in P$ with $\tau \neq \tau'$*

$$\limsup_{n\to\infty} \mid \phi^n(\tau) - \phi^n(\tau') \mid > 0$$

(b) *for $\tau, \tau' \in S$ with $\tau \neq \tau'$*

$$\limsup_{n\to\infty} \mid \phi^n(\tau) - \phi^n(\tau') \mid > 0$$

(c) *for $\tau, \tau' \in S$ with $\tau \neq \tau'$*

$$\liminf_{n\to\infty} \mid \phi^n(\tau) - \phi^n(\tau') \mid = 0$$

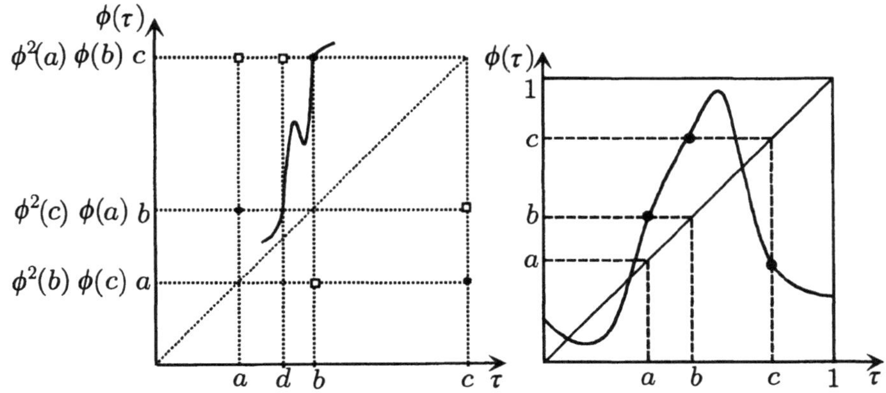

Figure A: *In the right hand panel a 3-cycle and in the left hand panel an illustration to* Theorem C. *Only part of the graph of ϕ is illustrated in the left hand panel*

By (a) there are points in S which induce solutions which do not converge to any cycle. Parts (b) and (c) tell us that also if we restrict attention to points in S which are arbitrarily close, the induced solutions will at some dates be close to each other and at other dates wide apart. This implies that it is very hard to predict the evolution for points in S since a small measurement error in the initial condition leads to a large prediction error at some future date. Dynamical systems, that are hard to predict in the sense of Theorem B, are referred to as *chaotic* dynamical systems.

The proofs of Theorem B and Theorem A are difficult, but it possible to get some intuition from the following corollary which concerns the construction of an n-cycle given that ϕ has a 3-cycle.

Theorem C *Assume that ϕ has a cycle of order 3 and let $n \in \mathbb{N}$. Then ϕ has a cycle of order n.*

Proof: Let the 3-cycle be a, b and c so that $\phi(a) = b, \phi(b) = c$ and $\phi(c) = a$. We will consider the case where $a < b < c$ and leave the only other case where $\phi(a) = c$ to the reader.

We will apply the following two observations.

Observation 1: *If I and J are closed intervals and $I \subset J$ and $\phi(I) \supset J$ then there is a point $\tau \in I$ such that $\phi(\tau) = \tau$.*

This is a simple consequence of the Intermediate Value Theorem. We also apply the following observation which we do not prove.

Observation 2: *Let A and B be closed intervals such that $\phi(A) \supset B$. Then there is a closed interval $I_A \subset A$ such that $\phi(I_A) = B$.*

For $n = 1$ there is, using Observation 1, a point $\hat{\tau} \in [b, c]$ such that $\phi(\hat{\tau}) = \hat{\tau}$, that is, a fixed point for ϕ. Cf. Figure A. If $n = 2$ then, as suggested by Figure A, there is a closed interval $[d, b]$ such that $\phi(\tau) \geq b > \tau$ for $\tau \in [d, b]$ and $\phi(d) = c$. Then ϕ does not have a fixed point in $[d, b]$. But since $\phi^2(d) = c$ and $\phi^2(b) = a$, Observation 1 implies that there is a point $\hat{\tau} \in [d, b]$ which is a fixed point of ϕ^2. Thus ϕ has cycles of order 1 and of order 2.

By assumption ϕ has a cycle of order 3 so we let $n = k+2$ with $k \geq 2$. We will prove that there is a point $\hat{\tau}$ which is a fixed point for ϕ^{k+2} and which is not a fixed point for ϕ^m for $m \in \{1, 2, \ldots, k+1\}$.

Let $J_0 = [b, c]$. Since $\phi([b, c]) \supset [b, c]$ there is, by Observation 2, a closed interval J_1 such that

$$J_1 \subset J_0 \text{ and } \phi(J_1) = J_0$$

Now Observation 2 can be applied to J_1. Thus there is a closed interval $J_2 \subset J_1$ such that

$$J_2 \subset J_1 \text{ and } \phi(J_2) = J_1 \text{ which implies } \phi^2(J_2) = J_0$$

Repeated application of Observation 2 yields a sequence of closed intervals such that

$$\begin{cases} J_3 \subset J_2 \text{ and } \phi(J_3) = J_2 \text{ which implies } \phi^3(J_3) = J_0 \\ \vdots \\ J_k \subset J_{k-1} \text{ and } \phi(J_k) = J_{k-1} \text{ which implies } \phi^k(J_k) = J_0 \end{cases}$$

Each of the intervals $J_0, J_1, \ldots, J_k$ is included in the interval $[b, c]$. Since $\phi([a, b]) \supset J_0$ there is, again by Observation 2, a closed interval J^*_{k+1} (the star is there to indicate that this is the only interval considered which is not included in $[b, c]$) such that

$$J^*_{k+1} \subset [a, b] \text{ and } \phi(J^*_{k+1}) = J_k \text{ which implies } \phi^{k+1}(J^*_{k+1}) = J_0$$

and since $\phi(J_0) \supset [a, b]$ there is a closed interval J_{k+2} such that

$$J_{k+2} \subset [b, c] \text{ and } \phi(J_{k+2}) = J^*_{k+1} \text{ which implies } \phi^{k+2}(J_{k+2}) = J_0$$

Hence ϕ^{k+2} maps the closed interval $J_{k+2} \subset J_0$ onto J_0 and by Observation 1, ϕ^{k+2} has a fixed point, $\hat{\tau}$, in J_{k+2} so that $\phi^{k+2}(\hat{\tau}) = \hat{\tau}$.

We have

$$\hat{\tau} \in J_{k+2}, \quad \phi(\hat{\tau}) \in J^*_{k+1}, \quad \phi^2(\hat{\tau}) \in J_k, \quad \ldots \quad , \phi^{k+2}(\hat{\tau}) \in J_0$$

Assume, in order to obtain a contradiction, that $\phi^m(\hat{\tau}) = \hat{\tau}$ for some $m < k+2$.

If $m = 1$ then $\hat{\tau} = \phi(\hat{\tau})$ belongs to $J_{k+2} \cap J^*_{k+1}$ and thus $\hat{\tau} = b$. However, then $\phi(\hat{\tau}) = c$. From this contradiction follows that $m > 1$. If this is the case then $\phi^{m-1}(\phi(\hat{\tau})) = \phi(\hat{\tau})$ and $\phi(\hat{\tau})$ belongs to $J_{k+2} \cap J^*_{k+1}$ which gives $\phi(\hat{\tau}) = b$. Then $\phi^2(\hat{\tau}) = c$ which shows that $m > 2$. But then $\phi^3(\hat{\tau}) = a$ contradicting that $\phi^3(\hat{\tau})$ belongs to $J_{k-1} \subset I_1$.

□

The structure of the real line is essential to Sarkovkii's theorem. The theorem does not generalize to several dimensions or to dynamical systems on a circle.

Bibliography

APOSTOL, T. M. (1974): *Mathematical Analysis.* Addison-Wesley.

ARROW, K. J. (1953): "Le Role Des Valeurs Boursieres Pour la Répartition la Meilleure Des Risques," in *Econométrie, Colloques Internationaux Du Centre National de la Recherhce Scientifique No. 40 (Paris 1952)*, pp. 41–47. CNRS, translation as Arrow, K.J. [1964].

——— (1964): "The Role of Securities in the the Optimal Allocation of Riskbearing," *Review of Economic Studies*, 31, 91–96, Translation of Arrow [1953].

ARROW, K. J., AND G. DEBREU (1954): "Existence of an Equilibrium for a Competitive Economy," *Econometrica*, 22, 265–292.

AZARIADIS, C. (1993): *Intertemporal Macroeconomics.* Blackwell, Oxford, UK.

AZARIADIS, C., AND R. GUESNERIE (1986): "Sunspots and Cycles," *Review of Economic Studies*, 53, 725–737.

BALASKO, Y. (1988): *Foundations of the Theory of General Equilibrium.* Academic Press.

BALASKO, Y., AND D. CASS (1989): "The Structure of Financial Equilibrium I: Exogenous Yields and Unrestricted Participation," *Econometrica*, 57, 135–162.

BALASKO, Y., D. CASS, AND K. SHELL (1980): "Existence of Competitive Equilibrium in a General Overlapping Generations Model," *Journal of Economic Theory*, 23, 307–322.

BALASKO, Y., AND C. GHIGLINO (1996): "On the Existence of Endogenous Cycles," *Journal of Economic Theory*, 67, 566–577.

BALASKO, Y., AND K. SHELL (1980): "The Overlapping Generations Model, I: The Case of Pure Exchange without Money," *Journal of Economic Theory*, 23, 281–306.

BENHABIB, J., AND R. H. DAY (1982): "A Characterization of Erratic Dynamics in the Overlapping Generations Model," *Journal of Economic Dynamics and Control*, 4, 37–55.

BILLINGSLEY, P. (1986): *Probability and Measure.* John Wiley & Sons, New York.

BLACK, F., AND M. SCHOLES (1973): "The Pricing of Options and Corporate Liabilites," *Journal of Political Economy*, 81, 637–654.

BORGLIN, A., AND H. KEIDING (1986): *Optimality in Infinite Horizon Economies.* Lecture Notes in Economics and Mathematical Systems 269, Springer Verlag.

BURKE, J. (1987): "Inactive Transfer Policies and Efficiency in General Overlapping-Generations Economies," *Journal of Mathematical Economics*, 16, 201–222.

CASS, D. (1972): "On Capital Overaccumulation in the Aggregative Neoclassical Model of Economic Growth: A Complete Characterization," *Journal of Economic Theory*, 4, 200–223.

——— (1985): "Competitive Equilibria in Incomplete Financial Markets," CARESS Working Paper No. 84-09, University of Pennsylvania.

DEBREU, G. (1952): "Definite and Semidefinite Quadratic Forms," *Econometrica*, 20, 295–300.

——— (1953): "Un economie de l'incertain," mimeographed manucscript, Electricite de France, translated in Debreu [1983].

——— (1959): *Theory of Value.* Wiley.

——— (1970): "Economies with a Finite Set of Equilibria," *Econometrica*, 38, 387–392.

——— (1983): "Economics of Uncertainty," in *Mathematical Economics*, chap. 8. Cambridge University Press, Translation of a mimeographed manuscript from 1953: "Un economie de l'incertain", see Debreu [1953].

DEMARZO, P. (1988): "An Extension of the Modigliani-Miller Theorem to Stochastic Economies with Incomplete Markets," *Journal of Economic Theory*, 45, 353–369.

DEVANEY, R. L. (1986): *Chaotic Dynamical Systems*. Benjamin Cummings Publishing, Menlo Park, CA.

DIERKER, E. (1974): *Topological Methods in Walrasian Economics*. Springer Verlag.

DUFFIE, D., AND W. SHAFER (1985): "Equilibrium in Incomplete Markets I: Basic Model of Generic Existence," *Journal of Mathematical Economics*, 14, 285–300.

——— (1986): "Equilibrium in Incomplete Markets II: Generic Existence in Stochastic Economies," *Journal of Mathematical Economics*, 15, 199–216.

FARMER, R. E. (1993): *The Macroeconomics of Self-Fulfilling Prophecies*. MIT Press, Cambridge, MA.

GALE, D. (1973): "Pure Exchange Equilibrium of Dynamic Economic Models," *Journal of Economic Theory*, 6, 12–36.

GALE, D., AND A. MAS-COLELL (1975): "An Equilibrium Existence Theorem for a General Model without Ordered Preferences," *Journal of Mathematical Economics*, 6, 9–15.

GEANAKOPLOS, J., AND A. MAS-COLELL (1989): "Real Indeterminacy with Financial Assets," *Journal of Economic Theory*, 47, 22–38.

GEANAKOPLOS, J. D., AND H. POLEMARCHAKIS (1991): "Overlapping Generations," in *Handbook of Mathematical Economics, vol. IV*, ed. by W. Hildenbrand, and H. Sonnenschein. Elsevier, Amsterdam.

GHIGLINO, C., AND M. TVEDE (1995): "Endowments, Stability and Fluctuations in OG models," *Journal of Economic Dynamics and Control*, 19, 621–653.

GRANDMONT, J. M. (1985): "On Endogenous Competitive Business Cycles," *Econometrica*, 82, 995–1045.

GROSSMAN, S. (1977): "A Characterization of Optimality of Equilibrium in Incomplete Markets," *Journal of Economic Theory*, 15, 1–15.

HART, O. (1975): "On the Optimality of Equilibrium When the Market Structure is Incomplete," *Journal of Econmic Theory*, 11, 381–408.

HIRSCH, M. D., M. MAGILL, AND A. MAS-COLELL (1990): "A Geometric Approach to a Class of Equilibrium Existence Theorems," *Journal of Mathematical Economics*, 19, 95–106.

KOOPMANS, T. C. (1957): *Three Essays on the State of Economic Science.* McGraw-Hill.

KRASA, S. (1989): "Existence of Competitive Equilibria with Option Markets," *Journal of Economic Theory*, 45, 353–369.

KREPS, D. (1982): "Multiperiod Securities and the Efficient Allocation of Risk: A Comment on the Black-Scholes Option Pricing Model," in *The Economics of Information and Uncertainty*, ed. by J. McCall. University of Chicago Press.

LEROY, S., AND J. WERNER (2001): *Principles of Financial Economics.* Cambridge University Press.

LI, T. Y., AND J. A. YORKE (1975): "Period Three implies Chaos," *American Mathematical Monthly*, 82, 985–992.

LUCAS, R. E. (1978): "Asset Prices in an Exchange Economy," *Econometrica*, 46, 1429–1445.

MAGILL, M., AND M. QUINZEE (1996): *The Theory of Incomplete Markets.* The MIT Press.

MAGILL, M., AND W. SHAFER (1991): "Incomplete Markets," in *Handbook of Mathematical Economics, vol. IV*, ed. by W. Hildenbrand, and H. Sonnenschein. Elsevier, Amsterdam.

MAS-COLELL, A. (1974): "An Equilibrium Existence Theorem without Complete or Transitive Preferences," *Journal of Mathematical Economics*, 1, 237–246.

——— (1985): *The Theory of General Economic Equilibrium: A differentiable approach.* Cambridge University Press.

MAS-COLELL, A., D. WHINSTON, AND J. GREEN (1995): *Microeconomic Theory.* Oxford University Press.

MCCANDLESS JR., G. T., AND N. WALLACE (1991): *Dynamic Macroeconomic Theory.* Harvard University Press, Cambridge, MA.

MODIGLIANI, F., AND M. H. MILLER (1958): "The Cost of Capital, Corporate Finance and the Theory of Investment," *American Economic Review*, 48, 261–297.

PLISKA, S. R. (1997): *Introduction to Mathematical Finance.* Blackwell.

POLEMARCHAKIS, H., AND B. I. KU (1990): "Options and Equilibrium," *Journal of Mathematical Economics*, 19, 107–112.

RADNER, R. (1968): "Competitive Equilibrium under Uncertainty," *Econometrica*, 36, 31–58.

——— (1972): "Existence of Equilibrium of Plans, Prices and Price Expectations in a Sequence of Markets," *Econometrica*, 40, 289–304.

——— (1979): "Rational Expectations Equilibrium: Generic Existence and the Information Revealed by Prices," *Econometrica*, 47, 655–678.

SAMUELSON, P. A. (1958): "An Exact Consumption-Loan Model of Interest with or without the Social Contrivance of Money," *Journal of Political Economy*, 66, 467–482.

SHAFER, W. J., AND H. SONNENSCHEIN (1975): "Some Theorems on the Existence of Competitive Equilibrium," *Journal of Economic Theory*, 11, 83–93.

SHELL, K. (1971): "Notes on the Economics of Infinity," *Journal of Political Economy*, 79, 1002–1011.

——— (1987): "Sunspot Equilibrium," in *The New Palgrave Dictionary of Economics*, ed. by J. Eatwell, M. Milgate, and P.Newman. MacMillan, London.

SYDSAETER, K., AND P. HAMMOND (1995): *Mathematics for Economic Analysis*. Prentice Hall, New Jersey.

WERNER, J. (1985): "Equilibrium in Economies with Incomplete Financial Markets," *Journal of Economic Theory*, 36, 110–119.

——— (1987): "Arbitrage and the Existence of Competitive Equilibrium," *Journal of Economic Theory*, 36, 110–119.

Index

LIST OF NOTATION

The page number indicates the page where the notation is first used.

A	real dividend matrix, 92
$(a_t)_{t\in\mathbb{Z}}$	reallocation, 225
β	vector of discount factors; state prices, 35, 136
$B(p,w)$	budget set at (p,w), 8
$(b_t)_{t\in\mathbb{Z}}$	improvement, 254
$\circ$	composition of sets, 257
c	consumption vector, 2
(c_t^y, c_t^o)	consumption, 224
C	consumption set, 2
$\mathbb{C}^2(C,\mathbb{R})$	twice differentiable functions with domain C and range $\mathbb{R}$, 2
D	domain of indirect utility function, 39
$Du(c)$	gradient of u at c, 2
$D^2u(c)$	Hessian of u at c, 2
$E(p)$	domain for $v(p,\cdot)$, 39
$\tilde{e}$	support function, 290
$\mathcal{E}$	economy, 13
$\mathcal{E}^s$	stationary economy, 275
$\|\cdot\|$	Euclidean norm, 8
$f(r,e)$	Walrasian demand at (r,e), 277
$\hat{f}(p,w)$	Walrasian demand at (p,w), 8
G_e	equilibrium locus, 313
$g(p,r)$	spot-market demand at (p,r), 42
$\hat{g}(r,e)$	real savings at (r,e), 286
$\operatorname{grad} u(c)$	gradient of u at c, 4
h	demand function under uncertainty, 350
$\mathbb{I}$	set of consumers, 13
J_e	interval $[\min(\rho(e),1), \max(\rho(e),1)]$, 311
$\mathbb{J}$	set of assets, 79
$\mathbb{L}$	set of goods, 30
$\mathbb{M}$	subspace of net income vectors, 51
m	nominal savings, 228
M	nominal savings, 228
M_e	maximal invariant set, 322
$\mathbb{N}$	the natural numbers, 46
P	future prices, 32
$\mathcal{P}$	partition, 140
p	price vector, 8
$(p_t)_{t\in\mathbb{Z}}$	price system, 222
pr_i	projection on the i'th factor, 258
q	date 0 prices of assets, 83
$\mathbb{R}^n$	Euclidean n-space, 2
$\mathbb{R}^n_{++}$	positive vectors of $\mathbb{R}^n$, 2
$\mathbb{R}^n_+$	non-negative vectors of $\mathbb{R}^n$, 6

$R_t(\cdot)$	translated upper contour set, 252
R_t	real interest rate, 287
r	net income vector Ch. 2-6, 35
r	relative price, Ch. 9-11, 277
$\rho(c)$	marginal rate of substitution, first component of normalized gradient, 283
$\mathbb{S}_1$	set of future elementary events, 129
S_e	invariant set, 318
$(S_t)_{t\in\mathbb{Z}}$	reduced model, 252
$S_H(a,q)$	hyperbola set for (a,q), 262
σ	event, 140
$\mathbb{T}$	set of dates, 30
$\mathbb{T}_1$	set of future dates, 30
θ	portfolio vector, 78
u	utility function, 2
V	value dividend matrix, 78
v	indirect utility function, 42
w	wealth, 8
W	extended dividend matrix, 80
Y	production set, 103
y	net production vector, 103
$\mathbb{Z}$	set of integers; dates, 222
z	net trade, 32
$z(r,e)$	excess demand at (r,e), 283

Druck: Strauss Offsetdruck, Mörlenbach
Verarbeitung: Schäffer, Grünstadt

MIX
Papier aus verantwortungsvollen Quellen
Paper from responsible sources
FSC® C105338

If you have any concerns about our products,
you can contact us on
ProductSafety@springernature.com

In case Publisher is established outside the EU,
the EU authorized representative is:
Springer Nature Customer Service Center GmbH
Europaplatz 3, 69115 Heidelberg, Germany

Printed by Libri Plureos GmbH
in Hamburg, Germany